Low-Power High-Level Synthesis for Nanoscale
CMOS Circuits

Saraju P. Mohanty · Nagarajan Ranganathan
Elias Kougianos · Priyadarsan Patra

Low-Power High-Level Synthesis for Nanoscale CMOS Circuits

 Springer

Saraju P. Mohanty
University of North Texas
Denton, TX
USA

Nagarajan Ranganathan
University of South Florida
Tampa, FL
USA

Elias Kougianos
University of North Texas
Denton, TX
USA

Priyadarsan Patra
Intel Corporation
Hillsboro, OR
USA

ISBN: 978-0-387-76473-3   e-ISBN: 978-0-387-76474-0
DOI: 10.1007/978-0-387-76474-0

Library of Congress Control Number: 2008920380

To Uma, my sisters and my parents.

To Radhika, Sheela, Shiva and my parents.

To Diane and my family.

*To Anu, Ashish, Adarsh and my parents
Harihar and Padmabati.*

Preface

Low-Power High-Level Synthesis for Nanoscale CMOS Circuits shows very-large-scale integration (VLSI) researchers and engineers how to minimize the different types of power consumption of digital circuits. This text provides the fundamentals of high-level (also called behavioral- or architectural-level) synthesis of digital circuits. In addition, this text provides adequate knowledge in the design of low-power CMOS circuits. Students will acquire a sufficient understanding of the different methodologies of power reduction while taking into account process variations during characterization and modeling.

This self-contained text is directed to nanoscale VLSI design engineers, graduate students in electrical engineering and computer scientists who are about to start their research in high-level VLSI design. It is also an important comprehensive source for microelectronics engineers who would like to understand the different aspects of power reduction. The text assumes basic knowledge of algorithms and familiarity with digital circuit design. Since the research focus in this field is continually progressing, students will acquire sufficient knowledge after studying this book so as to be able to understand more specialized literature on their own. Each chapter has simple relevant examples for a better grasp of the principles presented. Several algorithms are given to provide a better understanding of the underlying concepts. However, the goal of this book is to train the students to develop algorithms rationally.

The text provides a sufficient amount of fundamentals to become familiar with the terminology of the field. The main objective is to achieve in-depth knowledge in a few topics instead of a shallow, broader presentation. The purpose and objective of low-power high-level synthesis is explained in Chapter 1. The initial chapters deal with the basics of high-level synthesis (Chapter 2), power (Chapter 3) and power estimation (Chapter 4). In subsequent parts of the text, a detailed discussion of methodologies for the reduction of different types of power is presented: power reduction in Chapter 5, energy or average power reduction in Chapter 6, peak power reduction in Chapter 7 and transient power reduction in Chapter 8. As the dimensions of CMOS devices decrease to nanometer scale to adhere to Moore's law, different leakage mechanisms are observed. Different schemes to minimize various forms of leakage in ICs are presented in Chapter 9.

S.P. Mohanty
Denton, TX
USA

Acknowledgments

Saraju P. Mohanty acknowledges the full support and encouragement from the Department of Computer Science and Engineering at the University of North Texas. This book derives a lot of material from the very-large-scale integration (VLSI) design courses that he teaches there as well as from his doctoral dissertation. The author is very thankful to his students for providing constant valuable feedback. Saraju P. Mohanty also thanks Dr. Krishna Kavi, the Department Chair and Dr. Oscar García, Founding Dean of the College of Engineering. He acknowledges the faculty and staff of the Department of Computer Science and Engineering for their generous support.

Nagarajan Ranganathan would like to express his gratitude to the Department of Computer Science and Engineering at the University of South Florida. He acknowledges with gratitude the research contributions of all his current and past students.

Elias Kougianos expresses his gratitude to the Department of Engineering Technology at the University of North Texas. In particular, he thanks the Department Chair, Dr. Nourredine Boubekri, for the collegial environment that he fosters in the department.

Priyadarsan Patra is grateful to Intel Corporation for fully supporting his authorship and for providing a world-class environment. He thanks his colleagues at the Strategic CAD Labs and the Microprocessor Technology Labs of Intel, where he has worked, and other colleagues from the leading product groups, for providing perspectives that have influenced what is good in this book.

The authors are grateful to Dr. Uma Choppali who has helped immensely in the preparation of figures in the book as well as Ms. Otilia Sánchez, M.A. Both of them have read the entire manuscript thoroughly and provided excellent editorial comments.

The publication of this book would not have been possible without the support of the Springer staff. We would like to thank Alex Greene and Katelyn Stanne for their promptness and patience in answering all our questions.

The authors are greatly indebted to their families for their patience, understanding and support during the innumerable hours of preoccupation with the writing of this book. Authors Saraju P. Mohanty and Elias Kougianos would like to acknowledge NSF award number 0702361.

Contents

List of Figures

List of Tables

Acronym Definition

ACO	Ant colony optimization
ACS	Ant colony scheduling
AF	Activity factor
AHD	Average Hamming distance
ALAP	As late as possible
ALU	Arithmetic logic unit
AMPL	A mathematical programming language
ARF	Autoregressive filter
ARMA	Autoregressive moving average
ASAP	As soon as possible
ASIC	Application-specific integrated circuit
ATG	Automatic test generation
BFS	Breadth first search
BN	Boolean network
BPF	Band-pass filter
BSIM	Berkeley short channel insulated gate model
BTBT	Band-to-band tunneling
CAD	Computer-aided design
CCLC	Cross-coupled level converter
CDFG	Control data flow graph
CFG	Control flow graph
CG	Compatibility graph
CLT	Central limit theorem
CMOS	Complementary metal oxide semiconductor
CMP	Chemical mechanical polishing
CPF	Cycle power function
CPLD	Complex programmable logic device
CPU	Central processing unit
CVD	Chemical vapor deposition
DCN	Diffusion-connected network
DCT	Discrete cosine transform
DCU	Dynamic clocking unit
DCVS	Differential cascade voltage switch
DFC	Dynamic frequency clocking
DFG	Data flow graph

DFM	Design for manufacturing
DG	Distribution graph
DIBL	Drain-induced barrier lowering
DPCD	Dynamic programmable clock divider
DPM	Dynamic power management
DSP	Digital signal processing
DT	Dynamic threshold
DVS	Dynamic voltage scaling
DWT	Discrete wavelet transform
EDA	Electronic design automation
EDP	Energy delay product
ESL	Electronic system level
ESTG	Extended state transition graph
EWF	Elliptic wave filter
FDS	Force directed scheduling
FET	Field effect transistor
FFT	Fast Fourier transform
FIFO	First in first out
FIR	Finite impulse response
FN	Fowler–Nordheim
FPGA	Field programmable gate array
FSM	Finite state machine
FU	Functional unit
GAS	Genetic algorithm scheduling
GDS	Generalized data stream
GIDL	Gate-induced drain leakage
GIS	Geographic information system
GTS	Game-theory (based) scheduling
HCDG	Hierarchical conditional dependency graph
HDL	Hardware description language
HL	High to low
IC	Integrated circuit
IDCT	Inverse discrete cosine transform
IEEE	Institute of Electrical and Electronics Engineers
IIR	Infinite impulse response
ILP	Integer linear programming
I/O	Input/output
IP	Intellectual property
IR	Infrared
ITRS	International Technology Roadmap for Semiconductors
LBS	List-based scheduling
LECTOR	Leakage control transistor
LCT	Leakage control transistor
LH	Low to high
LP	Linear programming

LUT	Look-up table
MC	Multicycling
MCM	Monte Carlo method
MESVS	Minimum energy schedule with voltage selection
Mκ	Multiple dielectric
MMV	MPEG motion vectors
MOSFET	Metal oxide semiconductor field effect transistor
MOVER	Multiple operating voltage energy reduction
MOX	Multiple oxide
MPEG	Moving Picture Experts Group
MT	Multiple threshold
MU	Memory unit
MV	Multiple supply voltage
MVDFC	Multiple supply voltage and dynamic frequency clocking
MVMC	Multiple supply voltage and multicycling
MVSF	Multiple supply voltage and single frequency
MWIS	Maximum weight independent set
Nano-CMOS	Nanoscale CMOS
NFA	Nondeterministic finite automata
NBTI	Negative bias temperature instability
NMOS	Metal oxide semiconductor, n type
NoC	Network on a chip
NOP	No operation
NP	Nondeterministic polynomial
ODT	On-die termination
PC	Personal computer
PCB	Printed circuit board
PDA	Personal digital assistant
PDP	Power delay product
PDSS	Profile-driven synthesis system
PI	Primary input
PLA	Programmable logic array
PLL	Phase-locked loop
PMOS	Metal oxide semiconductor, p type
PSB	Processor side bus
PSF	Process scaling factor
RAM	Random access memory
RC	Resource constrained (scheduling)
RF	Radio frequency
ROM	Read-only memory
RTL	Register transfer level
SAS	Simulated annealing scheduling
SDL	Set dominant latches
SoC	system on a chip
SOI	Silicon on insulator
SP	Signal probability

SRAM	Static random access memory
STG	State transition graph
SVDFC	Single supply voltage and dynamic frequency clocking
SVMC	Single supply voltage and multicycling
SVSF	Single supply voltage and single frequency
TC	Time constrained
TDP	Thermal design power
TLB	Translation look-aside buffer
TLM	Transaction-level modeling
TSS	Tabu search scheduling
UDFG	Unscheduled data flow graph
VHDL	VHSIC Hardware Description Language
VHSIC	Very-high-speed-integrated circuit
VLSI	Very-large-scale integration
VT	Variable threshold
WDF	Wave digital filter
YACC	"Yet another compiler" compiler

Chapter 1
Introduction

The market demand for portability, performance and high functional integration density of digital devices has made the scaling of complementary metal oxide semiconductor (CMOS) devices inevitable. Such devices are susceptible to new leakage mechanisms because of tunneling through the gate dielectric as well as the junctions of the transistors. These leakage components, exacerbated by process and environmental variations, become nearly as prominent as the switching power dissipation of a chip and thus have a severe impact on the power and power-density budgets available for useful, reliable computation. Total power greatly influences power delivery and heat removal complexity, product cost, form factor and portability. Thus, there is a critical need for the analysis, characterization, estimation and optimization of the various forms of power dissipation, especially in the presence of process variations of nanoscale CMOS (or nano-CMOS) technologies. The amount of power consumption, as well as the fluctuation in power consumption of a circuit, affects its operational attributes. Increase in power consumption is detrimental to battery life and degrades the reliability of the device because of the reduced efficiency in electrochemical conversion and heat-induced anomalies. Power fluctuation also leads to larger power supply noise due to self-inductance and can introduce significant noise in signal lines due to mutual inductance and capacitance (cross-talk). High current peaks in short time spans can cause high heat dissipation in a localized area of the die and may lead to failures. The magnitude of each leakage component (gate oxide, subthreshold leakage, etc.) of the device is mostly dependent on the device geometry, doping profiles, oxide thickness, voltage and temperature, whereas the algorithms and design structures used to effect computation greatly influence switching power. At nanometer dimensions, the effects of variations in these process and design factors grow prominently. This growth necessitates any methodology for power and power fluctuation reduction to account for process variation during characterization and modeling and also in design and synthesis frameworks.

Various techniques are available for power estimation and optimization at different levels of abstraction: from circuit level to system level. This book primarily deals with the high level (*also known as* architectural or behavioral) because the behavioral level is neither as highly abstracted as the system level nor as com-

S.P. Mohanty et al., *Low-Power High-Level Synthesis for Nanoscale CMOS Circuits*, DOI: 10.1007/978-0-387-76474-0_1, © Springer Science+Business Media, LLC 2008

plexly detailed as the gate/transistor level. At the behavioral level, there is a balanced degree of freedom to explore power reduction mechanisms, power reduction opportunities are greater and one can cost-effectively explore lower power design alternatives prior to actual circuit layout or silicon implementation. This book is a self-contained low-power, high-level synthesis text for engineers and researchers interested in nanoscale very large scale integration (VLSI) design. The goal is to provide power modeling, characterization, estimation and optimization in the framework of high-level synthesis to provide a global perspective to the reader. The following topics are covered in detail in the following chapters:

- Chapter 2, *High-Level Synthesis Fundamentals*: This chapter provides a brief overview of the integrated circuit (*aka* IC) design flow and a summary of the high-level (architectural, behavioral or algorithmic) synthesis. Various algorithms of scheduling and binding are presented in a simplified manner.
- Chapter 3, *Power Modeling and Estimation at Transistor and Logic Gate Levels*: This chapter addresses the problem of modeling and estimating the various mechanisms of power (current) dissipation specific to nanoscale CMOS transistors, with the 45 nm technology node as a case study.
- Chapter 4, *Architectural Power Modeling and Estimation*: Modeling, estimation and analysis of power and thermal dissipation form an integral and important part of VLSI design. They go hand in hand with power optimization and design for low power.
- Chapter 5, *Power Reduction Fundamentals*: Low-power circuit design is a three-dimensional problem involving area, performance and power trade-offs. Power dissipation occurs in various forms, such as dynamic, subthreshold leakage and gate leakage and there is need to reduce each of these forms of dissipation. This chapter discusses various techniques for reducing these different sources of power dissipation. In addition, it is essential to study the dissipation profile of CMOS circuits and formulate the optimization problem accordingly for more effective power optimization. The chapter also highlights metrics for capturing the power profile.
- Chapter 6, *Energy or Average Power Reduction*: In this chapter, data path scheduling algorithms that use multiple supply voltages and dynamic clocking in a coordinated manner to reduce energy and energy delay products are discussed. The strategy is to schedule high-energy units, such as multipliers, at lower frequencies so that they can be operated at lower voltages to reduce energy consumption and low-energy units, such as adders, at higher frequencies to compensate for speed. The chapter also discusses techniques for switching activity reduction during scheduling and binding.
- Chapter 7, *Peak Power Reduction*: The use of multiple supply voltages for energy and average power reduction is well researched, and several works have appeared in the literature. However, in low-power design for battery-driven nanoscale circuits, peak power, peak power differential, average power and total energy are equally critical design constraints. Data path scheduling algorithms for peak power and simultaneous peak and average power minimization are presented in this chapter.

- Chapter 8, *Transient Power Reduction*: In this chapter, a framework for simultaneous reduction of the energy and transient power during behavioral synthesis is presented. Metrics are provided that capture transience or fluctuation in the power dissipation of a data path circuit. Then, heuristic and integer linear programming-based algorithms are proposed for their optimization.

- Chapter 9, *Leakage Power Reduction*: In a short-channel nano-CMOS transistor, several forms of leakage current exist, such as reverse biased diode leakage, subthreshold leakage, gate-oxide tunneling current, hot-carrier gate current, gate-induced drain leakage and channel punch-through current. Of all these leakage mechanisms, gate-oxide (direct) tunneling current that flows during both active and sleep modes of a device is the most significant component for low-end nano-CMOS technology at 65 nm and below. Thus, the major sources of power dissipation in a nano-CMOS circuit can be summarized as dynamic, short-circuit and subthreshold and gate-oxide leakage. In this chapter, reduction of gate-oxide leakage and subthreshold leakage in CMOS data path circuits during high-level synthesis is discussed.

- Chapter 10, *Conclusions and Future Directions*: This chapter summarizes the models, metrics and techniques presented in this book for power optimization in nano-CMOS circuits. The chapter then provides suggestions and guidelines for future research.

Chapter 2
High-Level Synthesis Fundamentals

2.1 Introduction

This chapter provides a brief overview of circuit design flow and a summary of high-level (*aka* architectural, behavioral or algorithmic) synthesis. High-level synthesis is the translation process from a behavioral description to a structural description. This is analogous to "compilation" that translates a high-level language program in C/C++ to an assembly language program. High-level synthesis is also known as behavioral-level synthesis or algorithmic-level synthesis. The constraints to be considered in high-level synthesis are area, performance, power consumption, reliability, testability and cost. This synthesis allows a design engineer to make decisions at an early stage of the design cycle, thus ensuring correct design. Typical steps involved are scheduling, binding, allocation, etc.

2.2 The Complete Chip Story: From Customers' Requirements to Silicon Chips for Customers

An integrated circuit (IC, informally called a chip) is a silicon semiconductor crystal containing electronic active components and interconnects. Since the invention of the first point contact transistor in 1947 and the first IC in 1960 [285, 364], the very-large-scale integration (VLSI) industry has been a driving factor in the growth of civilization. Starting from a mobile phone in our pocket, television at home, desktop computer to the space shuttle, everywhere, chips are performing their operations faithfully. Each of these chips is quite diverse, designed for different specifications to perform various operations under different operating conditions. The process of realizing a user specification or requirement in silicon is quite involved. Experts in physics, electronics, chemistry, etc. have worked together to develop the circuits. A broad perspective of this process is schematically presented in Fig. 2.1 [269, 370, 210, 289, 285, 364].

The customer or user requirement is first translated to user specifications that a VLSI design engineer can understand. Once the specification is determined, the

S.P. Mohanty et al., *Low-Power High-Level Synthesis for Nanoscale CMOS Circuits*,
DOI: 10.1007/978-0-387-76474-0_2, © Springer Science+Business Media, LLC 2008

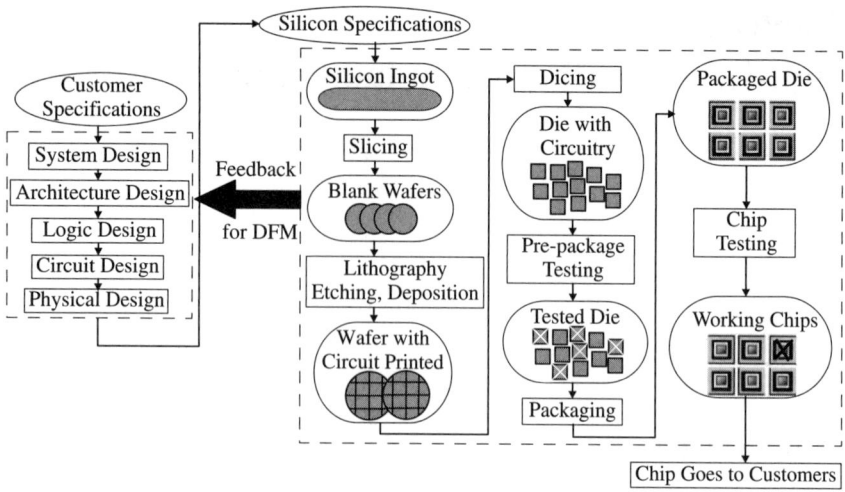

Fig. 2.1 The complete chip story: from user specification to user's chip

design process involving various levels of circuit abstraction is performed. It includes system design, architectural (high-level or behavioral) design, logic design,
circuit design and physical design. At each level of circuit abstraction, the circuit
is equivalent and performs the same target operation, but its structural components
(and hence the component's granularity) are different, and the design issues may be
different. Once the chip layout is finalized after rigorous design and verification, it
is taped out to the manufacturing plant for its realization as a chip.

The starting point of chip fabrication is a silicon ingot. The ingot is first sliced
to obtain blank wafers (several circuits will eventually be produced on each blank
wafer). The blank wafer goes through several lithographic, mechanical and chemical
processes to become imprinted with the desired circuits. To facilitate fabrication of
circuits using nano-CMOS technology, more and more sophisticated lithographic,
chemical and mechanical processing steps are adopted. There is uncertainty in the
process parameters, such as ion implantation, chemical mechanical polishing (CMP)
and chemical vapor deposition (CVD). In nano-CMOS fabrication has caused variations in process parameters [42, 327, 95] such as channel length, gate-oxide thickness, threshold voltage, metal wire thickness, contact and resistance. This has
become a major challenge for a VLSI designer to handle in order to ensure the
maximum possible yield. Thus, in the context of VLSI design, nano-CMOS technology is a feedback loop from the process to the design phase to ensure design
for manufacturing (DFM) [303, 314]. Once the wafer is imprinted with circuits, it
is then diced to obtain individual die with circuits. The individual die that passes
through the prepackaging test are then bonded with the package. The packaged chip
then goes through testing, and healthy chips are collected. The healthy chips go to
the user or customer whose specifications they implement.

2.3 Various Phases of Circuit Design and Synthesis

A chip under design is represented in various forms to support hierarchical design and synthesis [285, 364, 333, 210, 289, 209, 102]. A hierarchical design approach is essential to support modern systems on a chip (SoC), which are made of several millions of transistors (predicted to reach the billions soon). The circuit of a chip can be expressed at various levels of design abstraction as shown in Fig. 2.2. They can be distinguished from each other as follows:

- *System level*: The system level is the highest level circuit abstraction, at which the system is specified as processes and tasks with a mix of hardware and software. The system level is concerned with overall system structure and information flow. Computer systems are described as an interconnected set of processors, memories and switches.
- *Behavioral level, algorithmic level or high level*: This level is also called as instruction set level or algorithmic level. At this level, the focus is on the computations performed by an individual processor; i.e., the way it maps sequences of inputs to sequences of outputs.
- *Architecture, microarchitecture, RTL*: The system is viewed as a set of interconnected storage elements and functional blocks. The behavior of the system is described as a series of data transfers and transformations between the storage elements. Microarchitectural-level representation of the chip resources, such as adders and subtractors, is determined along with decisions such as single-cycle, multicycle, pipelined or superscalar implementation.

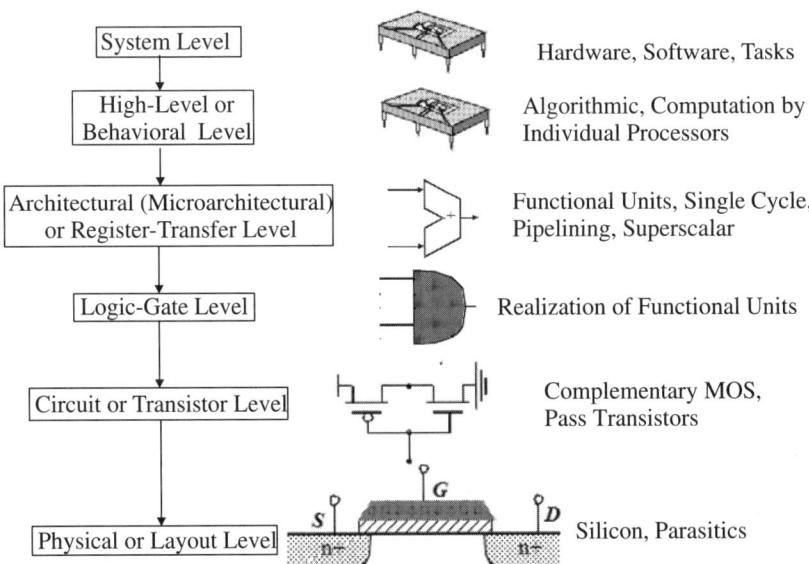

Fig. 2.2 Different levels of design abstractions

- *Logic level*: Below the RTL is the logic level. The system is described as a network of gates and flip-flops, and the behavior is specified by logic equations. At the logic level, the circuit is represented in the form of a netlist at which level logic realizations of functional blocks are determined.
- *Circuit or transistor level*: At the transistor level, the circuit is a netlist of transistors. Decisions such as how and what types of transistors will be used, complementary CMOS, pass transistors and domino logic are the main issues.
- *Physical or layout level*: In this level, the system is specified in terms of the individual transistors of which it is composed. The behavior of the system can be described in terms of the network equations. Physical or chip layout level is the lowest level of circuit abstraction at which the chip is a sequence of layers (masks), each layer of which is composed of polygons. It is this level that is transferred to the manufacturing process.

VLSI design flow is evolutionary. It consists of a number of design and test levels to conform to the user or design specifications. A typical VLSI design flow is best represented by a Y-chart that describes the design flow using design activities on three different axes (domains) resembling the letter Y. Figure 2.3(a) describes the design automation terminology, such as optimization, synthesis, analysis and optimization in the hardware representation domain. The axes in the Y-chart (Fig. 2.3(b)) represent three different domains of description, such as behavioral, structural and physical. Each concentric circle intersects the axes at a particular level of representation within a domain. In circuit analysis, the behavior or characteristics of a circuit are studied. The synthesis process is the reverse of the analysis process. The task of synthesis is to take the specifications of the behavior required for a system and a set of constraints and goals to be satisfied and to find a structure that implements the behavior while satisfying the goals and constraints [187, 206, 81, 210, 289, 209, 102, 332]. The "behavior" of the system refers to the ways in which the system or its components interact with their environment (mapping from inputs to outputs). The "structure" refers to the set of interconnected components that constitute the system (described by a netlist). Finally, the structure must be mapped into a "physical" design. The behavior, structure and physical designs are considered the three domains in which hardware can be described (Fig. 2.3(a) and 2.3(b)). In the behavioral domain, interest is in what a design does, not in how it is built. The physical domain ignores what the design is supposed to do and binds its structure in space or to silicon. A structural representation bridges the behavioral and physical representations. It is a one-to-one mapping of a behavioral representation onto a set of components and connections under constraints, such as area, cost and delay. The design process proceeds from higher to lower levels of abstractions. This procession is shown in Fig. 2.3(b) – from outer to inner rings, starting from the algorithm and ending in the mask, the last step before fabrication.

The automatic design process of VLSI circuits is called synthesis. Corresponding to the different abstraction levels of circuit representation are different synthesis levels, as presented in Fig. 2.4 [210, 289, 7, 102, 364, 71, 206]. The system-synthesis process partitions the tasks into hardware, software and their communications [82, 289, 132]. The high-level synthesis process is the translation from behavioral

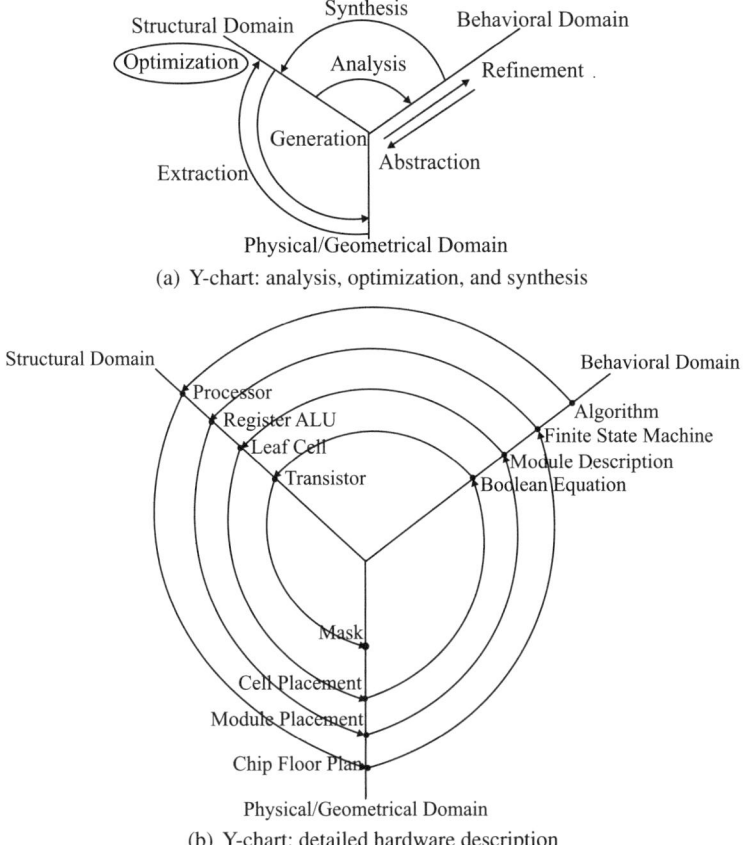

(a) Y-chart: analysis, optimization, and synthesis

(b) Y-chart: detailed hardware description

Fig. 2.3 Description of hardware in different domains and abstractions through the Gajski–Kuhn Y-chart

description to its equivalent structural description [206, 187, 102, 274, 210, 351]. Logic synthesis is the process of mapping from the design at the RTL to a gate-level representation that is suitable for input to physical design [336, 332, 152]. This synthesis precedes the circuit design, which addresses the creation of device and interconnect topologies (standard cells, full-custom analog, etc.) that attain the prescribed electrical and physical properties while remaining feasible with respect to process and manufacturability constraints. The physical design then addresses aspects of chip implementation (floor planning, placement, routing, extraction, performance analysis) related to the correct spatial embedding of devices and interconnects [112, 254, 372]. The output of physical design is the handoff ("tapeout") to manufacturing (currently centered around a generalized data stream, GDS II, stream file), along with verification of correctness (design rules, layout versus schematic, etc.) and constraints (timing, power, reliability, etc.). Together, logical, circuit and physical design comprises the implementation layer of a circuit that

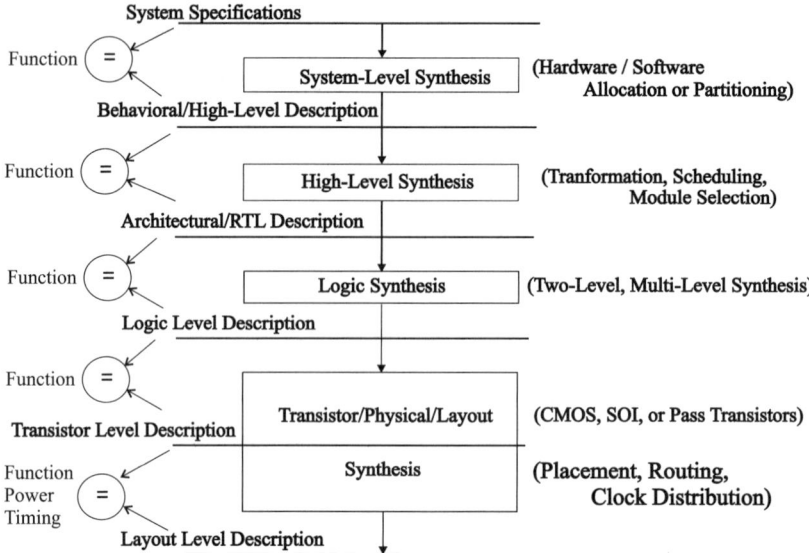

Fig. 2.4 Various synthesis processes that translate a system description to a physical description in silicon. Logic level to physical (layout) level synthesis occurs in one step through physical synthesis, but the transistor level or circuit level is shown because it is a form of description that can take place transparently

supports system-level design. During each phase of the synthesis process, a design engineer needs to check the functional equivalence of two consecutive phases to ensure that they are functionally the same. At the same time, a power and timing analysis study can be done by using compact models at the transistor level. At the physical level, more accurate power and time analysis is possible through the extraction of accurate parasitics.

2.4 High-Level or Behavioral Synthesis: What and Why

The high-level synthesis process can be defined as the translation of a behavioral description to a structural description [274, 206, 185, 102, 81, 33, 289, 209, 210, 284]. This process is analogous to a "compiler" that translates a high-level program in C/C++ to an assembly language program. High-level synthesis is also known as behavioral synthesis, algorithmic-level synthesis and architectural synthesis. The constraints to be considered in high-level synthesis are area, performance, power consumption, reliability, testability and cost. With the increasing demand for personal computing devices and wireless communications equipment, the demand for designing low-power-consuming circuits has increased. "Power" and "leakage" have become important parameters, along with area and throughput, for optimization during behavioral synthesis.

Behavioral or high-level synthesis has become popular because of several advantages it provides, as discussed below [206, 289, 210, 176, 86, 213, 34]:

- *Continuous and reliable design flow*: The high-level synthesis process provides a continuous and reliable flow from system-level abstraction to RTL abstraction automatically without manual handling. It is thus facilitating automatic translations from high-level specifications in the form of C or SystemC to RTL description of the circuit in the form of VHDL or Verilog.
- *Shorter design cycle*: If more of the design process is automated, faster products can be made available at cheaper prices. The shorter design cycle can reduce the number of man-hours used and time to market, and hence the overall cost of the chip.
- *Fewer errors*: Since the synthesis process can be verified easily, the chances of errors will be smaller. Correct design decisions at the higher levels of circuit abstraction can ensure that the errors are not propagated to the lower levels, which are too detailed and costly to correct.
- *Easy and flexible to search the design space*: Because a synthesis system can produce several designs in a short time, the designer has more flexibility to choose the proper design considering different trade-offs of power, leakage, area and delay.
- *Balanced degree of freedom for power optimization*: Power and performance optimization can be performed at any level of circuit abstraction, from system level to silicon. As the level of abstraction goes lower, the complexity of the circuit increases; additionally, the degrees of freedom, and thus power reduction opportunities, decrease. Hence, high level or behavioral level is an attractive level and provides a balanced degree of freedom for design space exploration. This feature is quite essential in nano-CMOS circuit design.
- *Documenting the design process*: An automated system can track design decisions and their effects. Thus, design debugging and continuation by third parties can be easily done. This will be useful for macrocell-based design and the sale of designs as intellectual property cores, one of the current market trends when fully custom design is very expensive.
- *Availability of circuit technology to more people*: As design expertise is moved into synthesis systems, it becomes easier for a non-expert to produce a chip that meets a given set of specifications. Hence, the designer can be hired at a lower price, which will reduce the non-recurring cost and overall design cost of the chip.

2.5 Various Phases of High-Level Synthesis

The high-level synthesis process takes a system in the form of a hardware description language (HDL) as input and generates an optimal RTL description. The various phases or tasks of high-level synthesis include compilation, transformation, scheduling, allocation and binding as detailed in Fig. 2.5 [187, 134, 102, 378, 150, 162, 266, 33, 206, 274, 210, 185, 176, 113, 60, 284]. Additional steps like power optimization, leakage optimization, register optimization and interconnect optimization also take place in synthesis either sequentially or along with the fundamental steps. Thus, various other inputs in the form of a data path component library and constraints are needed for trade-off and design space exploration study. There is no fixed sequence for performing various high-level synthesis tasks. Although they are independent of each other, these tasks should be performed simultaneously for effective optimization.

2.5.1 Compilation

The behavior of a system to be synthesized is usually specified at the algorithmic level using a high-level programming language like C/C++ or a hardware description language (HDL) such as VHDL and Verilog. The behavior of the system is then compiled into internal representations, which are usually data flow graphs (DFGs) and control flow graphs (CFGs). Each behavioral specification is transformed into a unique graphical representation. The DFG is a directed graph that represents data movement, whereas the CFG is a directed graph that indicates the sequence of operations.

2.5.2 Transformation

In the transformation step, the initial DFG is transformed so that the resultant DFG is more suitable for scheduling and allocation. These transformations include compiler-like optimizations such as dead-code elimination, common subexpression elimination, loop unrolling, constant propagation and code motion. In addition, some hardware-specific transformations like minimization of syntactic

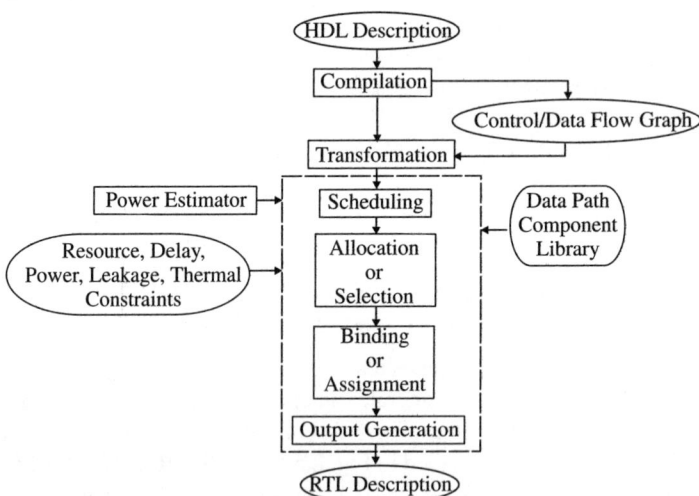

Fig. 2.5 Various phases of high-level synthesis

variances and retiming may be applied to take advantage of the associativity and commutativity of certain operations.

2.5.3 Scheduling

Scheduling is the process of partitioning the set of arithmetic and logical operations in the DFG into groups so that the operations in the same group can be executed concurrently, while taking into consideration possible trade-offs between the total execution cost and hardware cost. A group of concurrent computations to be executed simultaneously is referred to as a control step. The total number of control steps needed to execute all operations in the DFG, the minimum number of functional units of each type to be used in the design and the lifetimes of the variables generated during the computation of operations are determined in the scheduling step.

2.5.4 Selection or Allocation

Selection is the process of choosing resources from the library, which involves trade-offs according to different features like delay, area, power and leakage. Resource allocation is the process of determining the number of functional units of each type for performing operations, memory units (registers) for storing data values and interconnects for data transportation. Often, the selection and allocation processes are a single task. Allocation is further divided into sub-tasks, such as functional unit allocation, memory unit allocation and interconnect allocation. Resource allocation and binding may share resources so that the same hardware can be used to execute different operations or so that the same register can be used to store more than one variable.

2.5.5 Binding or Assignment

Binding or assignment is the process of assigning variables to memory units and data transfers to interconnections. Binding is further divided into several sub-tasks, such as functional unit binding, memory unit binding and interconnect binding. Functional unit binding involves the mapping of operations in the behavioral description into a set of selected functional units. Memory unit binding maps data carriers (constants, variables, arrays) in the behavioral description onto storage elements (read-only memories, registers, memory units) in the data path. The interconnect binding task maps every data transfer in the behavior onto a set of interconnection units for data routing.

2.5.6 Output Generation

In the output generation phase, design output is generated. The output should be in a form such that logic-level synthesis tools can optimize the combinational logic and layout synthesis tools can design the chip geometry. The generated output is generally in a low-level HDL, such as structural VHDL.

2.5.7 A Demonstrative Example

A small synthesis example is presented to demonstrate the various phases of behavioral synthesis in detail. Suppose that we want to synthesize hardware to perform the operation $y = (a+b) * (c-d)$ [210]. The steps can be seen in the self-explanatory Fig. 2.6 and 2.7.

2.6 Behavioral HDL to CDFG Translation or Compilation

The translation or compilation phase of high-level synthesis is the process by which a behavioral HDL (VHDL or Verilog) is translated to an intermediate representation [266, 33, 206, 274, 210]. This is the first step of high-level synthesis; however, this step often does not attract much attention because tools perform the task without the intervention of the user. This step is crucial and results in an intermediate representation that is essential for subsequent synthesis. There are many such intermediate representations including DFG, control flow graph (CFG) and control data flow graph (CDFG) [166, 250, 299, 43, 307, 30]. Each of these has its own advantages; for example, although the DFG is simple, the CDFG provides better options for design space exploration. The need for a CDFG and a DFG in high-level synthesis and hardware/software co-synthesis has been emphasized in [88, 353, 25, 91]. A representation called sequencing DFG, which is a directed acyclic graph, is a simpler form of CDFG [209, 210]. For circuits with a balanced mix of data and control flow, a hierarchical conditional dependency graph (HCDG) is often used [167].

An example of a CDFG is presented in Fig. 2.8 [250, 30]. The formal definitions of DFG and CFG are as follows [206, 209, 210]:

A *DFG* is a directed graph $G = (V, E)$, where $V = v_1, v_2, ..., v_n$ is a finite set whose elements are "nodes" and $E = V \times V$ is an asymmetric "dataflow relation", whose elements are called "data edges".

A *CFG* is a directed graph $G = (V, E)$, where $V = v_1, v_2, ..., v_n$ is a finite set whose elements are "nodes" and $E = V \times V$ is a "control flow relation" whose elements are directed "sequence edges."

The steps for converting the VHDL description to a CDFG are presented in Fig. 2.9 [209, 250, 299, 30, 249]. The steps are similar to the steps of a high-language compiler used in standard computers. The source program is translated into a stream of tokens in the lexical analysis phase. Each token is a sequence of

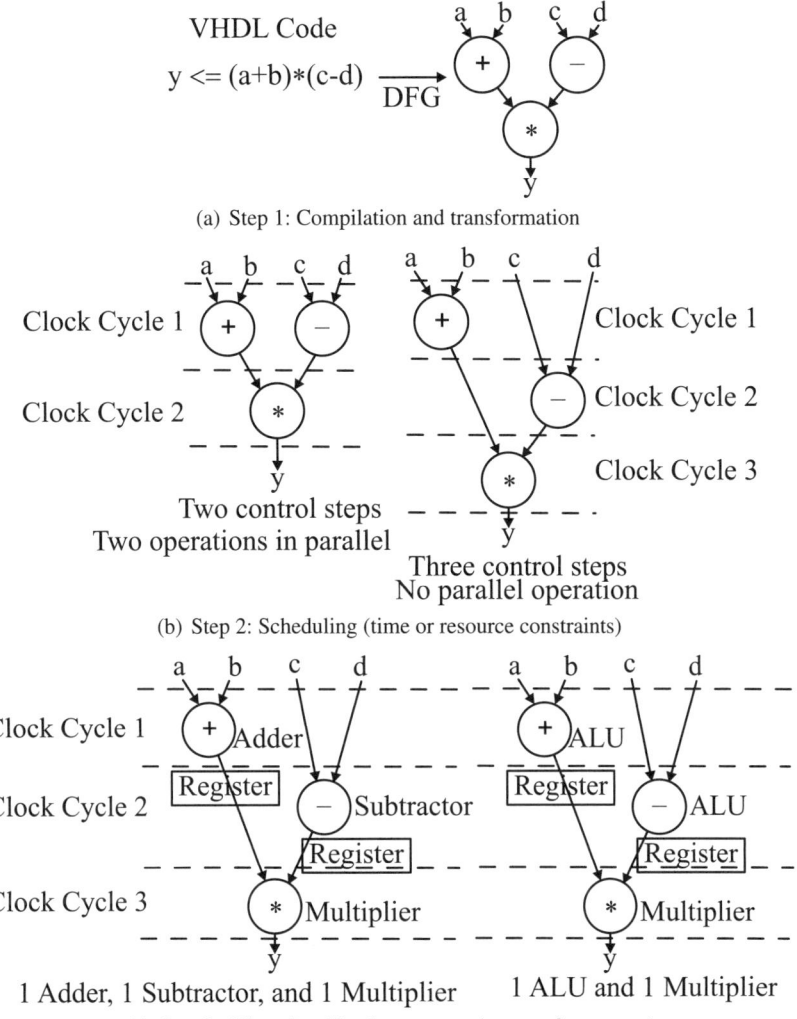

(a) Step 1: Compilation and transformation

(b) Step 2: Scheduling (time or resource constraints)

(c) Step 3: Allocation (fixed amount and types of resources)

Fig. 2.6 A synthesis example: step 1 to step 3

characters with collective meaning, such as an identifier, a keyword, an operator or a punctuation character. The syntax analysis phase verifies the syntax by imposing a hierarchical structure on the stream of tokens. These two steps use Lex and YACC tools for IEEE standard VHDL syntax. YACC code parses tokens and generates a parse tree using explicit codes [209, 250]. The parse tree is then compressed to obtain a syntax tree in which the operators appear as the interior nodes and the operands of an operator are the children of the node for that operator. Finally, the syntax tree is transformed to obtain the overall CDFG and DFG representation of the input VHDL. The example represented in Fig. 2.10 demonstrates the VHDL to

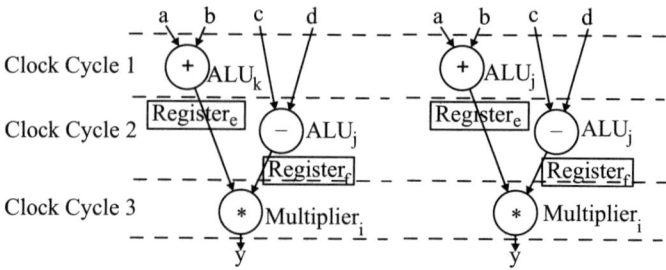

(a) Step 4: Binding (which resource will be used by which operation)

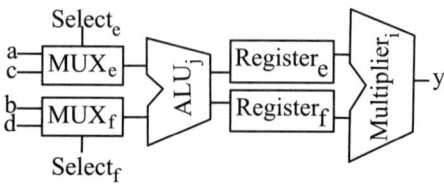

(b) Step 5: Connection allocation (communication between resources: bus, buffer, or MUX)

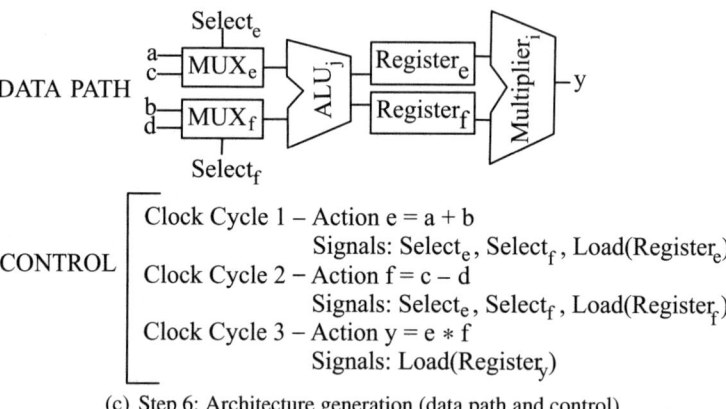

(c) Step 6: Architecture generation (data path and control)

Fig. 2.7 Synthesis example: step 4 to step 6

CDFG transformation. It may be noted that translators and compilers only support a subset and not all features of VHDL.

2.7 Scheduling Algorithms

The scheduling problem is a non-deterministic polynomial (NP) problem [83]. Behavioral scheduling algorithms may be of various types based on the constraints and optimization schemes, as shown in Fig. 2.11. Various scheduling algorithms are

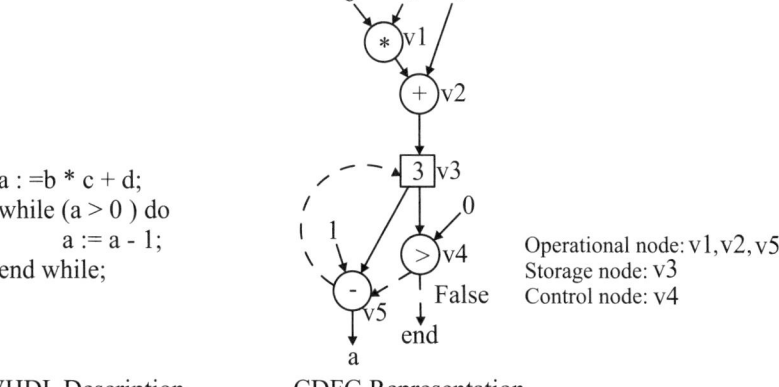

a : =b * c + d;
while (a > 0) do
    a := a - 1;
end while;

VHDL Description   CDFG Representation

Fig. 2.8 Control and DFG example

described in [378, 294, 272, 274, 134, 264, 162, 360, 121, 267, 138, 58, 124, 273, 200, 210, 22, 251, 282, 103]. The commonly used scheduling techniques are integer linear programming, as soon as possible (ASAP), as late as possible (ALAP), list-based scheduling, force-directed scheduling and freedom-based scheduling. The algorithms use power, leakage, area, etc. as objective functions and time, resources and area as constraints [366, 226, 221, 144, 320, 141]. In this section, selected algorithms are discussed generically.

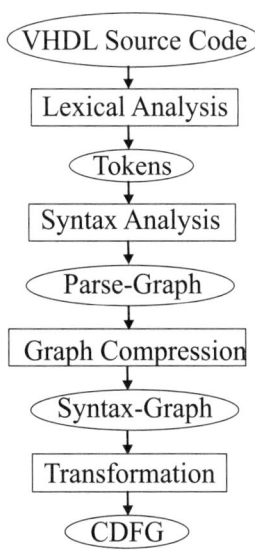

Fig. 2.9 Steps for converting VHDL to CDFG

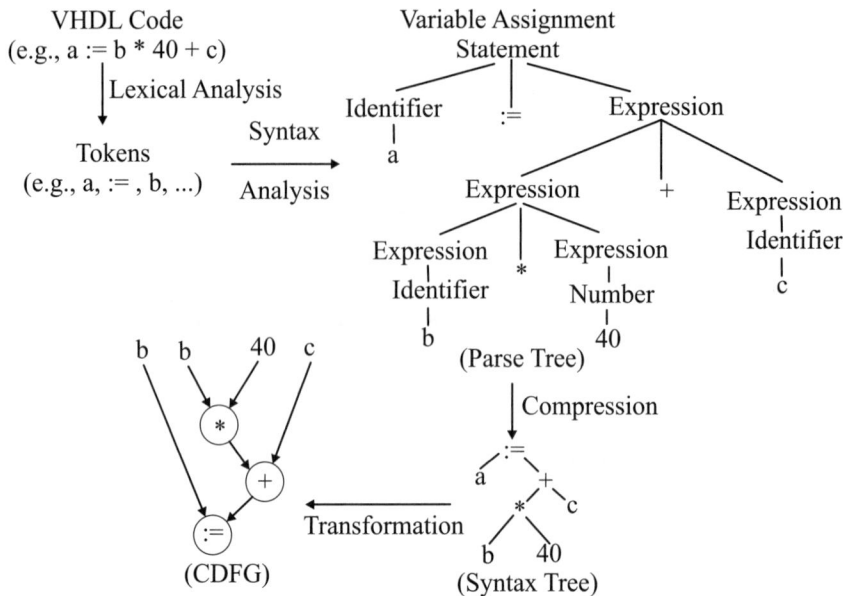

Fig. 2.10 VHDL to CDFG conversion example

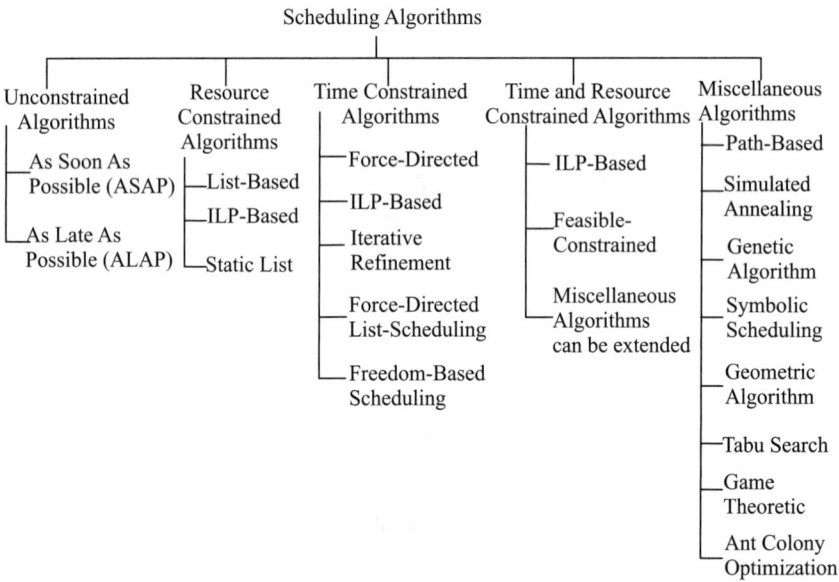

Fig. 2.11 Different types of scheduling algorithms

2.7.1 ASAP and ALAP Scheduling and Mobility

The ASAP scheduling process arranges the operations topologically according to their data or control flow. ASAP scheduling places the operations in the sorted order by stamping them in the earliest possible control step. The ALAP scheduling process is similar to ASAP; however, it is in the reverse order as that of ASAP. The ALAP scheduling places the operations in the latest possible control step. These two algorithms are simple, yet needed in most of the advanced scheduling algorithms. The ASAP algorithm is unconstrained, i.e., it considers neither resource nor time constraints for scheduling. The ALAP scheduling considers the number of steps resulting from the ASAP schedule as a latency constraint. The ASAP and ALAP time stamps provide lower and upper bounds on possible control steps for scheduling an operation. The difference between the ALAP and ASAP time stamp is called the mobility of an operation. The time complexity of these algorithms is $O(|V|+|E|)$ [83]. Algorithms 1 and 2, respectively, present the two algorithms. Figs. 2.12 and 2.13 demonstrate the algorithms with the help of the hardware abstraction layer (HAL) differential equation solver high-level synthesis benchmark [102, 209, 103].

2.7.2 Integer Linear Programming (ILP) Scheduling

ILP-based algorithms are the exact solutions of the scheduling problem [209, 102, 317, 230, 75, 368, 264, 144, 76, 186, 175, 210, 228]. The ILP-based scheduling algorithm is presented in Algorithm 3. The algorithm uses the ASAP and ALAP

Algorithm 1 As Soon As Possible (ASAP) Scheduling

1: For a sequencing DFG $G(V,E)$ time stamp source vertex v_0 with $c_0 = 0$.
2: **while** (Sink vertex v_N is not considered for time stamping) **do**
3:      Select a vertex v_i whose all predecessors are scheduled.
4:      Schedule v_i by setting its time stamp as c_i maximum start time of all its predecessors plus its delay.
5: **end while**
6: Return schedule of the DFG $\{C_S = c_0, c_1, ..., c_N\}$.

Algorithm 2 As Late As Possible (ALAP) Scheduling

1: For a sequencing DFG $G(V,E)$ time stamp sink vertex v_N with $c_N = $ ASAP Latency Bound $+1$.
2: **while** (Source vertex v_0 is not considered for time stamping) **do**
3:      Select a vertex v_i whose all successors are scheduled.
4:      Schedule v_i by setting its time stamp as c_i minimum start time of all its successor minus its delay.
5: **end while**
6: Return schedule of the DFG $\{C_L = c_0, c_1, ..., c_N\}$.

$$\frac{d^2y}{dx^2} + 3\frac{dy}{dx} - x + 3y = 0$$

while (x < a) do
    x1 := x + dx;
    u1 := u - (3 * x * u * dx) - (3 * y * dx);
    y1 := y + (u * dx);
    c = x < a;
    x := x1;
    u := u1;
    y := y1;
end while

(a) HAL solver

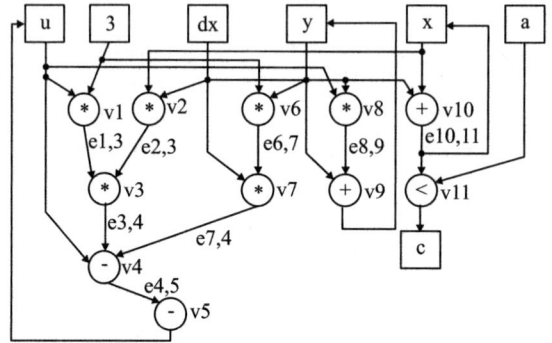

(b) HAL DFG showing edges and variables

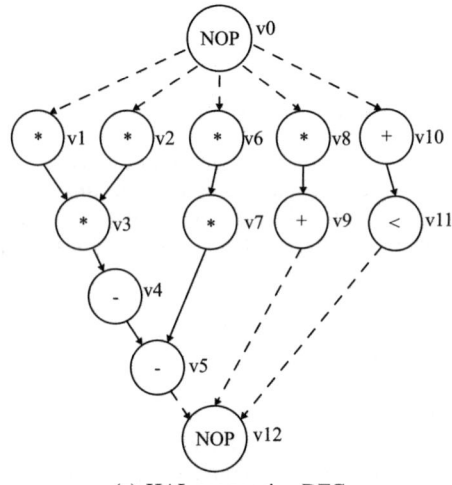

(c) HAL sequencing DFG

Fig. 2.12 DFG and sequencing DFG of HAL differential equation solver

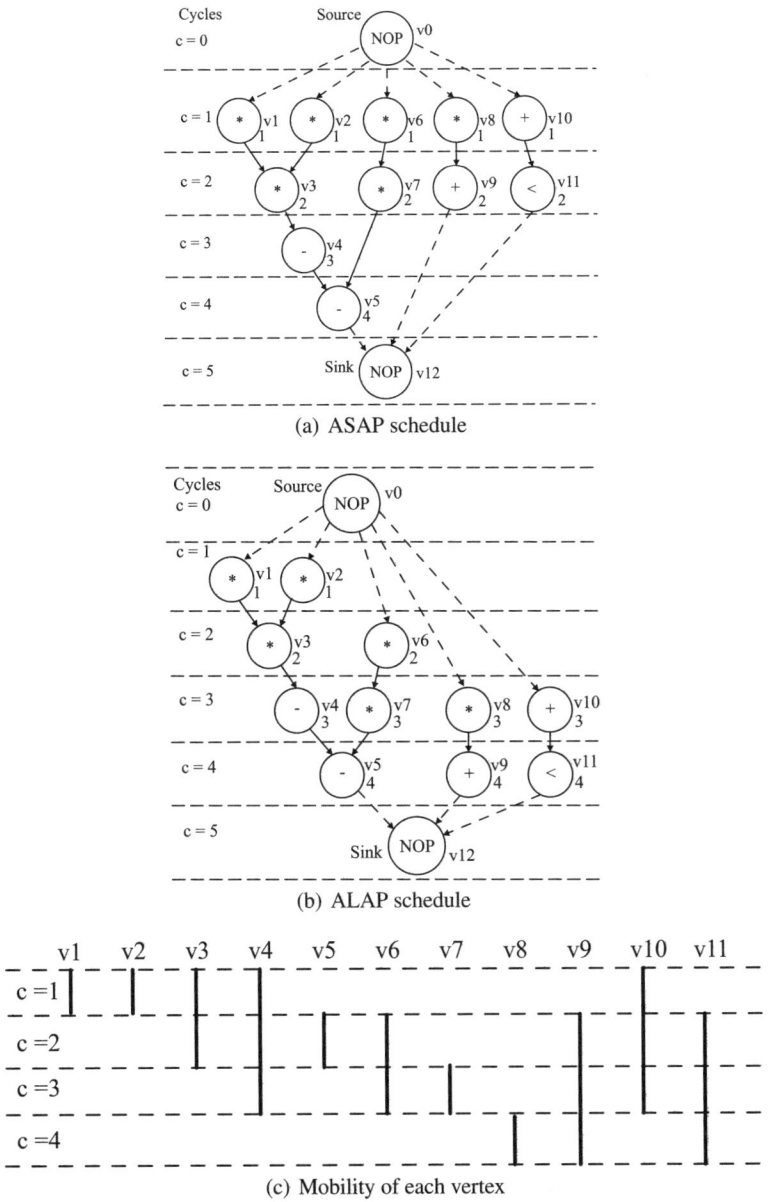

Fig. 2.13 ASAP and ALAP schedules with mobility graph

schedules to impose bounds on the ILP solution space to reduce the search space. ILP formulations can be modeled using the AMPL language and solved using various solvers such as LP-Solve [101, 260, 14, 302, 205].

Algorithm 3 Resource-Constrained ILP-Based Scheduling

1: Find ASAP and ALAP schedule of the sequencing DFG.
2: Determine the mobility graph of each vertex.
3: Construct the ILP formulations for the DFG.
4: Solve the ILP formulations.
5: Find the scheduled DFG.
6: Return schedule of the DFG $\{C_{\text{ILP}} = c_0, c_1, ..., c_N\}$.

ILP minimizes cost functions in the form of power, area, delay under resource, time or power constraints. The formulation needs the help of a binary decision variable that is either 0 or 1. The following notations will be used to formulate the ILP:

1. N_v is the total number of vertex operations in the sequencing DFG, excluding the source and sink nodes (NO–OPs).
2. v_i is any vertex in the DFG performing certain operations and $1 \leq i \leq N_v$.
3. FU_k is the functional unit of type k.
4. M_k is the maximum number of functional units of type FU_k.
5. $Cost_k$ is cost of functional unit of type FU_k may be power, area or delay.
6. $C_S[i]$ is the ASAP time stamp for the operation v_i.
7. $C_L[i]$ is the ALAP time stamp for the operation v_i.
8. $x_{i,c}$ is the binary decision variable which is 1 if vertex v_i starts in control step c, else 0.

1. *Objective function*: The objective is to minimize the total cost corresponding to the DFG over all control steps simultaneously. Using decision variables, one can rewrite the objective function as follows:

$$\text{Minimize} : \sum_c \text{Cost}_{k,v} \times M_k. \tag{2.1}$$

2. *Uniqueness constraints*: These constraints ensure that each operation v_i is scheduled to one unique control step within the mobility range $(C_S[i], C_L[i])$. They are represented as $\forall i, 1 \leq i \leq N_v$,

$$\sum_c x_{i,c} = 1. \tag{2.2}$$

3. *Precedence constraints*: These constraints ascertain that for a vertex v_i, all its predecessors are scheduled at an earlier control step and that its successors are scheduled at a later control step. These are modeled as $\forall i, j, v_i \in \text{Pred}_{o_j}$,

$$\sum_{d=C_S[i]}^{C_L[i]} d \times x_{i,d} - \sum_{e=C_S[j]}^{C_L[j]} e \times x_{j,e} \leq -1. \tag{2.3}$$

4. *Resource constraints*: These constraints establish that no control step contains more than M_k operations of type k. These can be enforced as $\forall c, 1 \leq c \leq N_c$,

$$\sum_{i \in F_k} x_{i,c} \leq M_k, \tag{2.4}$$

where N_c is the number of clock cycles.

As an example, let us assume four types of resources: multiplier, adder, subtractor and comparator. Let us also assume that the cost of a multiplier is 2 units and that the cost of the other units is 1 unit. Let us assume that two instances of a multiplier and one instance of the other resources are available. Let us consider the ASAP schedule, ALAP schedule and the mobility graph shown in Fig. 2.13(a), (b), (c), respectively, for the HAL DFG. The following solution is obtained: $x_{1,1} = x_{2,1} = x_{3,2} = x_{4,3} = x_{5,2} = x_{6,3} = x_{7,3} = x_{8,4} = x_{9,4} = x_{10,2} = x_{11,4} = 1$ and the rest $x_{i,j} = 0$. The corresponding scheduled DFG is presented in Fig. 2.14.

The ILP-based algorithm provides an exact solution, but it is slow and has an exponential worst-time time complexity. It is difficult to use it for large and practical circuits as the formulation grows exponentially with the number of vertices. When the number of variables in an ILP formulation reaches around 200, a result is not obtained even after several days of computation [186, 175].

2.7.3 List-Based Scheduling (LBS)

This algorithm is primarily resource-constrained but can be modified to accommodate time constraints [102, 209, 103, 330]. The problem of minimizing latency under resource constraints or minimizing resources under latency constraints is intractable [83]. List-based scheduling is essentially a heuristic approach to solve the scheduling problem.

The list-based algorithm takes a sequencing DFG and resource constraints as inputs and generates a scheduled sequencing DFG as output. In list scheduling, the

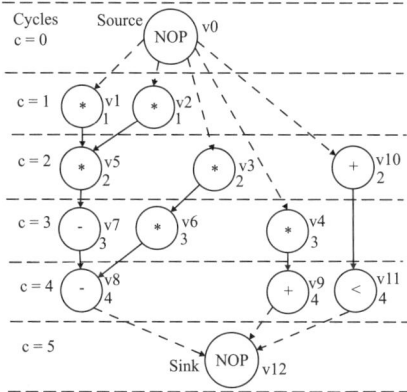

Fig. 2.14 Schedule resulting using the ILP-based algorithm

Algorithm 4 Resource-Constrained List-Based Scheduling to Minimize Latency

---

1: Initial clock cycle $c = 1$.
2: **while** (For a sequencing DFG $G(V, E)$ sink vertex v_n is not considered for time stamping) **do**
3:      **for all** (Resource of type k) **do**
4:          Determine list of vertices of type k whose predecessors are already scheduled to
            finish execution in c, $L_{F_{c,k}}$.
5:          Determine list of vertices of type k that started at an earlier cycle but whose exe-
            cution is not finished in c, $L_{U_{c,k}}$.
6:          Select the vertices from list $L_{F_{c,k}}$ such that the resource needed by these vertices
            and $L_{U_{c,k}}$ do not violate resource constraints.
7:          Time stamp the above selected vertices at clock cycle c, i.e., $c_i = c$.
8:      **end for**
9:      Increment the clock cycle c.
10: **end while**
11: Return schedule of the DFG $\{C_{\text{LBS}} = c_0, c_1, ..., c_N\}$.

---

operations available for scheduling are kept in a list for each control step. This list
is ordered by some priority function: mobility of the vertex or the length of path
from the operation to the sink while ranking the vertices in decreasing order. An
operation on the list is scheduled one by one if the resource needed by the operation is free; otherwise, it is deferred to the next clock cycle. Algorithm 4 shows the
pseudocode of resource-constrained list-based scheduling that minimizes latency
[102, 209, 103]. Figure 2.15 shows the scheduled DFG result for a resource constraint of three multipliers and one ALU.

The time complexity of this algorithm is $O(|E| + |V| \log(|V|) + |V| T_0)$, where
$O(|V| T_0)$ is the worst-case time for creating the list and selecting its first element.
This algorithm has better time complexity compared to force-directed scheduling

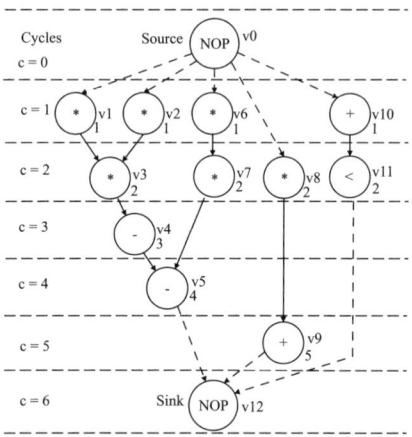

Fig. 2.15 Schedule resulting by using a list-based scheduling algorithm assuming unit delay for
each operation with a resource constraint of three multipliers and one ALU

and guarantees results [103, 209]. The time complexity of force-directed scheduling is $\left(O\left(|V|\right)^3 C_{max}\right)$, where C_{max} is the time constraint expressed in terms of the number of clock cycles [272, 209]. Thus, list scheduling is a better choice for large and practical designs.

2.7.4 Force-Directed Scheduling (FDS)

Force-directed scheduling is a heuristic algorithm that can consider both resource and time constraints [114, 272, 271, 274, 355, 102, 209, 103, 151]. The basic idea of this algorithm is to balance the concurrency of operations without increasing the total execution time to maximize the utilization of resources such that the number of required resources is minimal. The algorithm is iterative; that is, one operation is time-stamped in each iteration, and selection of the time stamp is made to achieve balance of distribution of operations. The term "force" has originated from a mechanical analogy; i.e., force is the product of a spring constant and displacement. This is motivated by Hooke's law, Force = Spring constant × Displacement. The algorithm is presented with the help of pseudocode in Algorithm 5.

The algorithm starts with the computation of the mobility of each operation based on its ASAP and ALAP time stamps. An operation is time-stamped for a specific clock cycle for which the total force is minimum. The total force is the sum of two forces, the self-force and mutual force, due to predecessors and successors. For vertex v_i to be considered for control step or clock cycle c has the total force:

$$\text{Total Force}(i,c) = \text{Self-Force}(i,c) + \text{Mutual Force}(i,c). \tag{2.5}$$

The self-force of an operation or vertex is the set of forces the operation is subjected to while tentatively scheduled in a control step in its mobility range. For a vertex v_i needing a resource type k in clock cycle c, the self-force can be expressed as follows:

Algorithm 5 Force-Directed Scheduling (FDS)

1: **while** (All vertices v_i in V are not considered for time stamping) **do**
2:     Find ASAP schedule C_S.
3:     Find ALAP schedule C_L.
4:     Determine mobility of each operation from C_S and C_L.
5:     Update distribution graphs (DG) for each resource type.
6:     Calculate self-forces for every feasible clock cycle.
7:     Add predecessor and successor forces to self-forces to calculate total force.
8:     Schedule the operation with lowest total force by setting the time stamp c_i as the selected clock cycle c and adjust the mobility.
9: **end while**
10: Return schedule of the DFG $\{C_{FDS} = c_0, c_1, ..., c_N\}$.

$$\text{Self-Force}\,(i,c) = \sum_{k} \text{Probability}\,(k,c)$$

$$- \frac{1}{m_i + 1} \sum_{c^* = C_{\mathrm{S}}[i]}^{C_{\mathrm{L}}[i]} \sum_{k} \text{Probability}\,(k,c^*). \qquad (2.6)$$

Here Probability (k,c) is the sum of probabilities of the operations that can be implemented by a specific resource of type k in the type distribution. Type distribution is indicative of concurrency of similar types of operations, the plot of which is called distribution graph (DG). The mobility of each operation v_i is $m_i = C_{\mathrm{L}}[i] - C_{\mathrm{S}}[i]$ and mobility range is $[C_{\mathrm{S}}[i], C_{\mathrm{L}}[i]]$.

The mutual force due to predecessor/successor forces is due to the operation dependencies. This force arises when a tentative assignment of an operation to a control step or clock cycle restricts the mobility of its predecessors and successors. For a vertex v_i needing a resource of type k in clock cycle c, the mutual force can be expressed as follows:

$$\text{Mutual Force}\,(i,c) = \frac{1}{m_i^+ + 1} \sum_{c^* = C_{\mathrm{S}}^+[i]}^{C_{\mathrm{L}}^+[i]} \sum_{k} \text{Probability}\,(k,c^*)$$

$$- \frac{1}{m_i + 1} \sum_{c^* = C_{\mathrm{S}}[i]}^{C_{\mathrm{L}}[i]} \sum_{k} \text{Probability}\,(k,c^*), \qquad (2.7)$$

where $[C_{\mathrm{S}}[i], C_{\mathrm{L}}[i]]$ is the initial mobility range and $[C_{\mathrm{S}}^+[i], C_{\mathrm{L}}^+[i]]$ is the restricted mobility range.

The time complexity of force-directed scheduling is $\left(\mathrm{O}\,(|V|)^3 C_{\max}\right)$, where $C_{\max}$ is the time constraint expressed in terms of number of clock cycles [272, 209]. The cubic time complexity makes it more time consuming for large designs and hence has reduced its role in solving practical designs which are often large. Thus, list-based scheduling may be a better choice.

2.7.5 Game Theory Scheduling (GTS)

Recently, game theory has been explored as a solution approach for behavioral scheduling [298, 245, 244]. Game theory has been used by economists to understand the action of economic agents by analyzing the interaction of decision makers with conflicting objectives [323, 298, 253]. In game theory, a game can be either cooperative or non-cooperative depending on whether an agreement has been made before [323, 298, 253]. The notable and unique features of game theory that can facilitate associated problems of high-level synthesis are as follows [298]:

1. *Rationality*: Each player always selfishly tries to maximize its gain, which can be suitable to meet conflicting objectives of synthesis.
2. *Coalition*: An important feature that can assist resource allocation during high-level synthesis is coalition formation when a subset of players having the same agenda in terms of strategies can pool their resources.
3. *Competition*: Multiple decision makers control a specified set of variables and seek to optimize their conflicting objectives, a natural fit for behavioral synthesis.
4. *Equilibrium*: When all the objectives of all players are met, a solution is in an equilibrium state, thus ensuring the convergence of solutions.

The scheduling problem is modeled as a first-bid sealed auction and solved using a game theory approach. The auctioning of items by a seller through bidding can be extended to the auctioning of operations in the sequencing unscheduled DFG to the available resources with an operation being bought by only one resource and the sale entity being the cost of execution, with cost being area, power or delay. In the presence of multiple operations and resources, an equilibrium point needs to be achieved to converge to a solution. An auction consists of a set of available resources M_k (of type k) and a set of interested buyers V (operations or vertices). A buyer ($v_i \in V$) may also have some preference for some resources over the others, and this can be specified in terms of a cost function Cost_k, in which v_i needs an instance of resource of type k. If the cost of two instances of the same type of resource is needed by a operation in vertex v_i, then obviously the instance is chosen for which the cost is lower. In other words, if FU_k^a and FU_k^b are two instances of FU_k and the corresponding costs are Cost_k^a and Cost_k^b, respectively, then FU_k^a is chosen if $\text{Cost}_k^a < \text{Cost}_k^b$; otherwise, FU_k^b is chosen. A scheduling algorithm using game theory is presented in Algorithm 6 [298, 153].

For N_P players, a game with S strategies and a number of clock cycles N_c, the time complexity of the game theoretic-based scheduling algorithm is $\text{O}\left(N_c S N_P{}^S\right)$ [207, 298]. The game theory-based formulations can be solved by the Gambit software [207].

Algorithm 6 Game Theory Scheduling (GTS) Algorithm

1: Initialize control step $c = 0$.
2: Assume the initial scheduled as the ASAP schedule C_S.
3: **while** (Sink vertex v_n is not scheduled) **do**
4:         Perform breadth first search (BFS) of current DFG.
5:         Create a list of all vertices List_v from current DFG having same time stamp.
6:         **for all** (Each set of vertices needing type k resource in cycle c, i.e., List_v^k) **do**
7:                 Calculate cost of the schedule using payoff matrix for List_v^k and M_k at step c.
8:                 Find Nash equilibrium solution at control step c.
9:                 Assign vertices in step c that form the Nash equilibrium.
10:                 $c = c + 1$.
11:         **end for**
12: **end while**
13: Return schedule of the DFG $\{C_{\text{GTS}} = c_0, c_1, ..., c_N\}$.

2.7.6 Tabu Search Scheduling (TSS)

Tabu search is a meta-heuristic algorithm that takes a more aggressive approach than other search algorithms. The algorithm skips inferior solutions other than the cases when it needs to exit out of the local optimum. It provides useful solutions to the scheduling problem in a reasonable amount of time [299, 250, 24, 363, 108, 128, 26, 342, 238]. The pseudocode of the TSS algorithms is presented in Algorithm 7 [299, 250, 363].

In Algorithm 7, the ASAP and ALAP algorithms are used to obtain the lower and upper bounds on possible control steps in which a vertex can be scheduled. Then, an initial schedule can be either the ASAP or ALAP schedule. Based on penalty weights, the algorithm evaluates the solutions to reach a final solution. The penalty weights are the factors that could affect the number of resources, such as operation type, number of non-mutually exclusive vertices, life time of operations and number of buses or inputs needed. The algorithm undergoes iterations in which a tentative solution is selected and compared against a current solution. If a solution is not visited previously and has lower weight factor than the current one, then a new solution is accepted. A neighborhood of a current solution is obtained by moving vertices from their present clock cycle to another clock cycle within the mobility range. The vertices considered for movement are chosen from the clock cycles that contribute the most to the penalty weights in the present solution. Such vertices are moved to clock cycles where there are fewer number of vertices of the same operation type and the same equivalence class.

The TSS algorithm is able to perform global optimization of the number of resources used, even with the large solution space explored. Thus, this scheduling scheme can be attractive for large designs.

Algorithm 7 Tabu Search Scheduling (TSS) Algorithm

1: Initialize iteration counter Counter $= 0$.
2: Determine ASAP and ALAP schedules.
3: Generate initial feasible solution S_i.
4: **while** (Counter $<$ Max_Counter) **do**
5:      Counter $=$ Counter $+ 1$.
6:      Generate neighborhood S_i^*.
7:      **if** (S_i is not visited in the previous iterations) **then**
8:          **if** (S_i^* is a better solution than S_i) **then**
9:              **return** $S_i \leftarrow S_i^*$.
10:         **else**
11:             **return** Discard the solution S_i^*.
12:         **end if**
13:     **end if**
14: **end while**
15: Return schedule of the DFG $\{C_{\text{TSS}} = c_0, c_1, ..., c_N\}$.

2.7.7 *Simulated Annealing Scheduling (SAS)*

Simulated annealing optimization, proposed in [161, 59, 248], is an extension of Monte Carlo [208]. Simulated annealing algorithms borrow ideas from Materials Science. Annealing is the process of heating and cooling a material slowly until it crystallizes. The atoms of this material have higher energies at very high temperatures. This gives the atoms a great deal of freedom in their ability to reconstruct themselves. As the temperature decreases, the energy of the atoms decreases. The simulated annealing optimization approach needs three attributes: an appropriate cooling schedule, a suitable objective function and a well-defined neighborhood structure.

Analogous to the annealing process, the mobility of nodes in a DFG is dependent on the total available resources. Here, the nodes of a DFG are analogous to the atoms, and temperature is analogous to the total number of available resources. The mobility of the nodes/vertices is dependent on the total number of available low-cost resources. The SAS approach has been proposed to address scheduling problems to explore the trade-offs among power, performance and area [161, 234, 235, 110, 335, 90, 281]. The pseudocode of the SAS algorithm is presented in Algorithm 8 [234, 235, 110].

The simulated annealing algorithm flow is similar to that of the TSS algorithm presented before. However, because of different convergence criteria, the time complexity and convergence may be different.

Algorithm 8 Simulated Annealing Scheduling (SAS) Algorithm

1: Determine ASAP and ALSP schedules.
2: Determine the mobility graph.
3: Initialize iteration counter Counter $= 0$.
4: Initial feasible solution S as the ASAP schedule.
5: Determine initial Cost_i for the initial solution S_i.
6: Assume initial temperature T as T_i.
7: **while** (Cost is varying) **do**
8:     Counter $=$ Maximum number of iterations.
9:     **while** (Counter > 0) **do**
10:         Generate random transition from S to S^* for a neighborhood solution.
11:         Calculate change in cost as: $\Delta_{\text{Cost}} = \text{Cost}_S - \text{Cost}_S^*$.
12:         **if** ($\Delta_{\text{Cost}} < 0$ random$(0,1) < e^{\frac{\Delta_{\text{Cost}}}{T}}$) **then**
13:             Update the solution with new solution, $S \leftarrow S^*$.
14:         **end if**
15:         Counter $=$ Counter $- 1$.
16:     **end while**
17:     Decrease temperature as: $T = T * \text{Cooling\_Rate}$.
18: **end while**
19: Return schedule of the DFG $\{C_{\text{SAS}} = c_0, c_1, ..., c_N\}$.

2.7.8 Genetic Algorithm Scheduling (GAS)

Genetic algorithms are probabilistic search algorithms based on the principle of "survival of the fittest." Genetic algorithms create a collection of solutions that evolve according to a quality measure; the evolution works on a search space represented by a chromosome [97, 124, 105, 125, 111, 262, 98, 97, 123, 84, 169, 197]. A collection of chromosomes is called a population, and the algorithm improves the average fitness of a population by constructing a new population through selection and recombination. Operators performing recombination accept parent chromosomes and generate children chromosomes. Using a stochastic selection process, parent selection is performed from a population such that the better parent is preferred. Pseudocode for a generic genetic algorithm is shown in Algorithm 9.

Genetic algorithms are generic and can be applied to many types of problems. They have the problem, however, that they may not converge to the optimal solution. They may become trapped in local optimum solutions, thus leading to inefficiency.

2.7.9 Ant Colony Scheduling (ACS)

The ant colony optimization (ACO) heuristic has been used to solve a wide variety of intractable problems such as the traveling salesman problem, graph coloring, sequential ordering and quadratic assignment [94, 93]. However, it has recently been used to solve design automation problems that are NP-hard in nature [361, 163, 28, 94, 93, 362]. The advantages of ACO over traditional schemes are

---

Algorithm 9 Genetic Algorithm Scheduling (GAS) Algorithm

1: Initialize iteration counter $i = 0$.
2: Create initial population S_i.
3: **while** (Stopping criterion is not met) **do**
4:      $i = i + 1$.
5:      **repeat**
6:           Select parent's chromosomes from their population based on fitness criteria.
7:           Cross over with a high probability: the vertex and resource priority sub-lists for a portion of selected chromosomes, where chromosome is a list containing two sub-lists for vertices and resources in priority order.
8:           Mutate with low probability: the vertex and resource priority sub-lists for a portion of selected chromosomes containing vertices and resources sub-lists in priority orders.
9:           For all children chromosomes decode them to schedule of vertices using list-based heuristic and evaluate fitness.
10:          Create a population S_i by replacing least fit members of population with children chromosomes.
11:     **until** (New population size is smaller than the previous population size, i.e., Size(S_i) < Size(S_{i-1}))
12: **end while**
13: Return best solution S_i as $\{C_{GAS} = c_0, c_1, ..., c_N\}$.

---

near optimal solutions, reasonable run time and multiple solutions. The pseudocode of a scheduling algorithm is presented in Algorithm 10 [361, 362, 163].

In Algorithm 10, an evaporation ratio is needed to update the pheromone trail based on the quality of instruction lists. A vertex v_i is associated with the n pheromone trails τ_{ij} with $j = 0, 1, ...n$, which is an indicator of global favorableness of scheduling the v_i vertex in the jth position in the priority list. A vertex v_i is chosen for the jth position of the list probabilistically. An evolving bound is computed for each iteration to set a limit on the pheromone trails. The algorithm has been demonstrated capable of solving large designs.

2.7.10 Automata-Based Symbolic Scheduling

An exact approach for scheduling looping DFGs with implicit support of functional pipelining and loop winding is given by Haynal and Brewer [121]. This algorithm uses a set of local, non-deterministic finite automata (NFA) to represent the sequencing constraints. The algorithm derives all optimal solutions and can handle very large designs.

2.7.11 Chaining, Multicycling and Pipelining Data Paths

Because of data dependencies, maximum utilization of resources is not possible. Data dependencies place bottlenecks on concurrent execution of operations even when abundant resources or functional units are available. Thus, techniques like chaining, multicycling or pipelining and their combinations are used to improve the

---

Algorithm 10 Ant Colony Scheduling (ACS) Algorithm

1: Initialize parameters such as evaporation ratio, pheromone trails, conditional probability, and evolving bounds.
2: Construct n_{ant} number of ants.
3: Assume an initial solution S.
4: **while** (Stopping criterion is not met) **do**
5:      **for** $i = 0$ to n_{ant} **do**
6:          ith ant constructs a list of vertices $List_i$ using pheromone trails and mobility.
7:          Generate a new schedule S^* using List-based or Force-directed approaches for graph $G(V, E)$, resource constraints, and $List_i$.
8:          **if** (S^* is better than S) **then**
9:              Update solution with new solution, i.e., $S \leftarrow S^*$.
10:         **end if**
11:     **end for**
12:     Update evolving bounds of the pheromone trails.
13:     Update mobility and pheromone trails.
14: **end while**
15: Return best solution S as $\{C_{ACS} = c_0, c_1, ..., c_N\}$.

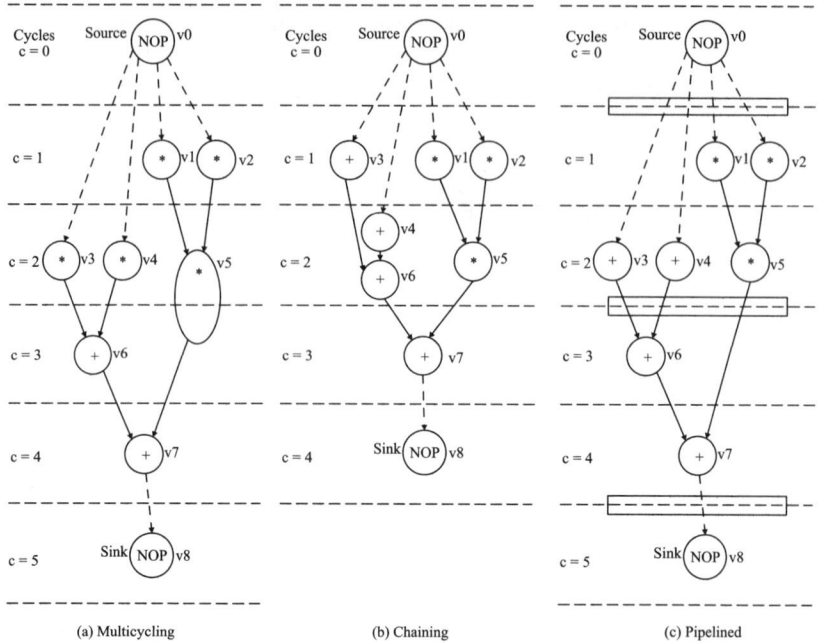

Fig. 2.16 Sequencing DFGs for chaining, multicycling or pipelining data paths

optimization of resource utilization. A data path can be designed for single cycling and multicycling executions. In a single-cycle data path operation, each operation is executed in a single cycle with clock width determined by the slowest instruction. In multicycling data path, one operation can take more than one clock cycle for its execution. In chaining, more than one operation are executed together in a clock cycle. In a pipelined scenario, a data path is segmented into several stages, where each stage is made available for the resources to execute the relevant operations of that stage. Figure 2.16 represents the three different data path scenarios. The scheduling and binding problems are more complex for pipelined data paths. The above scheduling algorithms can handle these data path specifications.

2.8 Binding or Allocations Algorithms

Scheduling, binding and allocation tasks are related to each other and need to be solved simultaneously for optimal design [210, 209, 103, 102, 246]. However, in practice, they are typically solved sequentially: first, scheduling is solved and then binding (more conventional) or first binding and then scheduling. From a DFG perspective, scheduling is a temporal partitioning process, and binding is a spatial partitioning process. Several subproblems are associated with a binding problem, such as operation to resource binding, binding values to be stored to instances of storage elements and binding data transfers to bus instances (i.e., functional unit binding, storage binding and interconnection binding). Similarly, the allocation problem is divided into three tasks: functional unit, storage and interconnection allocation [103, 349, 90]. The optimization goals of the binding problem are to minimize total cost of functional units, register, bus driver and multiplexor, total interconnection length, critical path delay and power dissipation. The sub-tasks of binding can be solved by following graph theoretic approaches, such as clique partitioning, circular-arc graph coloring or left edge algorithm. In this section, the input graph is a scheduled sequencing DFG $G_s(V,E)$ in which source and sink vertices are not considered because they perform no operations (NOP) and do not need resources.

2.8.1 Clique Partitioning Approach

Two operations are compatible if they need resources of the same type and are not scheduled in the same clock cycle and thus can use the same resources. To analyze compatibility of vertices/operations, a data structure called "compatibility graph" is useful. This graph is defined as follows [209, 103, 334, 273, 293, 185]. A *resource compatibility graph* $G_s^{com}(V,E)$ is an undirected graph whose vertex set $V = \{v_i, i = 1, 2, ..., N_v\}$ is in one-to-one correspondence with the operations and whose edge set $E = \{(v_i, v_j), i = 1, 2, ..., N_v$ and $j = 1, 2, ..., N_v\}$ denotes the compatible vertex pairs. An example of such a graph is shown in Fig. 2.17(b) for a scheduled DFG of Fig. 2.17(a), for two types of resources, multipliers and ALUs [209]. As it is evident from the figure, the graph has two disjoint components, the same as the type of resources. A clique is a subset of vertices mutually connected by edges. For example, in Fig. 2.17(b), the cliques are $\{v_1, v_3, v_7\}$, $\{v_2, v_6, v_8\}$, $\{v_4, v_5, v_{10}, v_{11}\}$ and $\{v_9\}$. The optimal binding is then to partition the compatibility graph into the minimum number of cliques; this number is called the *clique cover number*. For the above example, the clique cover number is 4; thus, there is a need for four resources, two multipliers and two ALUs. The clique partitioning problem is NP-complete and, hence the binding problem is also NP-complete.

2.8.2 Graph Coloring Approach

Two operations have a conflict if they are not compatible. To analyze the conflicts of vertices/operations, a data structure called "conflict graph" is useful, which is defined as follows [209, 103, 334, 273, 293, 185]. A *resource conflict graph* $G_s^{con}(V,E)$ is an undirected graph whose vertex set $V = \{v_i, i = 1, 2, ..., N_v\}$ is in one-to-one correspondence with the operations and whose edge set $E = \{(v_i, v_j), i = 1, 2, ..., N_v$ and $j = 1, 2, ..., N_v\}$ denotes the conflicting vertex pairs. An example of such a graph is shown in Fig. 2.17(c) for the scheduled DFG of Fig. 2.17(a) for two

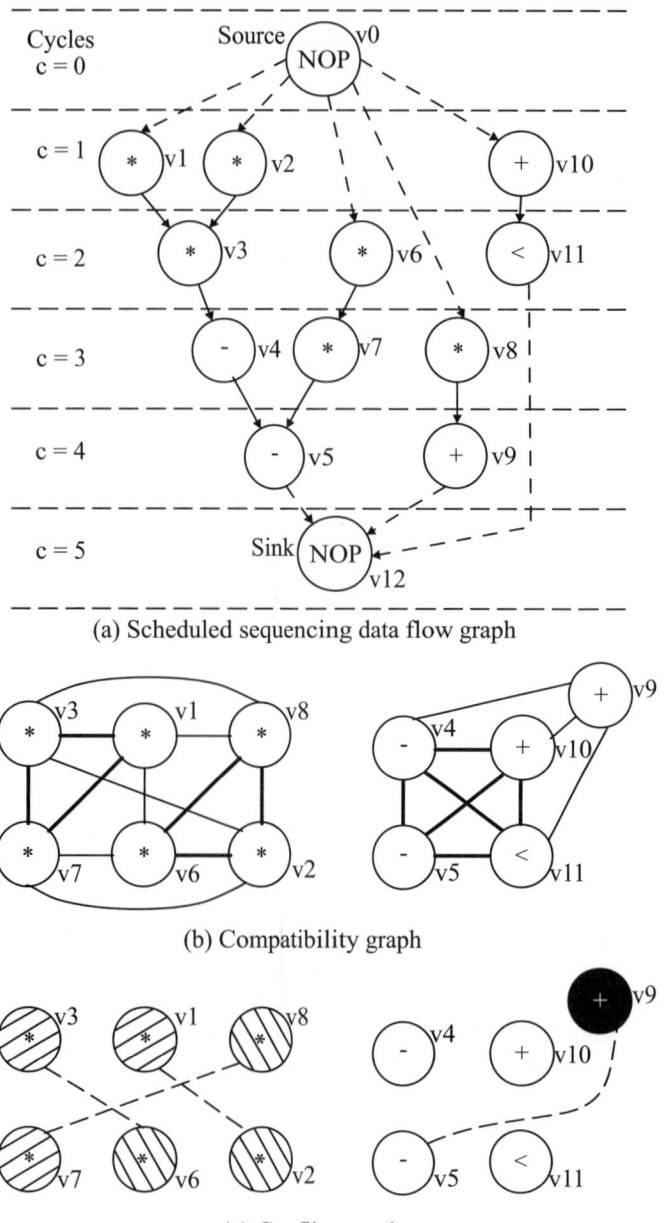

(a) Scheduled sequencing data flow graph

(b) Compatibility graph

(c) Conflict graph

Fig. 2.17 DFG and corresponding compatibility graph and conflict graph for two types of resources

types of resources, multipliers and ALUs [209]. In a conflict graph, an independent set is a subset of vertices that are not connected by edges representing a set of mutually compatible operations. For example, in Fig. 2.17(c), the independent sets are $\{v_1, v_3, v_7\}$, $\{v_2, v_6, v_8\}$, $\{v_4, v_5, v_{10}, v_{11}\}$ and $\{v_9\}$. The optimal binding is then to color the conflict graph with a minimum number of colors; this number is called the *chromatic number*. Each color corresponds to an instance of a resource. In the above example, each graph can be colored by two colors; thus, there is a need for four resources: two multipliers and two ALUs. The graph coloring problem is NP-complete; hence, the binding problem is also NP-complete.

The conflict graph and compatibility graph are complementary to each other. The choice between the use of compatibility is driven by the type of circuit; the graph which is sparse for a particular circuit is preferred for faster solution.

2.8.3 Left Edge Algorithm for Register Optimization

The register binding or assignment problem is similar to the resource (functional unit) binding. The register compatibility graph and register conflict graph are analogous to the resource compatibility graph and resource conflict graph, respectively. In the case of register binding, a conflict between two variables arises due to a lifetime overlap, in which the lifetime of a variable is the clock cycles in which the variable is active [266]. The register conflict graph is an interval graph and the complement of which is the compatibility graph. Thus, the register optimization problem can be cast as a clique partitioning problem of the compatibility graph or a minimum coloring problem of the conflict graph, both of which are NP-complete in nature [209, 103, 177, 86, 150]. However, the left edge algorithm can be used to solve the problem in polynomial time complexity [209, 103, 177].

In the left edge algorithm, the birth time of a variable is mapped to the left edge, and the death time of a variable is mapped to the right edge. The pseudocode of the algorithm is presented in Algorithm 11 [103]. In this algorithm, the variables are sorted in increasing order of their birth time. The first variable is then assigned to the first register. Then, the current register receives the next variable whose birth time is larger or equal to the death time of the previous variable.

The algorithm is greedy in nature and generates an optimal solution with a worst-case time complexity of $O(n \log n + dn)$ with n number of intervals and d number of rows. The left edge algorithm guarantees the allocation of the minimum number of registers but has disadvantages. In the case of iterative algorithms and conditional branches, some variables need to be alive across the boundaries of loops, thus requiring to be solved as circular-arc graph problems [103].

Algorithm 11 Left Edge Algorithm

1: Make a list of intervals, List.
2: Sort the intervals in List in order of increasing left coordinate.
3: Assume an initial solution S.
4: **while** (List is not empty) **do**
5:      Obtain the first element of the interval list List as first.
6:      Update List with first element removed.
7:      Initialize the row as row $\leftarrow 0$.
8:      **repeat**
9:          row $\leftarrow$ row $\cup$ {first}.
10:         first $\leftarrow$ First element in current *List* non-overlapping with first.
11:         Reconstruct List with original order removing elements equals to first.
12:     **until** (An element first can be found) $S \leftarrow S \cup$ {row}
13: **end while**
14: Return solution S.

2.8.4 Integer Linear Programming (ILP) Binding

For exact and optimal solutions to the binding problem, an ILP approach has been used [250, 209, 368, 319, 367, 304, 305]. The formulation is similar to the one done for ILP-based scheduling. Two decision variables are needed in this formulation as follows: $x_{i,c}$ is the binary decision variable which is 1 when vertex v_i starts in control step c, otherwise it is 0, and $b_{i,r}$ is the binary decision variable which is 1 when vertex v_i is bound to resource r, otherwise it is 0. Here $\{i = 1,2,...,N_v\}$, $\{i = 1,2,...,N_c\}$ and $\{r = 1,2,...,N_r\}$, with the number of vertices excluding NOP being N_v, the number of clock cycles being N_c and the number of resources of a particular type being N_r. Assuming a scheduled sequencing DFG, the following constraints are needed for each operation type:

1. *Uniqueness constraints*: These constraints ensure that each operation v_i is assigned to one unique resource. They are represented as $\forall i, 1 \leq i \leq N_v$,

$$\sum_r b_{i,r} = 1. \tag{2.8}$$

2. *Operation constraints*: These constraints ascertain that at most, one operation is executed among those assigned to resource r at any clock cycle. These are modeled as $\forall i, 1 \leq i \leq N_v$, $\forall i, 1 \leq c \leq N_c$ and $\forall i, 1 \leq r \leq N_r$,

$$\sum_i b_{i,r} \sum_{m=c-d_i+1}^{N_c} x_{i,m} \leq -1, \tag{2.9}$$

where d_i is the delay of operation associated with vertex v_i.

2.8.5 Heuristic Algorithm to Solve Clique Partitioning

To solve the clique partitioning problem, which is NP-complete, efficient polynomial time complexity heuristics have been developed that generate very good results [103, 349, 158]. In this algorithm, a super-graph is derived from the original compatibility graph. The super-graph contains super-vertices created by combining vertices of the compatibility graph. The algorithm then determines two connected super-vertices such that they have the maximum number of common neighbors. Then the two super-vertices are combined to form a new super-vertex. The algorithm terminates when no further merging of super-vertices can be performed. The pseudocode of the algorithm is presented in Algorithm 12 [103, 158]. Several other efficient clique partitioning heuristics are presented in [158].

2.8.6 GTS Algorithm

The pseudocode for the GTS binding algorithm is presented in Algorithm 13 [243, 250, 244, 299]. The algorithm assumes a scheduled DFG as an input and generates a binding matrix for the DFG. The Nash equilibrium is calculated for each clock cycle for each set of resources to ensure that the bidding strategy is applied separately for each clock cycle. The most time-consuming step of the algorithm is the Nash equilibrium determination. The existence of an equilibrium point is guaranteed when a number of alternative strategies are considered [345, 265].

Algorithm 12 Heuristic Algorithm to Solve Clique Partitioning

---

1: Initialize iteration step $k \leftarrow 0$.
2: Initialize an intermediate compatibility graph for step k as the input compatibility graph, i.e.,
   $G_s^{com_k}(V_k, E_k) \leftarrow G_s^{com}(V, E)$.
3: **while** (Set of edges E_k is not empty) **do**
4:     Find edge (v_i, v_j) in E_k that has largest set of common neighbors and let the set be
       $V_{neighbors}$. In case of a tie, select v_i and v_j such that sum of node degrees is maximum.
5:     Combine v_i and v_j and call it v_i.
6:     Delete edges from v_i and v_j that are not connected to common their common neighbors.
7:     Initialize a new set of edges; i.e., $E_{k+1} \leftarrow 0$.
8:     **for all** (Edge (v_m, v_n) in E_k) **do**
9:         **if** $((v_m, v_n)$ is different from $(v_i, v_j))$ **then**
10:            Add (v_m, v_n) to E_{k+1}.
11:        **end if**
12:    **end for**
13:    **for all** (v_n in $V_{neighbors}$) **do**
14:        Add (v_n, v_i) to E_{k+1}.
15:    **end for**
16:    $k \leftarrow k + 1$.
17: **end while**

Algorithm 13 GTS Binding Algorithm

1: **for all** (Clock Cycle c) **do**
2:       **for all** (Each set if compatible resources in clock cycle c) **do**
3:             Add (v_m, v_n) to E_{k+1}.
4:             Calculate the cost matrix.
5:             Determine Nash Equilibrium solution for cost matrix, vertex set and compatible
              resource set.
6:             Represent the Nash equilibrium solution as a binding matrix.
7:       **end for**
8: **end for**

2.9 Control Synthesis

The discussion in this book will be predominantly centered on data path synthesis. Control synthesis is beyond its scope. However, control synthesis is also an important high-level synthesis task [209, 77, 156]. High-level control synthesis is the process of generating a control unit that will drive the data path as required by the schedule using the values derived from the resource or interconnect allocation and binding. In a data path-dominated circuit, the design of the controller is simpler. In a control-dominated circuit, however, the design of the controller is very involved [117, 377]. In such a case, the controller needs to support conditional branch/looping, chaining, multicycling, data-dependent delay, multi-threading and multi-phase clocking [77]. The controller is typically a finite state machine that is either microcoded or hardwired.

2.10 High-Level Synthesis Benchmarks

Several high-level synthesis benchmarks are available in various forms to researchers to test the performance of their high-level synthesis-related research. The benchmarks can be obtained from various sources [10, 181, 182]. In addition, random CDFGs can be generated using a pseudorandom graph generator presented in [91]. Graphical representations of selected DSP benchmarks are provided in Figs. 2.18–2.22 that researchers can express in different intermediate forms for their research: autoregressive filter (ARF) [32, 31, 97], band-pass filter (BPF) [264], discrete cosine transformation (DCT) filter [100, 256, 255], discrete wavelet transformation (DWT) [139], elliptic wave filter (EWF) [162, 68], fast Fourier transformation (FFT) [33, 250], finite impulse response (FIR) filter [175, 221, 68, 97], inverse discrete cosine transformation (IDCT) filter [256, 255], MPEG motion vectors (MMV) [10] and wave digital filter (WDF) [123].

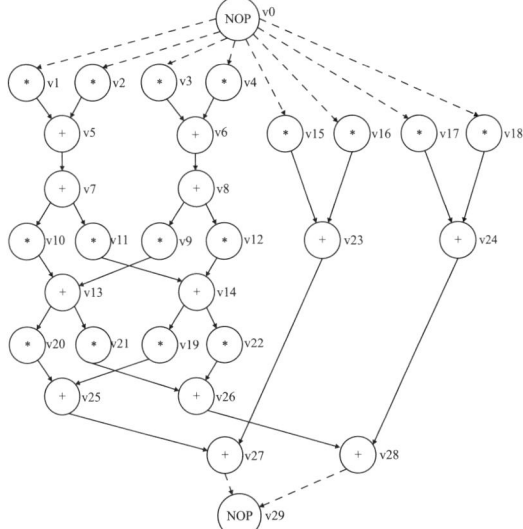

(a) Auto regressive filter (ARF)

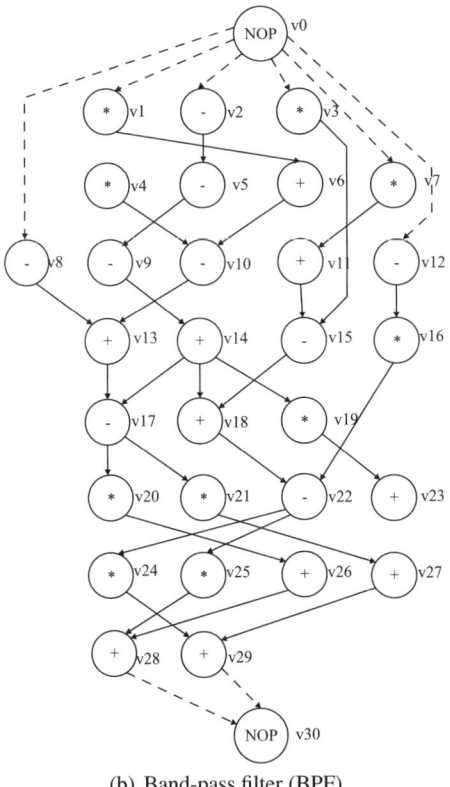

(b) Band-pass filter (BPF)

Fig. 2.18 Selected high-level synthesis benchmarks as sequencing DFG-1

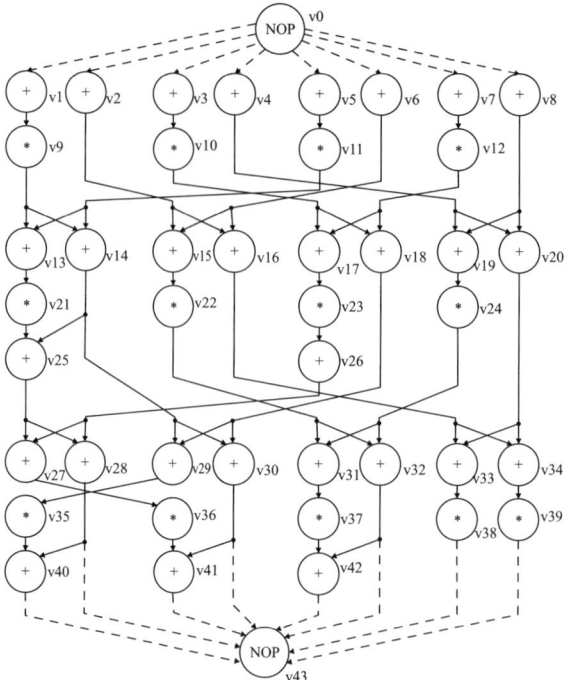

(a) Discrete cosine transformation (DCT)

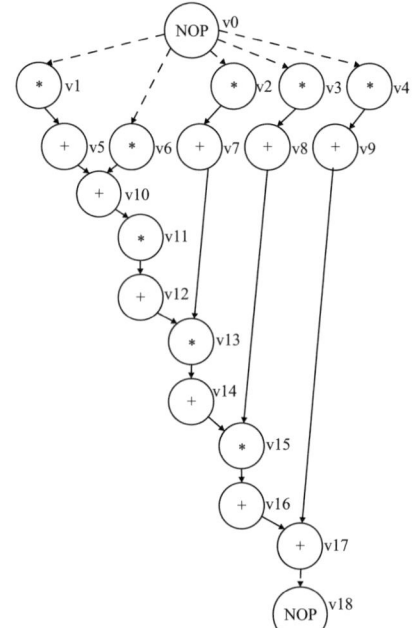

(b) Discrete wavelet transformation (DWT)

Fig. 2.19 Selected high-level synthesis benchmarks as sequencing DFG-2

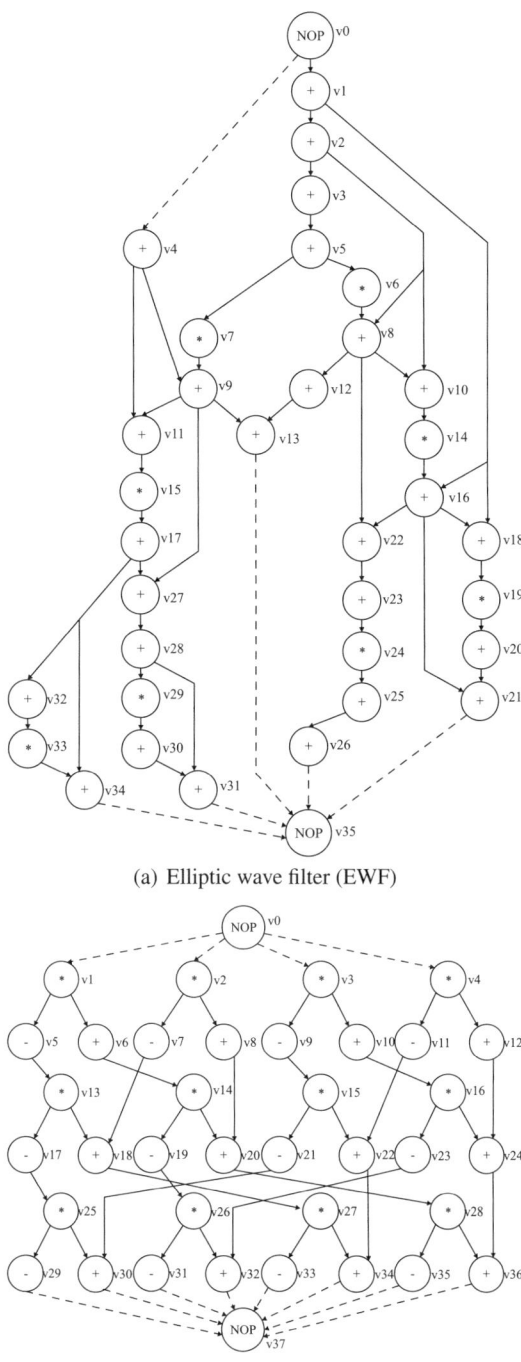

(a) Elliptic wave filter (EWF)

(b) Fast Fourier transformation (FFT)

Fig. 2.20 Selected high-level synthesis benchmarks as sequencing DFG-3

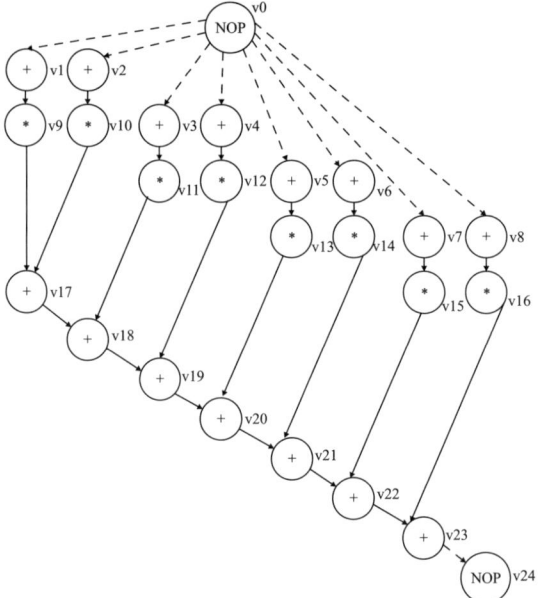

(a) Finite impulse response filter (FIR)

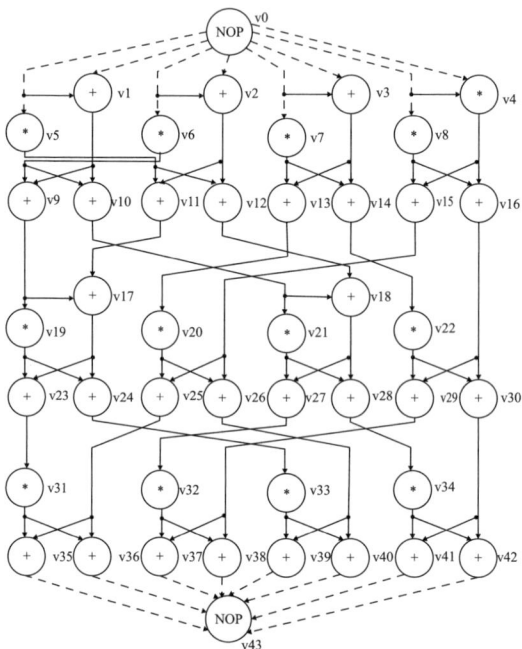

(b) Inverse discrete cosine transformation (IDCT)

Fig. 2.21 Selected high-level synthesis benchmarks as sequencing DFG-4

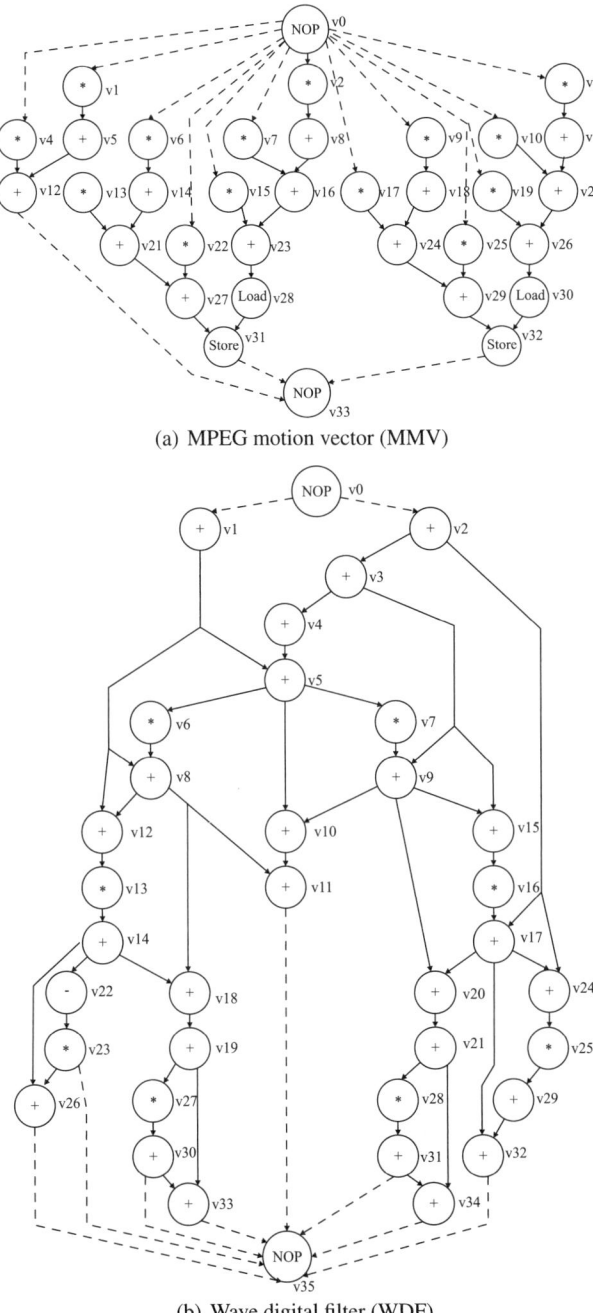

(a) MPEG motion vector (MMV)

(b) Wave digital filter (WDF)

Fig. 2.22 Selected high-level synthesis benchmarks as sequencing DFG-5

2.11 High-Level Synthesis Tools

Several industrial tools perform high-level or behavioral synthesis. In addition, all the tools that consider behavioral VHDL or Verilog use partial or full high-level synthesis process implicitly and transparently to the user. This section discusses selected high-level synthesis commercial tools.

2.11.1 CatapultC from Mentor Graphics

The first high-level synthesis tool, CatapultC from Mentor Graphics, automatically creates high-performance multiblock subsystems from ANSI C++ [11]. The CatapultC tool supports complex hierarchical designs and includes technologies that improve block-level performance and power analysis and reduction capabilities.

2.11.2 CyberWorkBench from NEC

CyberWorkBench from NEC supports behavioral synthesis [5]. It synthesizes algorithms described in C into RTL that meet various specifications in clock speed, chip area, power consumption, etc. It also supports multiple clocks, gated clocks, synchronous/asynchronous reset and synchronous/asynchronous/pipeline memory.

2.11.3 PICO Express from Synfora

PICO Express from Synfora compiles C to RTL exploiting parallelism [15]. The tool has verification and integration capabilities to ensure that verification and integration are minimized along with RTL generation. The tool also generates test benches, synthesis scripts, driver code and RTL.

2.11.4 Cynthesizer from Forte Design Systems

Cynthesizer from Forte Design Systems is a design environment that supports electronic system-level (ESL) design, transaction-level modeling (TLM) synthesis and verification, co-simulation, field-programmable gate array (FPGA) prototyping, power optimization, and dependency management in a single framework [6].

2.11.5 Cascade from Critical Blue

Cascade from Critical Blue synthesizes coprocessors. It generates a synthesizable RTL, a C functional model and a test bench [4]. The implementation then can be customized for SoC, FPGA, or structured application-specific integrated circuit (ASIC) design.

2.11.6 Agility Compiler from Celoxica

The Agility compiler from Celoxica provides behavioral design and synthesis for SystemC to RTL [2]. It is a unified framework for FPGA design and ASIC/SoC prototyping.

2.11.7 eXCite from Y Explorations

eXCite from Y Explorations synthesizes Verilog or VHDL RTL output from ISO/ANSI-C targeted for FPGA or ASIC technologies [9]. It allows software and hardware to communicate through channels and has algorithms for pipelining capabilities, automatic bit reduction and delay and area optimization.

2.11.8 ESEComp from BlueSpec

ESEComp from BlueSpec synthesizes SystemC descriptions of a design into highly efficient Verilog RTL [8]. It is a part of ESL Synthesis Extensions for SystemC, which is aimed at modeling and synthesis.

2.11.9 VCS from Synopsys

VCS from Synopsys is a comprehensive RTL verification solution [20]. It has a single compiler that supports several languages including Verilog, VHDL, SystemVerilog and SystemC.

2.11.10 NC-SC, NC-Verilog and NC-VHDL from Cadence

NC-SC, NC-Verilog and NC-VHDL from Cadence provide high-performance, high-capacity HDL simulation with integrated coverage viewing and analysis [12]. They have native support for Verilog, VHDL, SystemC Verification Library, PSL/Sugar assertions and acceleration on demand. They produce efficient native machine code directly from VHDL for high-speed execution.

2.11.11 Synplify from Synplicity

Synplify DSP from Synplicity can synthesize a digital signal processing (DSP) algorithm to FPGAs or ASICs [18]. The synthesis tool is faster and can create optimal circuits or pipelining, folding and multi-channelization.

2.11.12 ISE from Xilinx

The ISE tool from Xilinx generates intermediate RTL using a SmartCompile technology [13]. It can accept VHDL, Verilog or a mix of both and target designs for FPGA and complex programmable logic device (CPLD). It has built-in capability for power analysis, timing-driven design closure and HDL simulation.

2.11.13 Quartus from Altera

Quartus from Altera provides incremental compilation facility that can reduce the design cycle time [16]. It allows power analysis and optimization as well as memory compilation.

2.12 Summary and Conclusions

High-level synthesis is important for several reasons. In particular, reduction of design cycle time and rapid design space exploration at the higher level of abstraction will ensure that wrong decisions are not propagated to lower levels of design abstraction, which would be difficult and costly to correct at the lower levels. High-level synthesis involves several important steps, such as scheduling, allocation and binding. Several graph theoretical algorithms are available that can perform optimization while performing these tasks. Both data path and control synthesis need to be performed to ensure that the synthesis covers all types of real-life circuits. Industrial interest for the development of high-level synthesis tools has increased recently. There are existing tools to perform high-level synthesis explicitly, and some tools perform the behavioral to RTL compilation as an intermediate process.

Chapter 3
Power Modeling and Estimation at Transistor and Logic Gate Levels

3.1 Introduction

In this chapter, the current conduction and leakage mechanisms of the nano-CMOS transistor are discussed. First a brief overview of the expected trends of nano-CMOS technology in the near future is provided. The basic physics responsible for the behavior of the transistor as a switch is presented. Particular attention is paid to the leakage aspects of its operation. As a case study and to present the reader with concrete results, a bulk 45 nm state-of-the-art CMOS process is used. Even though the physics behind the conduction and leakage of the transistor are of interest, the primary focus will be that of a circuit designer, i.e., the modeling of the mechanisms in a manner suitable for circuit simulation and synthesis. The most widely used nano-CMOS model in the industry today is Berkeley's BSIM (Berkeley Short-Channel Insulated-Gate Model) [96, 3]. BSIM (currently at version 4.6.1) allows the process modeling groups to capture the complicated conduction and leakage physical mechanisms of the fabricated transistor in a *predictive* way and provide the designers with the ability to simulate their circuits with a high degree of confidence using standard SPICE engines. The predictive capability of the model (i.e., its validity for a wide range of lengths and widths) is of paramount importance for the success of the modeling and design efforts.

A more advanced version of the BSIM model (version 5, [122]) is currently under development. It is derived from the solution of the coupled system of current density and Poisson equations. One of the main shortcomings of earlier releases of BSIM addressed in this version is the asymmetry and discontinuity present in the calculation of derivatives based on these models. This can prevent the usage of BSIM in modeling distortion effects and other phenomena important in analog and radio-frequency (RF) design. Its use, however, is not yet wide in the digital community.

Good SPICE modeling and simulation at the transistor level permits the design process to climb to a higher level of complexity in the hierarchy of complex digital systems, namely, the logic gate. Although logic gates (such as NOT, NAND and NOR) are comprised of a small number of transistors, their function and power

S.P. Mohanty et al., *Low-Power High-Level Synthesis for Nanoscale CMOS Circuits*,
DOI: 10.1007/978-0-387-76474-0_3, © Springer Science+Business Media, LLC 2008

profile present new challenges in the nanometer regime. The NOT, NAND and NOR gates will be exhaustively presented. Particular attention will be paid to the new leakage mechanism called gate-oxide tunneling current for nano-CMOS technology. It will be seen that this power leakage mechanism is the dominant source of dissipation in the 45 nm and below regime and warrants special attention. An overview of all sources of power dissipation (static, dynamic and leakage) appropriate for logic gates is presented. The knowledge obtained from transistor modeling is extended to the problem of *estimation* of dissipative power from logic gate data. First the possibility of SPICE results is discussed but shown not tractable in medium to large complexity units and then HDL (Verilog and VHDL) simulation in tandem with probabilistic and/or statistical techniques is presented as a viable solution. Adder, subtractor, multiplier, divider, register, multiplexer and comparator data are given for comparative purposes and, finally, the transition is made to the next chapter, where the concepts, modeling and estimation techniques are extended to architectural subsystems comprised of hundreds or thousands of functional units.

3.2 CMOS Technology Trends

The primary source of information about the current state of semiconductor technology in general, and nano-CMOS technology in particular, is the International Technology Roadmap for Semiconductors (ITRS) [17]. The ITRS is a worldwide consortium composed of major semiconductor manufacturers, equipment suppliers, academic institutions and national laboratories. Part of the charter of the ITRS is to identify the major technological challenges facing the semiconductor industry in the upcoming years. It produces a report which presents, as a roadmap, the best estimates of the leading members of the semiconductor industry of the most significant challenges expected to be faced in the coming technology nodes. According to the ITRS, currently (2008) the working effective (printed) gate length for general high-performance digital nano-CMOS processes is approximately 45 nm with eventual reduction to as low as 9 nm by the year 2020. Two things become immediately obvious: nano-CMOS issues must be well understood for the technology to become successful and nano-CMOS process will be prominent for at least another decade.

Another very important observation from the ITRS is the anticipated range of power supply (V_{DD}): from approximately 1 V today to as low as 0.5 V by the year 2020. As will be seen in this chapter, the value of V_{DD} is extremely important as far as power dissipation is concerned. Even though 1 V may appear low, combined with gate-oxide thicknesses of 1–2 nm or lower, it may cause significant sources of unwanted dissipation to arise. For the purposes of this discussion, the most important information contained in the ITRS report is the redistribution of the various sources of power dissipation in nano-CMOS technologies, shown in Fig. 3.1. It is evident that the previously dominant source of power dissipation, dynamic power, is being eclipsed by gate-oxide leakage and subthreshold leakage which are comparable in magnitudes.

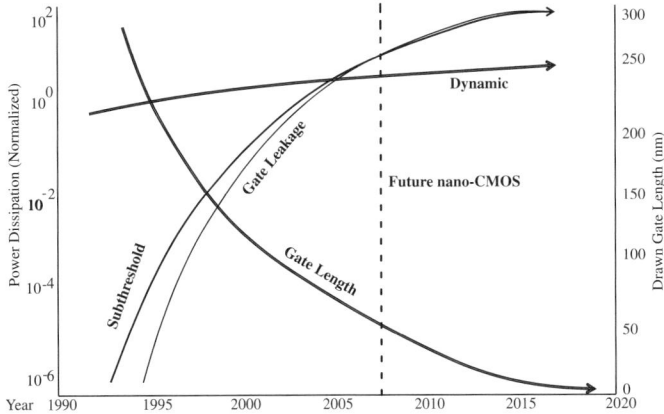

Fig. 3.1 Prediction of trends of major sources of power dissipation in nano-CMOS transistors and circuits [17]. It is predicted (area to the right of the dotted line) that gate-oxide leakage and subthreshold leakage will dominate the total power dissipation of nano-CMOS circuits. Note the logarithmic scale for power

3.3 Current Conduction Mechanisms in Nano-CMOS Devices: A Résumé

A nano-CMOS device in a complex digital circuit operates in three regions:

- In the ON state, a conducting channel forms that transforms the transistor into a voltage-controlled current source. In an ideal case, the transistor acts as a short circuit.
- In the OFF state, the transistor should act as an open circuit.
- In the transition from ON to OFF and OFF to ON states, the transistor should conduct minimal current and hence dissipate minimal power.

As will be seen in more detail in this section, the situation is more complex than the above-indicated scenario, particularly for nano-CMOS devices where a host of new phenomena come into play. A cross-section of a typical nano-CMOS n-type device is shown in Fig. 3.2 along with a pictorial representation of the different currents which may flow, depending on the region of operation.

3.3.1 The Ideal ON and OFF States

Ideally, the device should act like a switch: for an applied gate voltage V_G less than the threshold voltage V_{Th} the source and drain terminals are electrically disconnected. No current flows between the two terminals. If V_G exceeds V_{Th}, a

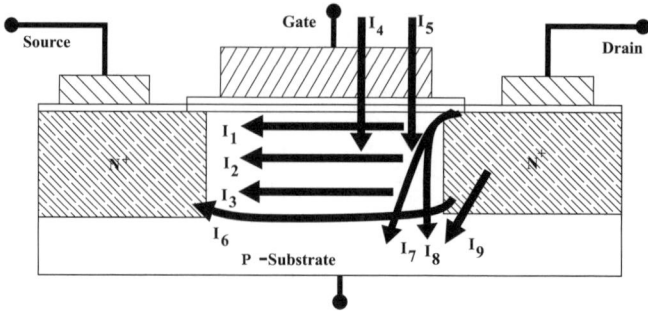

Fig. 3.2 Current flow paths in a nano-CMOS transistor during power dissipation in different states of its operation [308, 211, 55]: I_1 - drain-to-source active current (ON state), I_2 – drain-to-source short-circuit current (ON state), I_3 – subthreshold leakage (OFF state), I_4 – gate leakage (both ON and OFF states), I_5 – gate current due to hot-carrier injection (both ON and OFF states), I_6 – channel punch-through current (OFF state), I_7 – gate-induced drain leakage (OFF state), I_8 – band-to-band tunneling current (OFF state), I_9 – reverse bias PN junction leakage (both ON and OFF states)

well-defined channel forms and an active current flows between the drain and the source. This current is indicated as I_1 in Fig. 3.2. In the same figure, the origin of the short-circuit current I_2 is the same but the two currents arise in different contexts. The origin and impact of short-circuit effects will be discussed in Section 3.4. To obtain an accurate picture of the complex conduction processes in the nano-CMOS transistor, the $I-V$ characteristics of an actual device will be studied: an NMOS with drawn channel length $L = 45$ nm, width $W = 45$ nm, oxide thickness $T_{ox} = 1.4$ nm and threshold voltage $V_{Th} = 0.22$ V. A nominal power supply value of $V_{DD} = 0.7$ V is assumed. These values correspond to a state-of-the-art low-voltage 45 nm bulk nano-CMOS process. The corresponding data are shown in Fig. 3.3.

Several facts are obvious from these $I-V$ curves:

- The transistor conducts current even when in the OFF state. Several possible mechanisms for this current will be discussed in the course of this section.
- The device does not transition from the OFF to the ON state abruptly. Instead, there is a gradual increase as V_G varies from low to high values.
- The value of the current in the ON state depends on the applied drain voltage, V_D.
- Even at very low values of V_G, significant current can flow.

The picture presented above does not include another major source of leakage, gate-oxide tunneling current, as it manifests itself as a component of the gate current.

3.3.2 Junction Reverse Bias Current

In bulk CMOS processes, the wells are held at appropriate voltages (V_{DD} for the P-well and V_{SS} or ground for the N-well) to prevent the parasitic diodes, formed by the NMOS and PMOS source/drain diffusions and the wells, from conducting. This has the effect that when transistor terminals are held at high voltages, the parasitic diodes are strongly reverse biased and give rise to the junction reverse bias current [350], I_9 in Fig. 3.2.

This current is given by [373]:

$$I_9 = \mu_0 C_{ox} \frac{W}{L_{eff}} V_{therm}^2 e^{1.8}, \tag{3.1}$$

where μ_0 is the zero-bias mobility, C_{ox} is the gate-oxide capacitance per unit area, W is the device width, L_{eff} is the effective length of the transistor and V_{therm} is the thermal voltage. This current can be present even when the diodes are weakly reverse-biased and therefore is active in both the ON and OFF states.

3.3.3 Drain-Induced Barrier Lowering (DIBL)

Drain-induced barrier lowering (DIBL) is a direct consequence of the physical proximity of the source and drain regions in a nano-CMOS. As the name implies, the potential barrier between the source and drain is reduced due to the depletion region contact when high drain voltage is present [369]. The presence of DIBL is manifested in the $I-V$ curves by a vertical shift for high V_D. This shift is indicated in Fig. 3.3. DIBL is taken into account via the inclusion of a coefficient n in the calculation of the subthreshold current (Equation (3.3)). This coefficient is typically determined experimentally and is an input to the SPICE models. The actual lowering of the potential barrier ($\Delta\phi$) is given by [99]

$$\Delta\phi = \frac{\alpha_L}{L^2} \left(\psi_S + V_{SB}\right)^{1/2} V_{DS}, \tag{3.2}$$

where α_L is a technology-dependent constant, L is the channel length, ψ_S is the surface potential at the Si/SiO$_2$ interface, V_{SB} is the source to substrate voltage and V_{DS} is the drain-to-source voltage.

3.3.4 Subthreshold Leakage

In Fig. 3.3 the transition from the OFF to the ON states is clearly gradual and not abrupt. This is manifested by a slope of the $I-V$ curve in its linear region that has a finite and not infinite value. The physical mechanism responsible for this current (I_3 in Fig. 3.2) is the existence of a weak inversion region in the channel for small values of V_G. In long channel devices, the region of operation of the device, called *subthreshold*, can be controlled by keeping V_{Th} high. In nano-CMOS V_{Th} must be scaled down, hence the subthreshold effect becomes more dominant. The conduction mechanism in this case is dominated by diffusion due to the low electric

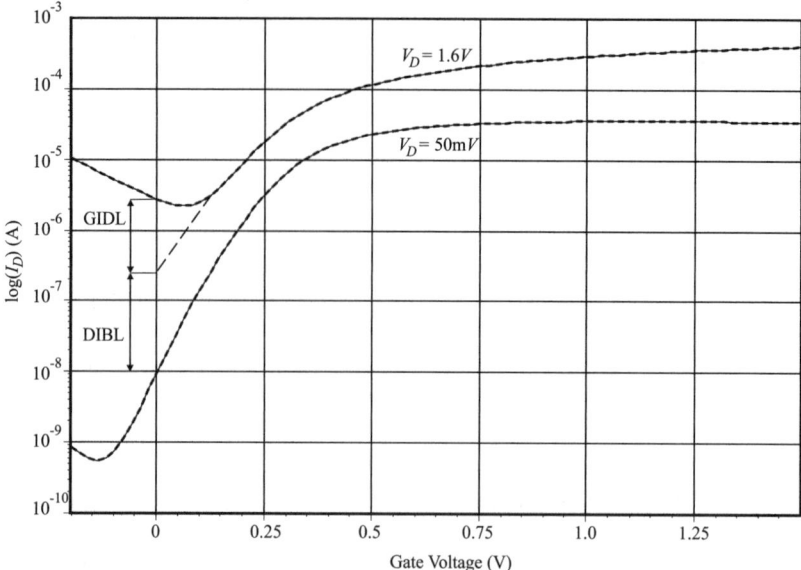

Fig. 3.3 Drain current as a function of gate voltage for an NMOS device with drawn channel length of $L = 45$ nm, width $W = 45$ nm, oxide thickness $T_{ox} = 1.4$ nm and threshold voltage $V_{Th} = 0.22$ V. The value of V_D in the upper curve is higher than typical operating values in order to exaggerate the DIBL and GIDL effects for illustrative purposes

field present in the channel [344, 350]. As can be seen from Fig. 3.1, subthreshold current conduction is one of the two dominant sources of power leakage in modern nano-CMOS devices. The BSIM model calculates the subthreshold current [96] accounting for DIBL as

$$I_3 = I_9 \left[1 - \exp\left(\frac{-V_{DS}}{V_{therm}} \right) \right] \exp\left(\frac{V_{GS} - V_{Th} - V_{off}}{nV_{therm}} \right), \qquad (3.3)$$

where I_9 is the junction reverse bias current, given by Equation (3.1), V_{DS} is the applied drain–source voltage, V_{GS} is the applied gate–substrate voltage, V_{Th} is the threshold voltage, n is the DIBL coefficient and V_{off} is the offset voltage calculated internally by BSIM to account for body effect and DIBL on threshold voltage. Due to its nature, this leakage mechanism is only active in the OFF state.

3.3.5 Gate-Induced Drain Leakage (GIDL)

Even though V_{DD} has scaled down significantly in nano-CMOS circuits, the gate-oxide thickness T_{ox} must scale down accordingly. The net result is that small applied voltages can give rise to large electric fields (which are proportional to V_{DD}/T_{ox}) in the gate-oxide region, near the drain junction. A direct consequence of these high fields is gate-induced drain leakage (GIDL) [344, 309, 373]. If the field is sufficiently high, a pair of carriers (electron–hole pair) can be generated in the drain–gate overlap region. The majority carrier recombines in the heavily doped drain region but the minority carrier is swept in the substrate, giving rise to the GIDL current component. The GIDL current is stronger when the gate voltage, V_G, becomes small or negative, as shown in Fig. 3.3. The GIDL contribution over the normal subthreshold component is clearly indicated. Since GIDL depends on a strong potential difference between V_G and V_D, it is present only in the OFF state. It is shown as I_7 in Fig. 3.2.

BSIM [96] calculates this current component from

$$I_7 = \text{AGIDL} \cdot W_{\text{effCJ}} \cdot N_{\text{f}} \cdot \frac{V_{DS} - V_{GS} - \text{EGIDL}}{3T_{\text{ox}}}$$

$$\exp\left(-\frac{3T_{\text{ox}} \cdot \text{BGIDL}}{V_{DS} - V_{GS} - \text{EGIDL}}\right) \frac{V_{DB}^3}{\text{CGIDL} \cdot V_{DB}^3}. \tag{3.4}$$

AGIDL, BGIDL, CGIDL and EGIDL are model parameters, obtained from experimental data. W_{effCJ} is the effective width of the drain diffusion, N_{f} is the number of fingers in the device and V_{DB} is the drain–bulk voltage.

3.3.6 Punch-Through

Another consequence of the physical proximity of the source and drain regions in a nano-CMOS (in addition to DIBL) is punch-through [339, 369]. When the depletion regions of the source/substrate and drain/substrate PN junctions touch, a conduction path is created from drain to source, in addition to the channel. The end result is a punch-through current that flows in the bulk of the channel region. It is shown as I_6 in Fig. 3.2. Punch-through current occurs in the OFF state of the transistor and contributes to subthreshold leakage.

3.3.7 Hot-Carrier Injection

One other effect caused by the high electric fields in the gate–drain overlap region (in addition to GIDL) is the injection of hot carriers from the substrate to the gate. For sufficiently high fields, electrons can become energetic enough (hot) to overcome the barrier height presented by the oxide and travel to the gate. This effect is known as hot-carrier injection. Holes usually do not participate in this effect due to their large effective mass and larger barrier height. This current is shown as I_5 in Fig. 3.2. BSIM calculates this current component from

$$I_5 = \frac{\text{ALPHA0} + \text{ALPHA1} \cdot L_{\text{eff}}}{L_{\text{eff}}} (V_{DS} - V_{DSeff})$$

$$\exp\left(\frac{\text{BETA0}}{V_{DS} - V_{DSeff}}\right) \cdot I_{\text{dsNoSCBE}}, \tag{3.5}$$

where ALPHA0, ALPHA1 and BETA0 are model parameters, V_{DSeff} is an internally computed effective drain–source voltage that ensures a smooth transition from the triode to the saturation region and $I_{dsNoSCBE}$ is the drain–source current ignoring the substrate current-induced body effect [96]. Typically hot-carrier injection occurs in the ON state but can also be present in the OFF state.

3.3.8 Band-to-Band Tunneling (BTBT)

Another leakage mechanism which can become important in nano-CMOS devices is band-to-band tunneling (BTBT) due to the gated drain/substrate diode [357]. BTBT takes place in the deeply depleted region between the gate oxide and the heavily doped (n+ for an NMOS) drain junction. If the electric field present in this region is large enough to cause band bending that exceeds the tunneling gap from conduction to valence band of the drain/substrate diode, a current flows from the drain–gate overlap region to the substrate. The energy band diagram for a reverse-biased diode exhibiting BTBT is shown in Fig. 3.4. This current is shown as I_8 in Fig. 3.2.

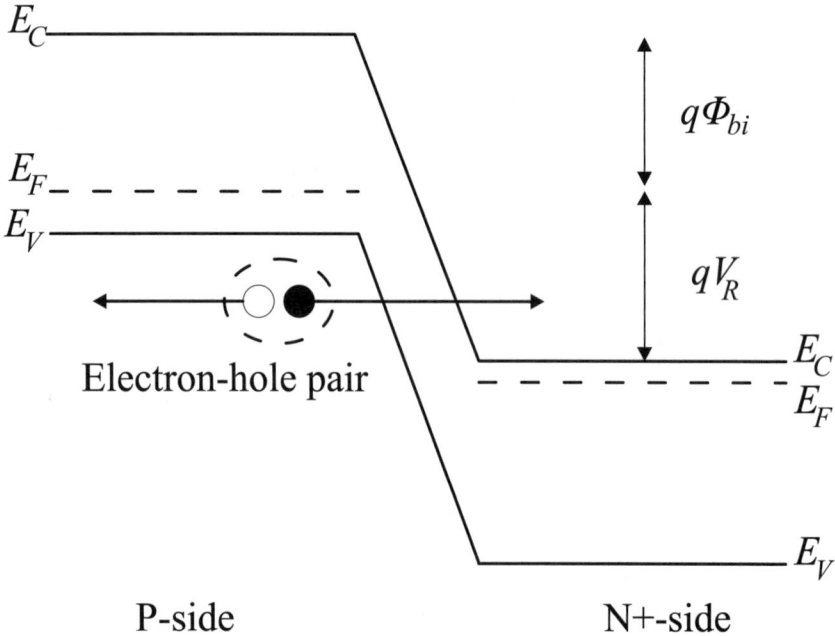

Fig. 3.4 Energy band diagram of a reverse-biased diode exhibiting BTBT current. E_C and E_V are the conduction and valence band edges, respectively, and E_F is the Fermi level. ϕ_{bi} is the built-in potential of the junction and V_R is the externally applied bias

BTBT current contributes to leakage in the OFF state and can be calculated from the following expression [344]:

$$J_{\text{BTBT}} = A \frac{EV_R}{\sqrt{E_g}} \exp\left(-B\frac{E_g^{3/2}}{E}\right),\tag{3.6}$$

where J_{BTBT} is the BTBT current density, A and B are constants that depend on the carrier-effective mass in silicon, E is the electric field across the junction, V_R is the applied voltage across the junction and E_g is the band gap at the junction. Equation (3.6) can be applied to a MOS transistor to calculate the overall contribution from both the source and the drain. A procedure that can yield numerical results for the calculation of the total BTBT is described in [242].

3.3.9 Gate-Oxide Tunneling

Gate-oxide tunneling current is the most adversely affected leakage mechanism resulting from aggressive nano-CMOS scaling. At the first two nanotechnology nodes, 90 nm and 65 nm, its impact is fairly small and can be easily accommodated by minor process changes. At 45 nm and below, however, it becomes equal to, and even surpasses, subthreshold leakage, as seen from Fig. 3.1. Of course, moving to high-κ dielectrics will provide some relief, but will it scale [74, 46]?

The process variable that causes this large leakage current is the oxide thickness, which in a typical 45 nm process is only 1.4 nm thick. In terms of actual physical scale, this is only a few monolayers thick. This fact has serious repercussions when process variation effects are considered, in Section 3.5. The direct result of this thin oxide is that a high electric field is present across the gate oxide, even with V_{DD} being scaled down to 0.7 V or lower. As a consequence of the small oxide thickness and large electric field, quantum mechanical tunneling occurs and a sizeable current can flow from/to the gate terminal. Classical CMOS devices are characterized by very large input impedance, due to the gate oxide, but nano-CMOS devices can exhibit very small impedance. This fact has implications for analog portions of large systems-on-chip (SoC). An energy band diagram of the gate-oxide-substrate structure is shown in Fig. 3.5(a) and (b) for different relative values of the potential drop across the oxide (V_{ox}) and the barrier height (ϕ_{ox}) between the oxide and silicon substrate. It is assumed that the gate is heavily doped polysilicon and that the substrate is P type.

The scenario shown in Fig. 3.5(a) corresponds to the case where $V_{\text{ox}} > \phi_{\text{ox}}$. As seen from the figure, the resulting conduction band profile is triangular. Tunneling through triangular barriers of this sort is known as Fowler–Nordheim (FN) tunneling [308]. This situation does not typically arise during the operation of nano-CMOS devices as the barrier height is significantly larger than the applied voltages to the device gates. On the other hand, thin oxides and high electric fields can result in the situation shown in Fig. 3.5(b) where $V_{\text{ox}} < \phi_{\text{ox}}$ and the electrons can tunnel through

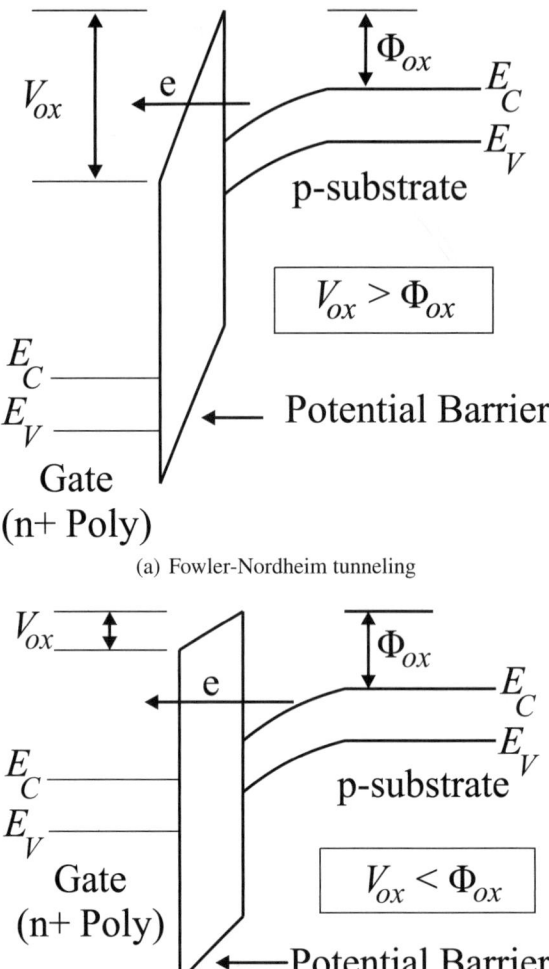

(a) Fowler-Nordheim tunneling

(b) Direct tunneling

Fig. 3.5 Tunneling mechanisms in the gate-oxide region of nano-CMOS [308]

a trapezoidal barrier. This is known as direct tunneling. This current component is indicated as I_9 in Fig. 3.2. Because of its significance for future nano-CMOS, it will be examined in more detail. Direct tunneling gate-oxide current can be present in both ON and OFF states. Hence its impact is more severe than subthreshold leakage which only occurs during the OFF state. The actual conducting paths are shown in Fig. 3.6(a) and (b) and discussed subsequently.

The general nature of the direct tunneling current density J_{DT} through the gate oxide is encapsulated in the following equation [313]:

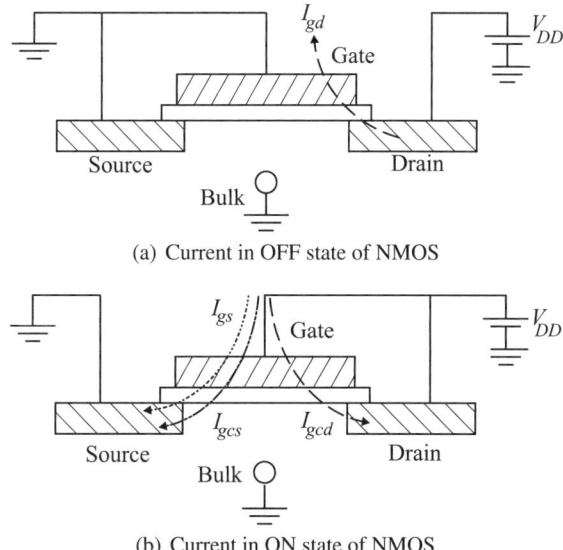

(a) Current in OFF state of NMOS

(b) Current in ON state of NMOS

Fig. 3.6 Gate-oxide tunneling current component flow in the various regions of operation of an NMOS transistor. I_{gd} and I_{gs} are the components from the diffusions (drain and source, respectively) directly to the gate. I_{gcd} and I_{gcs} are the components from the diffusions (drain and source, respectively) to the gate via the channel. The gate to bulk tunneling current (I_{gb}) is negligible and is not shown

$$J_{DT} = AE_{ox}^2 \exp\left\{ -\frac{B\left[1 - \left(1 - \frac{V_{ox}}{\phi_{ox}}\right)^{3/2}\right]}{E_{ox}} \right\}, \tag{3.7}$$

where E_{ox} is the electric field across the oxide and A and B are constants that depend on the effective mass of the carrier and the barrier height ϕ_{ox}. Since $E_{ox} = V_{ox}/T_{ox}$, the strong exponential dependence of the tunneling current on both power supply value and oxide thickness is evident from Equation (3.7). A quantitative appreciation of the magnitude of the tunneling currents involved during the entire switching cycle of a 45 nm NMOS can be obtained from Fig. 3.7(a) and (b), which shows the emergence of the various current components described above.

From this figure it can be observed that the gate to bulk component I_{gb} is negligible throughout all regions. It is also clear that different mechanisms contribute to the overall current during different phases of the switching cycle:

- *OFF region*: In the steady-state OFF region, both gate and source are at ground while the drain is at high (V_{DD}) voltage. Since no channel is formed in this condition, the only active component is I_{gd}, due to the overlap of the drain diffusion and the gate. The direction of the current flow is from diffusion to gate.

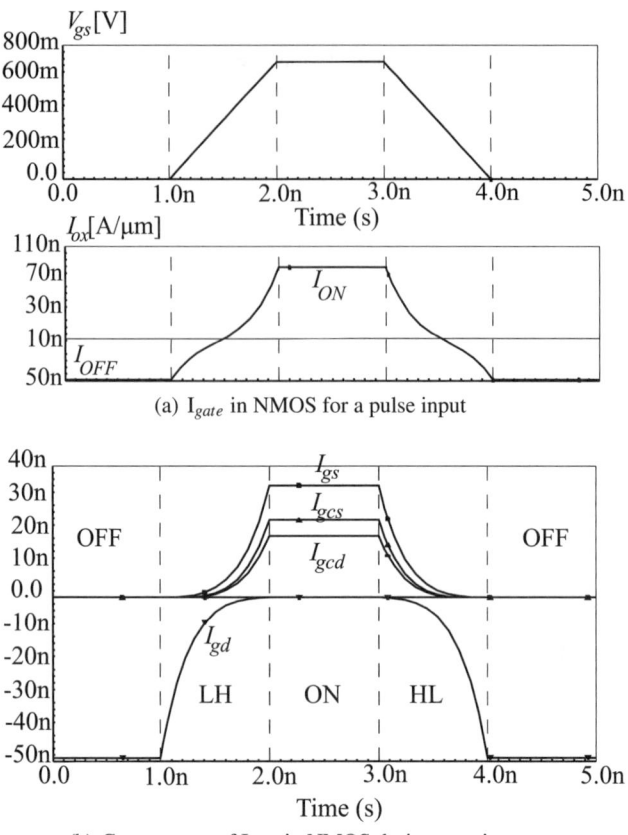

(a) I_{gate} in NMOS for a pulse input

(b) Components of I_{gate} in NMOS during transient state

Fig. 3.7 Study of gate-oxide tunneling current (I_{gate}) components in response to a pulse input. The input voltage V_{gs} is given for a complete cycle (low to high and then high to low) to bias the NMOS in the OFF, OFF-to-ON, ON and ON-to-OFF states

- *ON region*: In the steady-state ON region, both gate and drain are held at high with the source being grounded. A well-formed channel exists in this case and three separate components are active: (i) I_{gs} due to the overlap of the source diffusion with the gate, (ii) I_{gcs} due to the flow of current from the gate to the inversion layer in the channel and subsequently to the source and (iii) I_{gcd} due to the flow of current from the gate to the channel and subsequent collection at the drain. The component from gate to drain overlap (I_{gd}) has been extinguished due to the almost zero electric field in that region of the oxide. The overall current flow is from gate to source and channel, opposite to the flow in the OFF state.
- *Low-to-high (LH) or high-to-low (HL) transition region*: Finally, during the LH and HL transitions, all four components become active. In this case the source is at ground, the drain is at V_{DD} and the gate is switched from low to high or high to low. In the LH transition, the channel gradually forms from source to drain

and the components I_{gs}, I_{gcs} and I_{gcd} start becoming significant, in that order. Conversely, as the field across the oxide region over the drain is reduced, I_{gd} decreases to almost total extinction. The situation is reversed in the HL transition.

It should be noted that the total gate leakage component in the OFF state (I_{OFF}) is comparable to that in the ON state (I_{ON}).

3.4 Power Dissipation in Nano-CMOS Logic Gates

It becomes apparent from the discussion in Section 3.3 that the physics of current conduction and, particularly, of current leakage mechanisms in nano-CMOS devices is quite involved. Although the designer of nano-CMOS digital circuits must be aware of all the different sources of potential device degradation and power dissipation, it is impractical to handle them during the design process on a device-by-device basis. It is the function of the process and device modeling group to capture the device physics into a model that the designer can use with a high degree of confidence to predict not only system functionality but power consumption and dissipation as well. The most common means of transferring this knowledge is via a circuit-level (SPICE, SPICE derivative or SPICE equivalent) predictive model. As mentioned earlier, the most widely used model for nano-CMOS digital design is Berkeley's BSIM [3]. With the help of accurate and predictive models, such as BSIM, the designer is able to characterize exhaustively not only libraries of pre-built components but actual medium-to large-scale circuits with a high degree of confidence. Perhaps the most important library components to be characterized are the primitive gates (NOT, NAND and NOR) out of which larger functional units (adders, subtractors, multipliers, multiplexers, etc.) are built. These primitives will be examined in Section 3.4.2 but first a classification of the various types of power dissipation in digital circuits will be performed.

3.4.1 Static, Dynamic and Leakage Power Dissipation

A digital circuit comprised of CMOS devices is typically in two distinct states: at the highs and lows of the clock(s) it can be considered to be in a "static" state while during transition of the clock(s) it can be considered to be in a "dynamic" state. Accordingly, the power that a circuit consumes can be broken down to static and dynamic components:

$$P_{total} = P_{dyn} + P_{static}, \tag{3.8}$$

where P_{total} is the total power dissipation, P_{dyn} is the portion of power dissipation due to transitions and P_{static} is the power due to the static portion of the clock waveform. Each of these components, in turn, is due to several different mechanisms of conduction during the operation of the devices.

The *dynamic* portion of power dissipation will be considered first. It is dominated by three terms:

$$P_{\text{dyn}} = P_{\text{switch}} + P_{\text{short}} + P_{\text{gate}}, \tag{3.9}$$

where P_{switch} is due to the effort needed to switch a gate's output in the presence of loading from other gates in the circuit, P_{short} is the short-circuit current which flows during transitions and P_{gate} is the gate-oxide tunneling current that flows during the transition periods of the devices (areas indicated as HL and LH in Fig. 3.7).

An estimate of P_{switch} can be obtained by assuming that the average capacitive load per gate is C_{L} and by calculating the power needed to charge and discharge it under a clock of frequency f_c [365]:

$$P_{\text{switch}} = \alpha V_{\text{DD}}^2 C_{\text{L}} f_c, \tag{3.10}$$

where α represents an estimate of the switching activity of the device.

The short-circuit component occurs because the transition of NMOS and PMOS devices from OFF to ON, and vice versa, is not instantaneous but gradual. There are, therefore, short periods of time where both types are conducting at the same time. The net result is a current that flows briefly from the pull-up to the pull-down network of the circuit. This flow only happens during transitions; so it is a part of dynamic power consumption. The short-circuit power component is given by [89, 354]

$$P_{\text{short}} = \frac{\beta}{12}(V_{\text{DD}} - 2V_{\text{Th}})^3 \frac{t_{\text{r}}}{t_{\text{p}}}, \tag{3.11}$$

where β is the DC gain of the device, t_{r} is the rise time of the input waveform (assume equal to the fall time) and t_{p} is its period. The final contribution to Equation (3.9) is due to tunneling during transitions.

The *static* portion of power dissipation is the cumulative sum of all the leakage mechanisms active in the OFF *and* ON states and contains components I_3–I_9 in Fig. 3.2. The primary contributor in the ON state is direct tunneling current through the gate oxide. In the OFF state the tunneling current, with the addition of subthreshold leakage, dominates. A taxonomy of the different current components contributing to static and dynamic power is shown in Fig. 3.8.

3.4.2 Case Study: The 45 nm NOT, NAND, NOR CMOS Gates

From the discussions in Sections 3.3 and 3.4.1, it is anticipated that the estimation of power dissipation in a primitive logic gate (and, by extension, a functional unit) will exhibit a complicated dependence on the state of the individual transistors in the gate and hence on the inputs to the gate itself.

Since subthreshold leakage is present only during the OFF state, its impact is straightforward to assess. Direct gate-oxide tunneling current, on the other hand, is present during ON, OFF and transition states, which makes its evaluation and impact assessment possible only though circuit simulation.

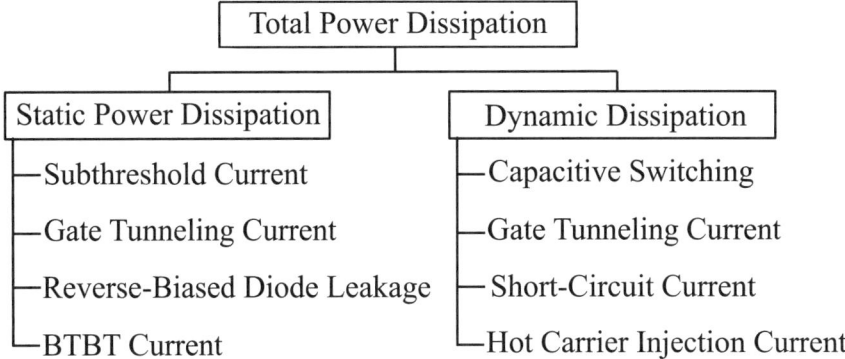

Fig. 3.8 Taxonomy of current components contributing to dynamic and static power dissipation

To obtain a quantitative and qualitative perspective on the tunneling current leakage mechanism in a logic gate, three prototype two-input gates (NOT, NAND and NOR) will be examined for a model 45 nm technology. Since both NAND and NOR gates are universal, it will be instructive to also determine which one performs better in terms of tunneling leakage.

Figure 3.9 shows the different tunneling conduction paths in a NOT gate (inverter) for the two possible input states.

With a gate that has two or more terminals, such as a NAND, the situation is considerably more involved. The actual leakage current paths for the different input combinations in a NAND gate are shown in Fig. 3.10.

To obtain a quantitative picture of the complex interrelation of tunneling current components, circuit-level simulation techniques must be used. Using the BSIM models for the prototype 45-nm process, a simulation of the NAND gate over all possible input states can be performed. Since this is a two-terminal gate, the possible states are "00," "01," "10" and "11." An interesting question that can also be

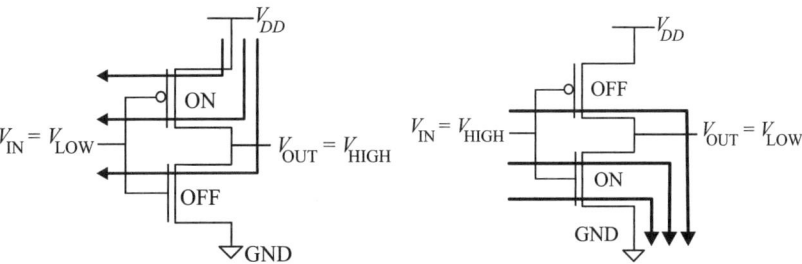

(a) Gate Oxide Leakage Path with input "0"   (b) Gate Oxide Leakage Path with input "1"

Fig. 3.9 Gate-oxide tunneling current paths in various switching states of an inverter logic gate. V_{Th} voltage drops have not been taken into account. The heavy arrows indicate the direction of the tunneling current flow in each device. In order to provide a clear perspective, individual components like I_{gd}, I_{gs}, I_{gcd} and I_{gcs} as in Fig. 3.6 are not shown

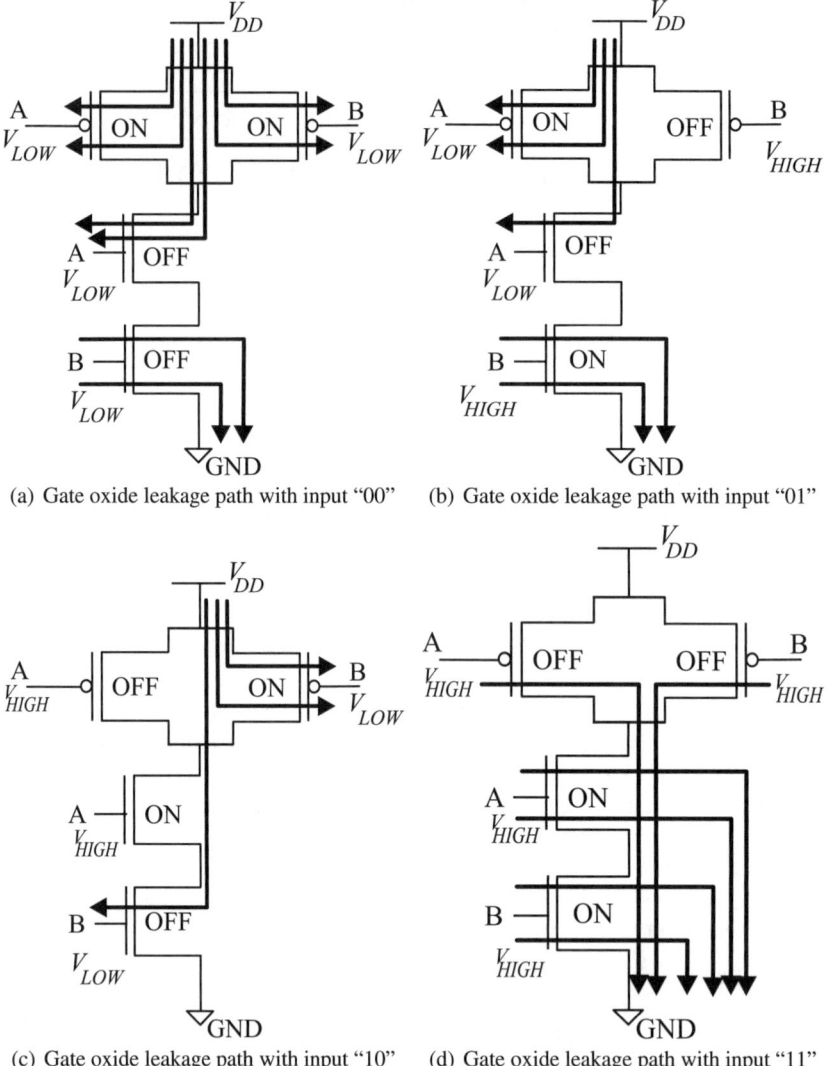

(a) Gate oxide leakage path with input "00"   (b) Gate oxide leakage path with input "01"

(c) Gate oxide leakage path with input "10"   (d) Gate oxide leakage path with input "11"

Fig. 3.10 Gate-oxide tunneling current paths in various switching states of a two-input NAND logic gate. V_{Th} voltage drops have not been taken into account. The heavy arrows indicate the direction of the tunneling current flow in each device. In order to provide a clear perspective, individual components like I_{gd}, I_{gs}, I_{gcd} and I_{gcs} as in Fig. 3.6 are not shown

answered with this type of characterization is whether a particular combination of inputs results in a worst-case scenario, as far as leakage is concerned. Examination of the simulation results in Fig. 3.11 demonstrates that the leakage current is strongly state dependent with state "11" providing the worst-case scenario.

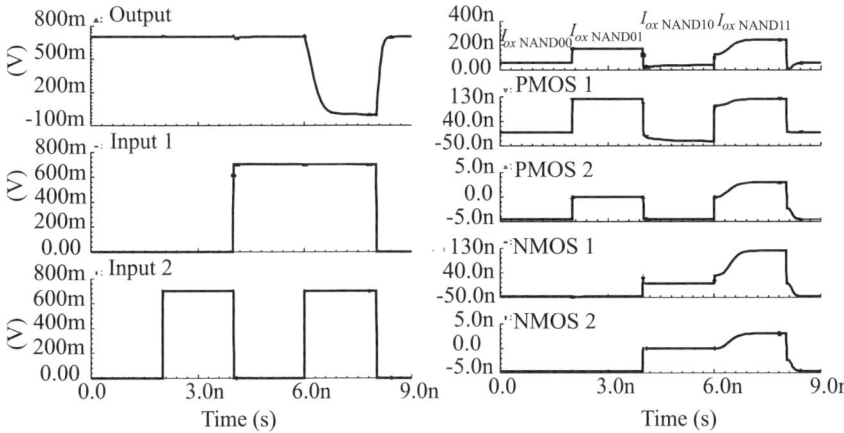

Fig. 3.11 Transient response of a two-input NAND gate for a fixed load. The left figures show voltages for two input voltages and the resulting output voltage. The gate-oxide tunneling current components in various individual transistors and the total value for the logic gate are presented on the right side figure

A similar analysis for the NOR gate is shown in Fig. 3.12 where it is seen that again the state "11" produces the highest leakage. Determination of optimal universal gates from graphs like these is not straightforward, and a more directly applicable figure of merit can be obtained by considering an average tunneling current. For a

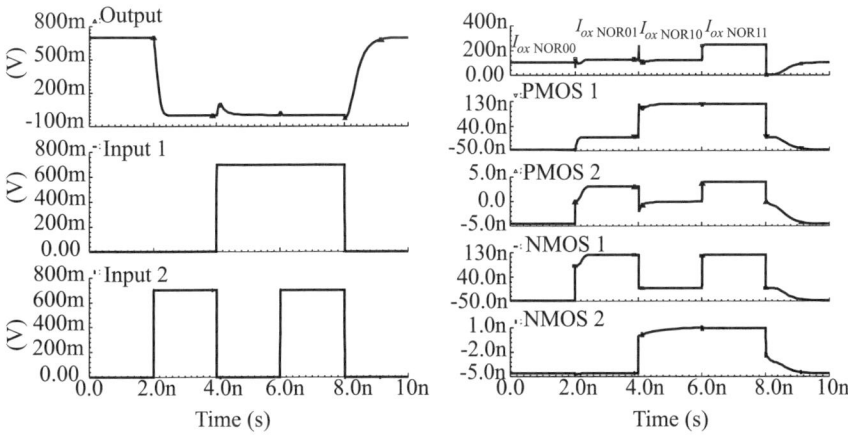

Fig. 3.12 Transient response of a two-input NOR gate for a fixed load. The left figures show voltages for two input voltages and the resulting output voltage. The gate-oxide tunneling current components in various individual transistors and the total value for the logic gate are presented on the right side figure

single terminal gate, there are only two possible inputs, "0" and "1." If I_0 and I_1 are the *total* leakage currents for each state, respectively, a current can then be defined:

$$I_{avg} = \frac{I_0 + I_1}{2},$$ (3.12)

where I_{avg} is the average leakage current flowing through the gate at a given time. Implicit in this definition is the assumption that the two states are equally probable. The definition of I_{avg} can be easily extended to a two-terminal gate, where the possible input states are "00," "01," "10" and "11," and the corresponding total leakage currents are I_{00}, I_{01}, I_{10} and I_{11}:

$$I_{avg} = \frac{I_{00} + I_{01} + I_{10} + I_{11}}{4}.$$ (3.13)

Again, it is assumed that all states occur with the same probability.

It is instructive to examine the dependence of these average tunneling leakage currents on the most important parameters appearing in Equation (3.7); namely, V_{DD} and T_{ox}, for the two universal logic gates NAND and NOR. Figure 3.13 demonstrates this dependence.

From Fig. 3.13, it can be seen that the NAND gate is preferred over the NOR gate because of its lower tunneling leakage current under equivalent V_{DD} and T_{ox} conditions.

Instructive as these graphs may be, they can be deceptive because they do not tell the whole story. Apart from the leakage current itself, another useful metric for estimating the impact of gate leakage on timing is the *effective tunneling capacitance* [164, 212, 165]. This metric, for a single transistor, is defined as follows:

$$C_{eff}^{tun} = \left| \frac{I_{ON} - I_{OFF}}{dV_g / dt} \right|,$$ (3.14)

where V_g is the voltage applied on the gate of the device and I_{ON} and I_{OFF} are the tunneling currents in the ON and OFF states, respectively, as indicated in Fig. 3.7(a). For simplicity, it is assumed that the rise (t_r) and fall (t_f) times of the gate input voltage are equal; thus, the two transition regions are symmetric with respect to their behavior during switching. This definition is then simplified to the following:

$$C_{eff}^{tun} = \frac{|I_{ON} - I_{OFF}|}{V_{DD}} t_r.$$ (3.15)

This definition can be extended to the case of a logic gate. However, the interaction of various ON and OFF devices in a logic gate complicates the analysis as seen in Figs. 3.11 and 3.12. A quantitative definition of the effective tunneling capacitance is formulated for a logic gate by considering the worst case among the transitions in the input switching states of the logic gate. Therefore, the variation

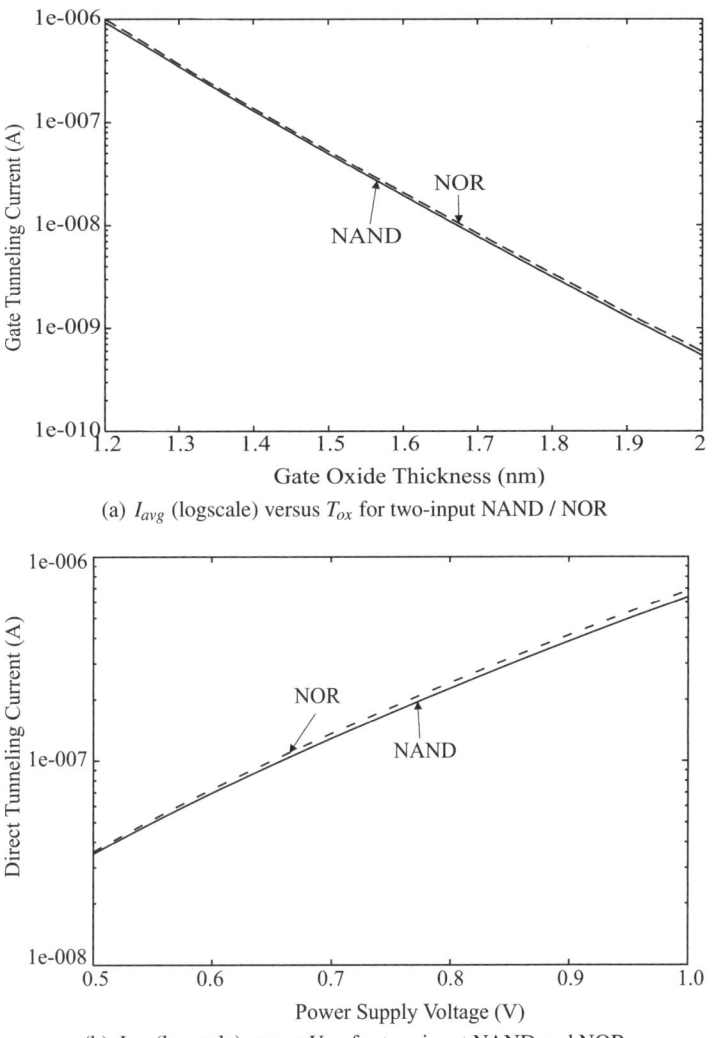

(a) I_{avg} (logscale) versus T_{ox} for two-input NAND / NOR

(b) I_{avg} (logscale) versus V_{DD} for two-input NAND and NOR

Fig. 3.13 Change in tunneling current due to oxide thickness and power supply variations. The NAND gate always exhibits lower leakage than the NOR gate

of the total tunneling current at this transition from I_{max}^{logic} to I_{min}^{logic} over a switching period can be used to define the worst-case effective tunneling capacitance of the logic gate as follows:

$$C_{eff}^{tun} = \left| \frac{I_{max}^{logic} - I_{min}^{logic}}{dV_{in}/dt} \right|, \tag{3.16}$$

where V_{in} is the input voltage waveform.

When the rise time and fall time are equal, the equation is simplified to the following:

$$C_{\text{eff}}^{\text{tun}} = \frac{\left| I_{\text{max}}^{\text{logic}} - I_{\text{min}}^{\text{logic}} \right|}{V_{\text{DD}}} t_r. \tag{3.17}$$

Using this definition for the effective tunneling capacitance of a *logic gate*, its dependence on V_{DD} and T_{ox} is plotted in Fig. 3.14.

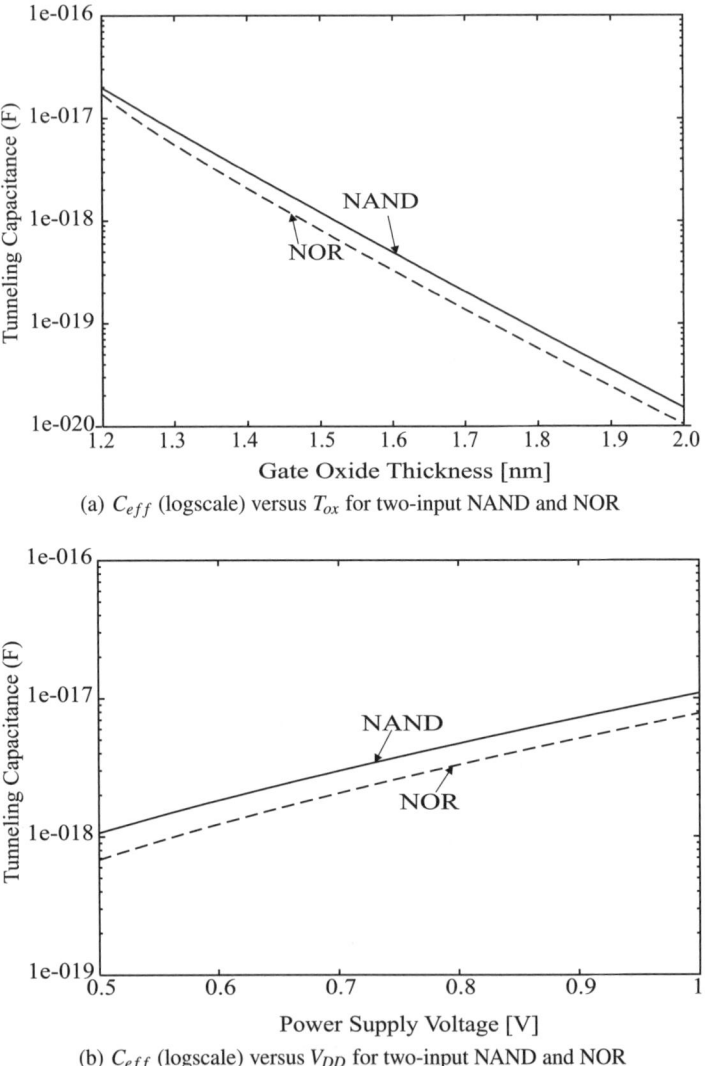

(a) C_{eff} (logscale) versus T_{ox} for two-input NAND and NOR

(b) C_{eff} (logscale) versus V_{DD} for two-input NAND and NOR

Fig. 3.14 Change in tunneling capacitance due to dielectric, oxide thickness and power supply variations. The NOR gate always exhibits lower capacitance than the NAND gate

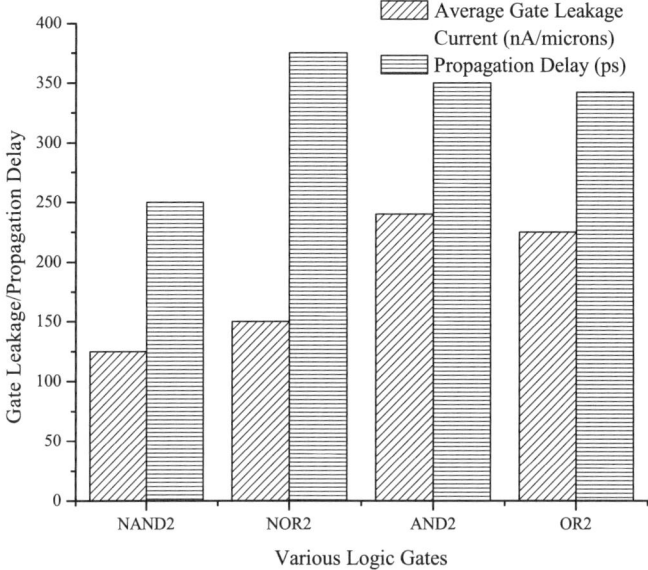

Fig. 3.15 Average gate-oxide leakage and propagation delay for various logic gates. It is evident from the figure that NAND2 has minimal average gate leakage and propagation delay, thus making it most suitable for functional unit realization for low gate leakage data path circuit synthesis

As is evident from the figure, even though the NOR gate produces more average gate leakage, it presents a lower capacitive fan-out load because of tunneling. The question of universal gates becomes now more complex when one tries to achieve balance between both leakage and timing.

Finally, a comparative view of various logic gates is presented, in addition to the NAND and NOR, in terms of average leakage current and *intrinsic* delay, i.e., the delay between input and output waveforms on the same gate, as shown in Fig. 3.15.

3.5 Process Variation Effects

3.5.1 Origins and Sources of Process Variation

In spite of the enormous technological progress that has given rise to nano-CMOS, some serious problems are pervasive to the technology. The most important of these problems, as far as design, synthesis and manufacturing of chips are concerned, is *process variation*.

Process variation is inherent in any process, and it is a function of quality engineers to bring the process in control and minimize the sources of extrinsic variation. The fact remains, however, that every process will have generic or intrinsic variation. The same is true for nano-CMOS processes. The situation is further aggravated by the fact that scaling often leads to process requirements beyond its intrinsic capability [42]. Variation in the process will invariably translate to variability in power and performance of the entire chip.

Examination of Equation (3.7) reveals the exponential dependence of direct tunneling current in the gate oxide, which can be a serious power dissipation source. When the oxide thickness itself is a few monolayers, it is very easy for T_{ox} to exhibit significant variation both within the same die (intra-die variation) and different dies (inter-die variation). When the power dissipation budget is very tight, the designer and the synthesis tools must consider this variation to produce working designs under a variety of process scenarios with acceptable yield. Even though T_{ox} is focused in this discussion, the same is true for numerous other process parameters such as threshold voltages, lithographically defined dimensions, etc., which directly affect the electrical performance of the devices.

To make the situation worse, the designer cannot usually count on very accurate power supply, not only off-chip but also on-chip, because of large parasitic IR drops.

3.5.2 Methodologies to Accommodate Process Variation

3.5.2.1 Corners Analysis

Presented with such variability, it is important to have design techniques (which translate to computer-aided design tools) that allow for exhaustive "what if" analysis in order to produce chips working in a variety of process capabilities and environments. One such technique is the "corners analysis" approach. To simplify the following discussion, it is assumed that only three process and design variables can affect the design, namely, gate-oxide thickness T_{ox}, threshold voltage V_{Th} and power supply V_{DD}. In a real environment, the number of variables can easily range in the tens or hundreds, but the methodology is essentially the same.

The corners analysis, which is an adaptation of the theory of design of experiments to circuit design, assumes that each of these variables has a minimum and a maximum value. The precise values are determined on experimental data from the wafer foundry. In fact, they are usually given in terms of statistical distributions, as can be seen later in the discussion. Therefore, if there are n variables, there are 2^n possible combinations ("corners") of their extremes. For the example of three variables, there will be a total of eight corners that can be visualized as a cube as shown in Fig. 3.16.

Since the cube represents the limits of the process, it is anticipated that any particular die will lie somewhere within this cube as far as the values of these particular process and design variables are concerned. If the design fulfills the specifications

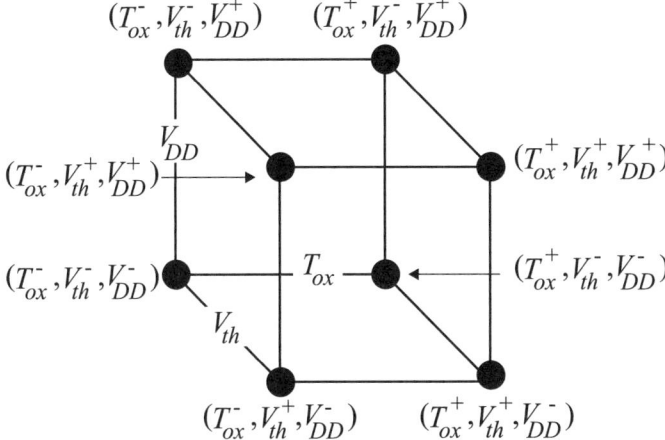

Fig. 3.16 Design cube for three process and design corners: gate-oxide thickness (T_{ox}), threshold voltage (V_{Th}) and power supply (V_{DD}). The superscripts $-$ and $+$ indicate the minimum and maximum values of the variable, respectively

in all eight corners, it will fulfill them anywhere in the cube. It is the task of the designer or synthesis tools to make the appropriate trade-offs to ensure that this is indeed the case. This problem is very difficult and in many cases automated synthesis algorithms may not be able to achieve design closure in all corners. Even in this case, knowing in which corners the design does not perform up to specification and by how much provides invaluable information about the anticipated yield of the design. This principle can be extended to arbitrary numbers of variables, and modern design frameworks available from electronic design automation (EDA) vendors provide these capabilities. Corners analysis has been a traditional tool for analog designers and for library characterization, but its usefulness for automatic synthesis cannot be overemphasized.

A variation of the corners technique is Monte Carlo simulation. The fundamental problem with corners analysis is that it lacks the "fuzziness" that real processes exhibit. When queried about the possible values of T_{ox} in a particular process, a wafer foundry will typically respond with statistical information, not minimum and maximum values. This statistical information is usually in the form of a probability density function (Gaussian, Uniform, Poisson, etc.) and its measures of central tendency (mean, μ) and dispersion (standard deviation, SD, σ). Armed with this information, the designer or synthesis tool must come up with a simulation strategy that will translate this statistical information about process and design variables to statistical information about design specifications. One could conceivably divide a region of the variable's range (say $\pm 3\sigma$ on each side of μ) into a number of points and then simulate the design at these points. The problem with this approach is that if there are n variables and each is sampled at N points, the total number of simulations needed to cover the process space is N^n. This number can reach astronomical values very quickly. An efficient alternative is to statistically sample the process space with a large number of samples. This is the basis of the Monte Carlo method (MCM).

3.5.2.2 Monte Carlo Method

In the MCM, a set of n random numbers (one for each process and design variable) is generated on the basis of the statistical properties of each variable. A simulation is then performed using these random numbers as the values of the process and design variables. This simulation is then repeated N times, where N is a fairly small number such as 100 or 1000. Thus, the process space is sampled with N points instead of N^n points. If the random number generators are of good quality, this approach typically provides excellent coverage of the process space with a fairly small number of simulations.

The input for the MCM is a set of statistical distributions for the process and design parameters, and the output is a set of statistical distributions for various design specifications of interest. The "engine" that transforms the inputs into outputs is typically SPICE; the process is pictured in Fig. 3.17. The output specification statistical distributions can then be used by the synthesis tools to make appropriate trade-offs for power, speed, etc.

An actual result of an MCM characterization of a 45 nm NAND gate is shown in Fig. 3.18. In this experiment, it was assumed that T_{ox}, V_{Th} and V_{DD} are normally distributed with means the nominal values for this process (1.4 nm, 0.22 and 0.7 V, respectively) and standard deviation equal to 10% of the mean. $N = 1000$ SPICE Monte Carlo runs are performed, and statistics on the specifications of interest are collected. For this experiment, total gate leakage (Fig. 3.18(a)), total subthreshold leakage (Fig. 3.18(b)), dynamic current (Fig. 3.18(c)) and propagation delay (Fig. 3.18(d)) are of interest. A load equal to 10 times the pMOS gate capacitance

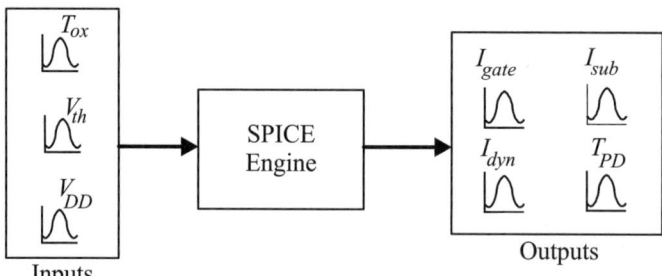

Fig. 3.17 Monte Carlo simulation methodology to account for process and power supply variation. The inputs are statistical distributions of the variables. The processing engine is SPICE executing a large number of Monte Carlo runs. The outputs are statistical distributions of specification parameters of interest for the entire circuit, such as total gate leakage I_{gate}, total subthreshold leakage I_{sub}, total dynamic current I_{dyn} and propagation delay T_{PD}

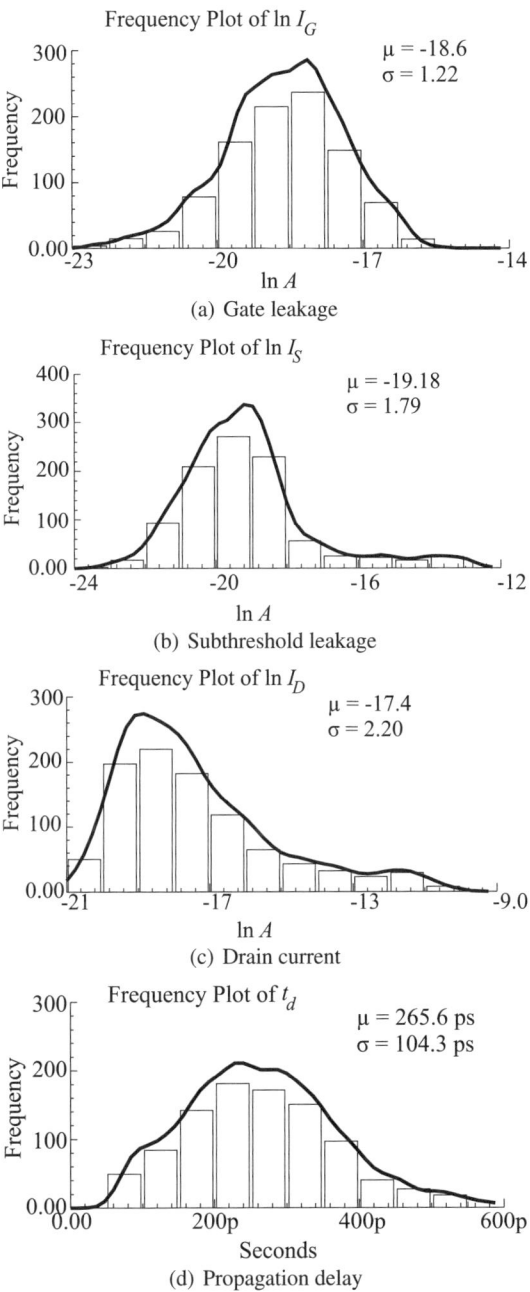

Fig. 3.18 Effects of statistical process variation on gate-oxide leakage, subthreshold leakage, dynamic current and propagation delay in a two-input NAND gate. It is observed that variability of gate-oxide leakage, subthreshold leakage, dynamic current is lognormal in nature and that variability of propagation delay is normal in nature. A load equal to 10 times the pMOS gate capacitance has been added to the output of the gate [120]

has been added to the output of the gate [120]. As discussed above, the outputs are actually statistical distributions characterized by means, standard deviations and distinct probability density function shapes. The propagation delay exhibits a normal distribution but the currents are lognormal in shape; this is anticipated because they depend exponentially on quantities having normal distributions.

3.6 From Gates to Functional Units: A Power Modeling and Estimation Perspective

The previous section shows how the basic device physics for nano-CMOS transistors gives rise to a multitude of current components that contribute to the power dissipation of the devices as they perform their essential switching functions. It is also observed that when the devices are assembled into logic gates, the interplay of current components becomes more complex and depends very strongly not only on state transitions but also on the states themselves. It is natural to attempt to extend the understanding of power modeling and estimation from single devices and gates to more complex subsystems, namely, *functional units*. Functional units are medium-sized digital blocks that perform standard logic functions. Examples are adders, subtractors, multipliers, comparators, dividers, registers, ROMs and small dynamic and static random access memory (RAM) blocks. Each functional unit can consist of tens up to thousands of gates. The question then arises: how can the power dissipation of such a functional unit be modeled and estimated, based on data from logic gates?

A pictorial view of such a functional unit, seen from three different levels of abstraction (behavioral, gate and transistor), is shown in Fig. 3.19.

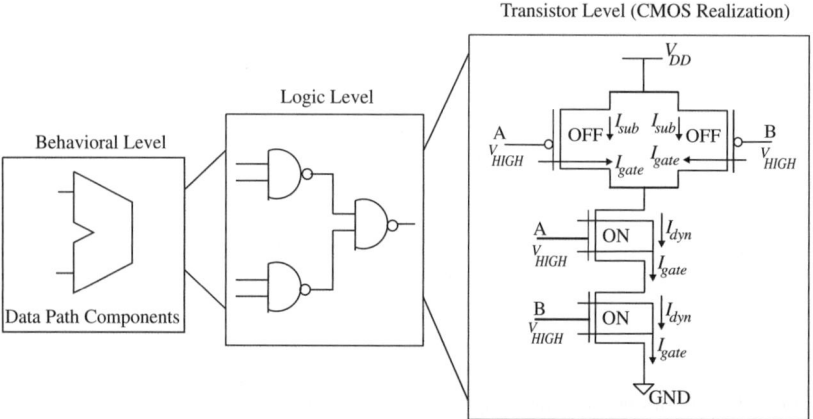

Fig. 3.19 Three levels of abstraction in which data path components are realized using two-input NAND gates

3.6.1 SPICE level

The most accurate means of profiling the power consumption of such functional units is through transistor-level circuit simulation, i.e., SPICE simulations. If the transistor models are accurate, the simulation results can be extremely accurate. The shortcoming of this approach is immediately apparent: large logic gate and transistor counts for each unit make the simulation time prohibitive. The situation is further aggravated by the fact that the power profile is strongly dependent on the state of the unit, which implies the need for complete characterization. Consider, for example, the case of a 16-bit adder with two inputs. A complete coverage of possible states requires input vectors for 2^{32} distinct cases. Such simulation vectors are possible with event-based, logic-level simulators but not with continuous-time tools like SPICE. A compromise solution to this problem is to use SPICE to characterize exhaustively the logic gates used in the construction of the functional unit and obtain a power profile of the unit by assuming a worst-case power dissipation for each gate. The total power consumption is then the sum of the profiles for each gate. This approach is often combined with corners analysis to provide a worst possible case power estimation for a functional unit. Results obtained for a variety of functional units using this technique are illustrated in Fig. 3.20. The corners referred to in this figure correspond to those of Fig. 3.16.

This approach is also amenable to statistical characterization of the functional unit performance metrics, e.g., by using the MCM outlined in Section 3.5. For instance, a library of data path components and functional units can be developed using universal NAND logic. At the architectural level, a state-independent approach can be followed by using the state average data derived from the SPICE-characterized NAND gate. The fact that some of the leakage currents do not follow normal distributions (as seen in Fig. 3.18) can be accounted for at the gate level via the use of the central limit theorem (CLT) [237]. Since a typical functional unit consists of hundreds of NAND gates, the leakage, dynamic and subthreshold currents for the total unit will be normally distributed even though the same currents are lognormally distributed for each individual gate. The currents and the delay for the functional units can thus be modeled by utilizing the characterized data for the two-input NAND gate. The total current in the functional unit (FU) can be defined as the sum of currents in the individual NAND gates comprising the unit. Assuming that the distributions for each gate are statistically independent of each other, the mean and variance of the currents can be derived as follows:

$$\mu_{FU} = N \, \mu_{NAND}, \tag{3.18}$$

$$\sigma_{FU} = \sqrt{N} \, \sigma_{NAND}, \tag{3.19}$$

where there are N NAND gates in the implementation of the FU. The assumption of statistical independence for all gates in a given FU implies that there are no statistical correlations between adjacent gates because of spatial effects. The approach presented here can be modified to account for such cases. From these equations, the

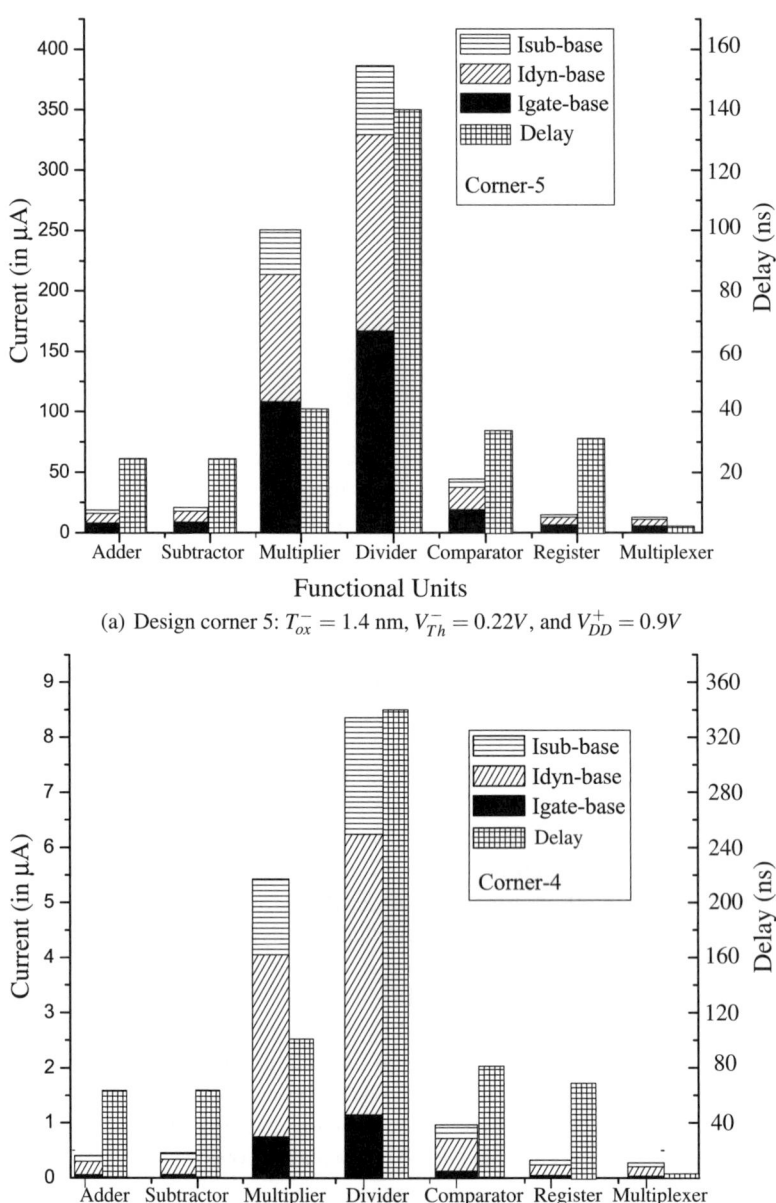

(a) Design corner 5: $T_{ox}^- = 1.4$ nm, $V_{Th}^- = 0.22V$, and $V_{DD}^+ = 0.9V$

(b) Design corner 4: $T_{ox}^+ = 1.7$ nm, $V_{Th}^+ = 0.25V$, and $V_{DD}^- = 0.7V$

Fig. 3.20 Nominal results showing individual components of power consumption for different output corners. It may be noted that the total current values are reduced and that the proportions of different components in the total current have changed. Only two corners are shown for brevity. The corners referred to in this figure correspond to those of Fig. 3.16

mean and variance of I_{gate}, I_{sub}, I_{dyn}, and delay T_{PD} for each of the FUs can be calculated. The use of universal NAND gates simplifies the construction of the cell library containing FUs like adder, subtractor, multiplier, etc. Other types of logic gates can be used to build data path and FU component libraries by applying the above statistical expressions provided that the number of individual logic gates in an FU is large enough to justify the use of the CLT, a realistic assumption for real-life designs.

Finally, this SPICE methodology allows for the characterization of specific power components as a function of process parameters, as shown in Fig. 3.21, where the dependence of total gate leakage current on oxide thickness is shown. Information of this kind is invaluable for synthesis tools attempting to optimize power–performance using resources of various characteristics (different oxide thicknesses in this case).

3.6.2 Probabilistic and Statistical Techniques

From the discussion in the previous section, it can be seen that complete and accurate power profiling of large FUs is not feasible using analog circuit simulation (SPICE) methods. In this section, statistical and probabilistic approaches are presented that can be used if detailed logic gate-level SPICE data is available. It is also assumed in the ensuing discussion that the NAND gate is used universally. The method can be applied to other cases with slight modifications.

It is assumed that there are total n_{total} NAND gates in the network of NAND gates constituting an n-bit FU. In this model, the effect of interconnects is not considered;

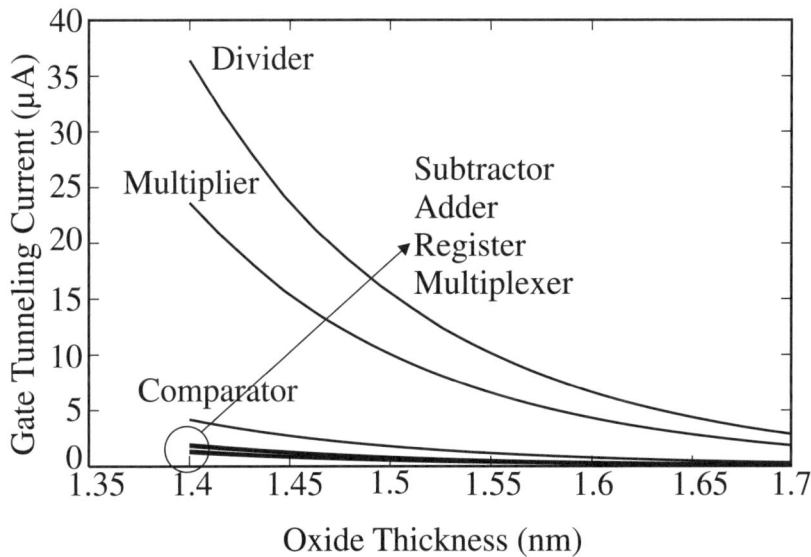

Fig. 3.21 Variation of gate-oxide leakage current with respect to gate-oxide thickness

the focus is on the power profiling of the FUs only. It may be noted that this assumption does not affect the dominant current leakage mechanisms as they happen only in the active devices but not in the interconnects. It does, however, ignore ohmic drops, which need to be accounted for in a complete methodology. In the following discussion, the notation p_a will be used to indicate the probability that node a in the FU is at logic level "1." Note that the node a is not necessarily an input node but can be an internal node. Similarly, $\bar{p}_a = (1 - p_a)$ is defined as the probability that the node a is at logic level "0". Even though the discussion assumes two inputs per gate, the theory can be easily generalized for multi-input gates.

From the SPICE characterization we described in Section 3.4.2, the total leakage current per gate is assumed known for each input combination (namely, the four currents I_{00}, I_{01}, I_{10} and I_{11}).

The average tunneling current for a logic gate with inputs a and b is then calculated as follows:

$$I_{\text{NAND}} = \bar{p}_a \bar{p}_b I_{00} + \bar{p}_a p_b I_{01} + p_a \bar{p}_b I_{10} + p_a p_b I_{11}. \tag{3.20}$$

Once the current per gate is known, the total tunneling current of an n-bit functional unit is calculated in the following manner:

$$I_{\text{FU}} = \sum_{j=1}^{n_{\text{total}}} I_{\text{NAND}} j, \tag{3.21}$$

where the index j runs over all the NAND gates in the functional unit. In order to use Equation 3.21 effectively, the probabilities p_a and p_b must be known. If a and b are primary inputs to the FU, then the assumption $p_a = p_b = \frac{1}{2}$ can be used. It can be seen shortly, however, that this can underestimate the total gate leakage current.

Consider the simple combinational logic circuit shown in Fig. 3.22. Simple probabilistic analysis indicates that if the two primary inputs a and b are equally probable ($p_a = p_b = \frac{1}{2}$), the internal node c has $p_c = \frac{3}{4}$ and not $\frac{1}{2}$. If it is assumed that all internal nodes are equally probable at logic level "1," then the total gate leakage current for this circuit would be as follows:

$$I_{\text{gate, circuit}} = 2 \times I_{\text{NAND}} = 2 \times \left(\frac{1}{4} \right) (I_{00} + I_{01} + I_{10} + I_{11}), \tag{3.22}$$

where I_{NAND} is the average leakage current per NAND gate. Using the data in Fig. 3.11 gives an estimate of $I_{\text{gate, circuit}} = 255.6 \frac{\text{nA}}{\mu\text{m}}$. On the other hand, using the probabilistic analysis mentioned above, the following is obtained:

$$I_{\text{gate, circuit}} = I_{\text{NAND}} + \bar{p}_c \bar{p}_b I_{00} + \bar{p}_c p_b I_{01} + p_c \bar{p}_b I_{10} + p_c p_b I_{11}, \tag{3.23}$$

which results in $I_{\text{gate, circuit}} = 296.6 \frac{\text{nA}}{\mu\text{m}}$. Clearly, there is a potential for *underestimating* the leakage current.

To address the problem of unknown a priori probabilities for the internal nodes of FUs, one approach is to simulate at transistor level the FU using SPICE. As it is

Fig. 3.22 A simple combinational logic circuit. The fractions indicate the probabilities that the indicated node will be at logic level "1"

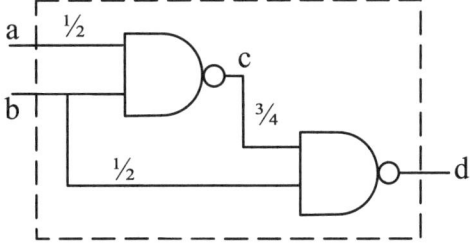

already seen, in order to obtain statistically meaningful probabilities, the simulations must assume random input vectors of very large length. Considering that each gate comprises four transistors and each FU may contain hundreds or thousands of gates, it becomes apparent that such simulations are not expected to complete in a realistic amount of time.

Another way to estimate the required probabilities is via the use of high-level HDL (such as Verilog or VHDL) simulations to perform probabilistic analysis of FU internal node probabilities. Consider the NAND implementation of a half-adder shown in Fig. 3.23. Structural Verilog simulations are performed for this combinational circuit assuming random input vectors a and b with 10^6 bits each. From these simulations, the probabilities indicated in the figure are determined. The total simulation time is 4.7 sec clock time on a standard PC. A similar SPICE simulation would have run times in the order of many hours.

If equal probabilities is assumed for all nets, the leakage current estimate for the circuit would be

$$I_{\text{gate, circuit}} = 5 \times I_{\text{NAND}} = 5 \times \left(\frac{1}{4}\right)(I_{00} + I_{01} + I_{10} + I_{11}) \qquad (3.24)$$

or, $I_{\text{gate, circuit}} = 639\frac{\text{nA}}{\mu\text{m}}$. On the other hand, using the probabilistic analysis mentioned above, $I_{\text{gate, circuit}} = 829\frac{\text{nA}}{\mu\text{m}}$ is obtained, an almost 23% underestimation of the leakage current. If a conservative approach is used and it is assumed that the NAND gate current is the worst case (which, from Fig. 3.11 is I_{11}), a leakage current of $I_{\text{gate, circuit}} = 1238\frac{\text{nA}}{\mu\text{m}}$ is obtained, an almost 50% *overestimation*! Clearly, without an HDL simulation for the probabilistic estimation of logic levels for the

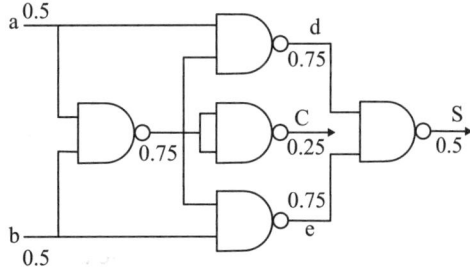

Fig. 3.23 A NAND implementation of a half-adder. The numbers indicate the probabilities that the indicated net will be at logic level "1"

FU, the chosen implementation can underestimate or overestimate the leakage current, particularly for large FUs with many NAND gates.

Since event-based HDL simulations have short execution times (compared to full-scale analog circuit simulation), their use in the probabilistic estimation of logic levels, in combination with accurate gate-level characterization data, can provide realistic and accurate power profiling.

3.7 Summary and Conclusions

The relentless drive for miniaturization has pushed CMOS technology deeply in the nanometer regime. As a consequence, a host of physical effects, previously unimportant, have come into play to produce sources of power dissipation in nano-CMOS devices that can dramatically affect their performance.

In this chapter, the sources and the relative magnitude of the most important of these sources were briefly reviewed during the three distinct operational regimes of the nano-CMOS in a digital circuit: the ON state, the OFF state and the transition state. A new leakage mechanism, direct tunneling current through the gate oxide, has come to dominate, along with subthreshold leakage and power dissipation and will be the major challenge in the integration of digital and analog components into an SoC. It also became apparent that controlling this leakage mechanism will require new technological breakthroughs (high-κ dielectrics, metal gate, etc. [74, 46]) and new ways of design because it is pervasive: it is important in all three modes of operation of a transistor. Solid understanding of the physical mechanisms responsible for leakage in nano-CMOS devices can be encapsulated in advanced device models for circuit simulation, such as BSIM. Armed with accurate models, the designer can explore and characterize exhaustively the performance of the building block of digital circuits, the logic gate. It is seen that even the smallest universal gate, a two-input NAND can exhibit surprisingly complex behavior, with regard to leakage, depending on its state. This has significant consequences for the characterization and power profiling of the next level of hierarchy, the FU. The concept of effective tunneling capacitance was also introduced, which helps quantify the impact of tunneling on delay estimation.

Closely related to the problem of gate and FU characterization is the issue of process and design parameter variation. Nano-CMOS processes have intrinsic variations that are large because of the small dimensions involved. Both on-chip and off-chip power distributions include variability. To guarantee the functionality of a design and to account for process and design variation according to DFM principles, the designer uses statistical techniques such as corners analysis and the MCM. Examples of their applicability to the design of nano-CMOS digital circuits were presented in this chapter.

The problem of extending the profiling and characterization methodologies in common usage from the realm of the logic gate to an FU that may consist of a few tens to a few hundreds or even thousands of gates was then tackled. Complete SPICE simulation would be desirable but is not practical for two reasons: the large transistor count of the units and the complexity of simulation vectors that must be considered. Both requirements lead to excessive simulation times. A compromise solution was presented where accurate SPICE data were combined with HDL simulation estimates of logic probabilities in the operation of the FUs. Accurate probabilistic estimates are extremely important because, as it was seen, equally probable assumptions can lead to serious *underestimation* of leakage, whereas worst-case data will invariably lead to serious *overestimation* and will unduly restrict the designer.

When making the transition from the FU to the next level of hierarchy, one must consider the new approaches in profiling and characterization of the subsystem or complete system. These new approaches are the subject of the next chapter.

Chapter 4
Architectural Power Modeling and Estimation

4.1 Introduction

Modeling, estimation and analysis of power and thermal dissipation form an integral and important part of VLSI design. They go hand in hand with power optimization and design for low power. Energy and power efficiency are now fundamental concerns from the ultra-mobile to server classes of computing applications. At any level of abstraction, a design improvement or power–performance trade-off will imply an ability to estimate or analyze power and its efficiency for the different options and choices in the design, synthesis or optimization processes. To enable a shorter development time and to make rational decisions about a design, one needs to rely on sufficiently efficient and accurate power estimation. However, the estimation, which relies on modeling at high level and the analysis that ensues it, is fairly complex with many subtleties. Moreover, the estimates may not be even a single number or a set of deterministic numbers but rather distributions, as will be seen later on. Some of these nuances and their contextual significance are introduced below.

Leading commercial processors have microarchitectural operating points called C states: C0–C3 are often the *active* states, whereas C4–C6 are *sleep* states. During active states (the lower C), in technology nodes 90 nm and beyond, processor leakage power tends to be as high as about 35% of the total power. On the other hand, in the higher C states, nearly no node in the circuit is toggled, and almost 100% of the power is leakage. *Active leakage*, which occurs during the normal activity of a processor, contributes toward the thermal design power (TDP) of the processor – an important parameter that defines its cooling system. System reliability, safety and performance critically depend on TDP. Maximum or *peak power* attempts to characterize the maximum power that a system, processor or core can produce in a specified voltage range. TDP is a fraction (approximately 80%) of the maximum power as it depends on real applications running on the target device. TDP is constrained by the cooling system in place and the junction temperatures allowable for safe and reliable operation of the transistors and other nanodevices.

S.P. Mohanty et al., *Low-Power High-Level Synthesis for Nanoscale CMOS Circuits,*
DOI: 10.1007/978-0-387-76474-0_4, © Springer Science+Business Media, LLC 2008

Average power, on the other hand, is a combination of the two broad categories of operating points mentioned above. Average power is highly important for mobile systems running on battery because the processors spend a fair amount of time in different sleep states. Contribution of leakage to average power may typically be as high as 55%. Also related is the *standby power* incurred in "shallow" sleep states where very little useful computation occurs; a few critical nodes are toggling and monitoring for possible wake-up of the rest of a chip or system. Hence, estimation of power does not correspond to a single number or a single attribute, nor does it lend itself to a single method of computation.

Designers have other reasons to model/estimate various kinds of power–such as burn-in power–often important to identify parametric losses and infant mortality through defect acceleration. This estimation is crucial to be able to insert the right circuit hooks to trade off between cache leakage and core leakage during burn-in. Note that exponential dependence of leakage on temperature is making burn-in increasingly less viable because power dissipated in the burn-in oven can burn the chip.

Sometimes energy is emphasized over power, which is the time derivative of the former, for special portable applications. Some of the most efficient batteries such as Li ion today provide an energy density of about 80 Watt-hours per pound. Energy efficiency is related not only to power management and power delivery issues but also to an environmental footprint of computing. The U.S. Environmental Protection Agency in its report to Congress on Server and Data Center Energy Efficiency Public Law (dated August 2, 2007) estimates that if current trends continue, an additional 10 power plants will be required by 2011 to meet the demands of U.S. data centers alone.

Estimation of power is important for effective thermal design of a device. Moreover, device thermal dissipation and power are becoming increasingly coupled because of the strong, exponential dependence of power on temperature and temperature's dependence on spatial and temporal distribution of power on a chip combined with the thermal resistances involved in the cooling system design of the device. A typical cooling setup of a processor is shown in Fig. 4.1.

Deriving the thermal map of a system involves using suitable die and package thermal models along with a power map of the die under its usage condition. Since power affects temperature and vice versa, iterative computation (temperature simulation) is usually performed to arrive at a converged map. A thermal map analysis also identifies *hot spots* (high-temperature areas) and reliability issues in a design.

The amount of effort for estimation and optimization should also be commensurate with the job at hand. In other words, it is important to know which part of the full system contributes how much power so that those components with the most room for improvement can be addressed. The power breakdown of two typical optimized industrial platforms are presented below: Fig. 4.2 represents an estimated power breakdown for a mobile platform, whereas Fig. 4.3 is for a desktop machine.

Significant loss in power supply for the desktop indicates a need to consider integrated voltage regulators, and prominent display power on a mobile machine indicates a need to consider smarter display management, for example.

It may be noted that power estimation often presupposes a Pareto-optimal design because sensitivity analysis for design choices is not very meaningful, nor is high-

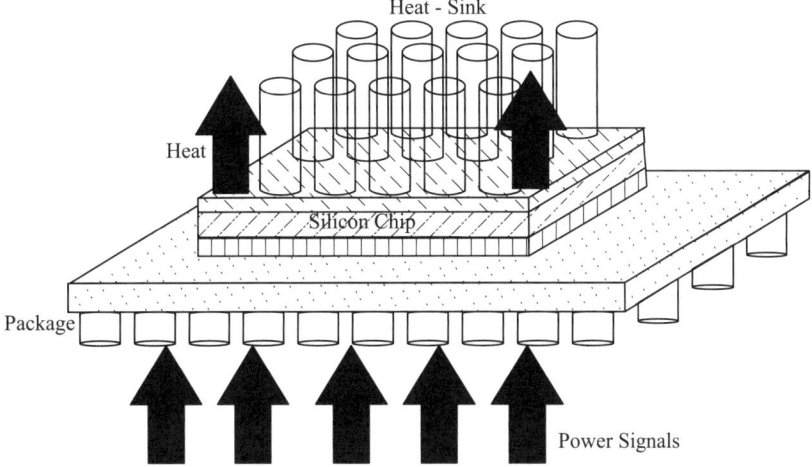

Fig. 4.1 Typical processor cooling setup

level estimation effective for suboptimal designs. A typical representation of energy delay trade-off is shown in Fig. 4.4. The figure provides a Pareto-optimal plot where for a given energy or power budget, the delay/latency of a circuit is optimized and vice versa.

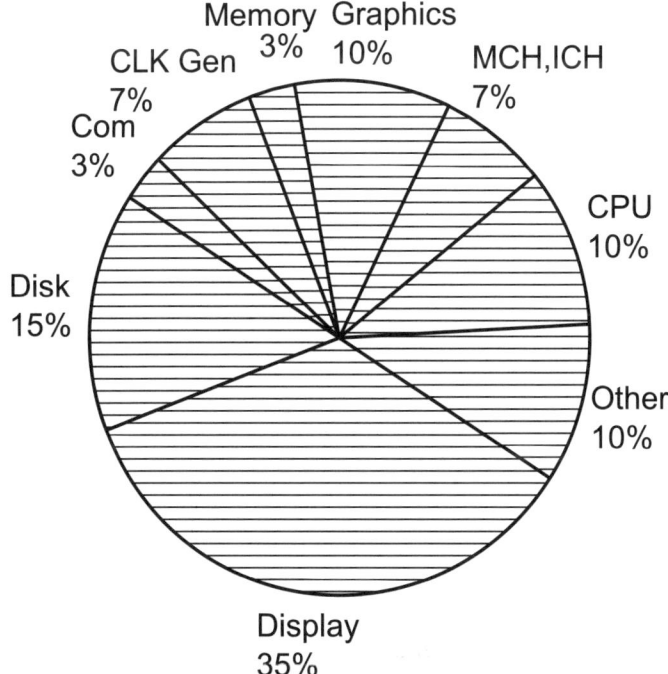

Fig. 4.2 Mobile platform power breakdown

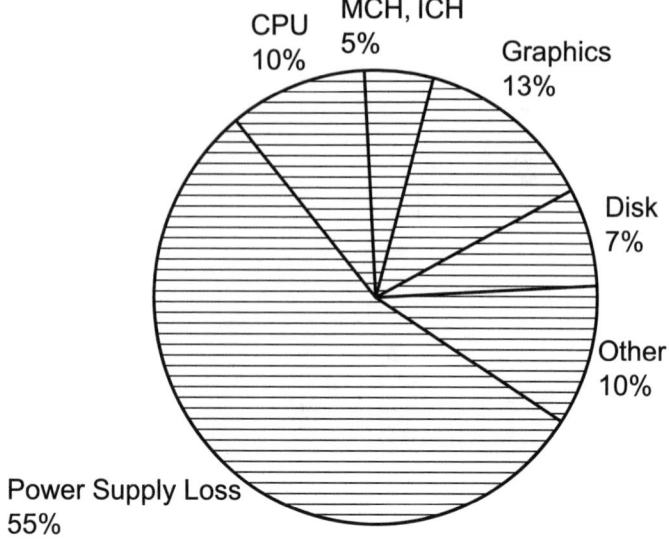

Fig. 4.3 Desktop platform power breakdown

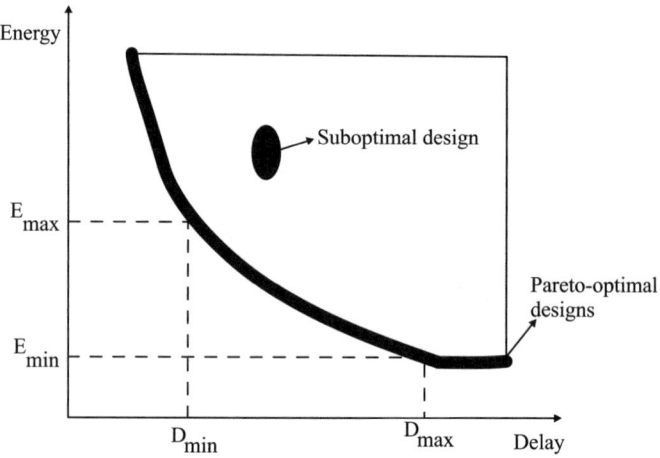

Fig. 4.4 Energy delay trade-off in design

4.2 Architecture-Level Estimation

Industrial and academic investigations have by now made amply clear the performance-limiting impacts of power dissipation. Hence, power estimation, an integral activity for power reduction, is done at different levels of abstraction: algorithmic and product-definition level, architectural, or RTL, and pre- and post-layout gate or circuit levels. The combination of advanced and complex architectures with the increasing complexity and cost of thermal controls integrated into these architectures has driven the importance of studying the design performance versus power trade-offs as early as possible in the design cycle, often at the architectural level. It is at this level that significant trade-offs are analyzed and performance decisions are made for commercially available microprocessors and SoC. The advantages of better exploration-based design decisions and faster analysis time are the hallmarks of estimation at the high levels of abstraction, but accuracy is a trade-off as indicated in Fig. 4.5. For application-specific designs, estimation can be done at an even higher level of abstraction – the algorithmic level. However, we restrict our focus to the architectural level in this book.

There are four major reasons and objectives for performing RTL power analysis and estimation:

- *Early prediction*: After initial RTL coding, the RTL power simulations can help determine power issues in the design and whether the design is likely to meet the requirements on power and thermal dissipation of the design. If the design is projected not to meet the power specifications, either more low-power design tricks can be implemented to decrease power consumption or potential thermal solutions, such as thermal throttling, can be identified and implemented. The availability of power estimates at the pre-synthesis stage is significant: It is early enough in the design cycle to make changes in the microarchitecture. Changes at the microarchitecture level have the most profound impact on power. The earlier the power issues are flagged in the design period, the easier it is to resolve issues without affecting the product schedule.
- *Power budgeting and hot-spot detection*: Designers and architects often need to know where in the chip the power goes. A major design activity is to allocate and

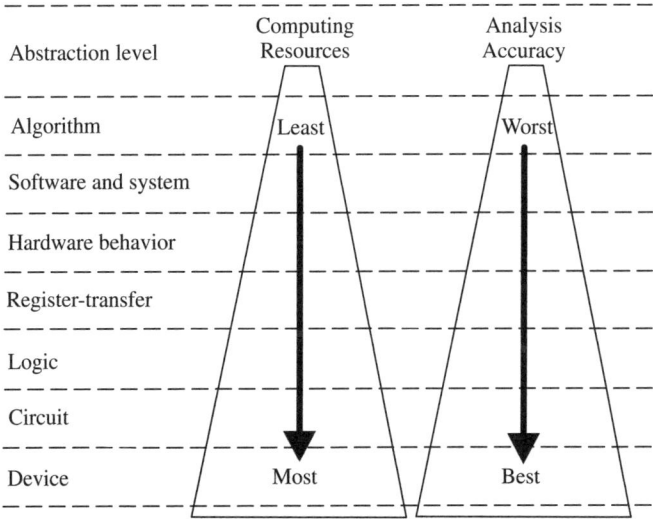

Fig. 4.5 Accuracy and analysis time trade-off for various design abstractions

subsequently verify power budgets for different components or the entire system. RTL power simulations can provide a detailed power breakdown for the design: FUs, clock network, I/O ring and embedded blocks. The simulation results point out the hot spots in the design. The designer can then focus attention on these hot spots and apply appropriate low-power techniques efficiently. Design effort can be prioritized to focus on the root causes of power consumption so that engineering resources may be allocated efficiently; this reduces time to market. For instance, if the power breakdown shows that clock power is a dominant part of the full-chip power, then either clock gating or power-optimized sequential cell elements can reduce the power significantly.

- *Priming the power optimization loop*: Estimation is in the "inner loop" of optimization at any level of abstraction. This enables power versus performance versus area trade-offs and the design of cooling and power delivery mechanisms. Speed of estimation is quite important here.
- *Evaluating low-power guidelines*: RTL power simulation or modeling can answer the question, how much power would be saved if a particular technique is implemented? For power-critical designs, design guidelines or rules may be applied to the design methodology to ensure that no power is wasted in a unit in idle state. To enforce the rules, power validation may be applied to identify any power bugs in the design that do not follow these guidelines by observing the units' relative power dissipation. One can also study the effect of clock buffer on the clock-distribution power through RTL power modeling and simulation.

High-level (architectural) estimation trades off accuracy for speed but copes with the lack of complete and concrete design data at the time of estimation. Even so, there is a growing concern about erosion in fidelity of classical power models due to the increasingly complex microarchitecture features and the rapidly changing technology and circuit landscape. However, designers demand at least relative (or *tracking*) accuracy of any estimation techniques and tools. Since design entry is often at an architectural level or RTL, estimation depends highly on design descriptions of the functional blocks of a CPU or the IP cores in an SoC, the target process technology and usage conditions or the software. Power estimation at any level of a CPU design requires the capability to estimate switching capacitance, with leakage properly approximated. One can solve the RTL power estimation problem using two different approaches: (a) The top-down approach tries to predict power using simple algorithmic models and does not consider circuit implementation styles. Analytical models are used that correlate design complexity with power consumption. (b) The bottom-up approach involves, for better accuracy, macro-modeling of blocks such as controllers, register files, programmable logic arrays (PLAs), cache and custom memories and the sequential and the clock buffers.

A typical commercial tool such as Sequence Design's WattWatcher has the following typical features: In terms of accuracy, simulated power is within about 25% of silicon measurements for all power benchmarks; in terms of speed, simulation time less than RTL simulation for a given benchmark; and for overall functionality, full chip with I/Os, clocks and embedded blocks like SRAMs. As for reporting, it shows results graphically and hierarchically with data and clock power reported separately. Figure 4.6 illustrates the flow of simulation for the WattWatcher tool.

The tool first performs a high-level synthesis of the RTL design, while using the information from the previously generated library. The library contains information about the process, logic cells, input/output (I/O) cells and embedded blocks like SRAM and phase-locked loop (PLL). Next, based on signal activity factors at all the RTL nodes generated from RTL simulations, internal algorithms in the WattWatcher engine estimate power using the additional information on clocking style used and capacitive loading for the design.

Typically, the user provides the information on process and technology, logic and I/O cells, analog and SRAM devices, capacitance loadings and clocking style. Process information includes numbers of layers of metal in the process, capacitance and maximum current density in each layer. The logic cells used in the design are characterized for power and are included in the library. For the WattWatcher tool to perform the high-level synthesis, one also needs to specify which cells are expected to be the default cells for cell categories like flip-flop, latch, NAND, inverter, XOR and MUX. I/O cells can be characterized by the circuit designers for power and included in the WattWatcher library. SRAMs can be modeled by the WattWatcher format specification, which assumes that total power dissipation is a sum of power dissipation for the smaller sub-tasks. The power for the sub-tasks can be estimated by circuit simulation or by hand calculations. Capacitive loading is the loading seen by the chip on the board. Clock tree design style is specified along with the names and the number of fan-outs of the root, branch and leaf drivers.

WattWatcher-like tools require availability of power benchmarks: Power benchmarks are tests that are developed and used specifically to estimate or measure power consumed by a component. Although power benchmarks have traditionally been used for post-Silicon power measurement for chips, they can also be used very effectively for pre-silicon power prediction. Typical benchmarks to estimate power consumption of core logic are as follows:

- Maximum-power benchmark (power virus)
- Maximum-application-power benchmarks
- Minimum-power benchmarks
- Power-on-suspend

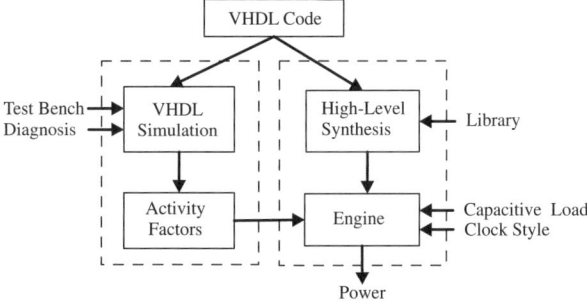

Fig. 4.6 Simulation flow for RTL power simulation by WattWatcher tool

- Suspend-to-RAM
- Suspend-to-disk
- Idle-power benchmark

Minimum-power benchmarks are used to estimate power consumed during low-power *suspend* modes. These benchmarks are developed based on the suspend-mode-entry methods described in the component specification. The idle-power benchmark is used to estimate component power consumption when it is not doing any useful work. This benchmark is useful for determining the efficiency of power-savings features such as clock gating.

Traditionally in the literature, high-level estimation implied modeling/analysis of macro-blocks such as ALUs, register files, vector operators, multiplexors and finite state machines (FSMs). But industrial central processing unit (CPU) experience shows that only about 16% of the active power is in logic gates, whereas roughly 20% is from die leakage. Most of the rest are in global and local clock distributions and in clock nodes inside sequential blocks.

Macro-modeling involves generating power models at the microarchitectural block level. There are two distinct approaches to incorporating these models: one that loosely couples pre-characterized power models in look-up tables (LUTs) to access estimated power of a component based on the Hamming distance between two consecutive input vectors [356]; and the other, which attempts better accuracy, closely couples with a performance simulator such as *SimpleScalar* [54] to obtain delay and switching power at simulation granularity (often at cycle level).

Major power model types include data path blocks, storage arrays, register file, PLA, sequential cells, clock-distribution network and other standard cells. Expected inputs to evaluate these (dynamic) power models are activity factors, signal probabilities and load capacitances of interface signals, which are available at RTL. Data path power models cover most of the custom-designed data path primitives such as wide rotators, comparators, adders, 2,3,4,5 to -1 MUXes (multiplexors), large decoders and shifters. Data path power models are built assuming that the logic implementation of data path blocks does not change drastically from the last design to the current; for example, MUXes can be implemented with pass gates, AND–OR gates or tri-state buffers.

Large blocks, like 64-bit Carry Lookahead adders, are divided into sub-blocks, and power models for sub-blocks are generated to cover a whole range of input bit widths and different circuit and logic implementation styles. Similarly, register file and PLA power models are also developed by functional partitioning of large schematic blocks into sub-blocks. This finer granularity level for large logic blocks helps to improve model accuracy, reduce characterization complexity and generalize the power model for different bit widths. For memory array power estimation, the power model is characterized by partitioning the cache block into many different sub-blocks to improve accuracy and to reduce characterization complexity. Power models for sub-blocks include sense amplifier, word-line drivers, bit-line precharge circuits, address decoders and timer circuits. All major power contributors in a typical memory block are covered. Power models are generalized for multi-port memories. Power models for sequential cells and other standard cells come from the standard library. This power characterization includes power consumption due to output switching, clock-driven nodes switching and leakage.

Power estimation of synthesizable control logic is more involved because the logic can be implemented in many different ways and each implementation will have a different power consumption. Power models also include empirical rules to count dissipation in data and clock buffers, which is a large percentage of chip power, as indicated before.

The RTL model is simulated by using a simulator and actual test vectors (for example, idle test, active test, high power test) as shown in Fig. 4.7. The output of an RTL simulation is a switching activity file, which has activity factor (AF) and signal probability (SP) for all nodes visible in RTL, where AF is toggle count normalized over the number of clock toggles and SP is probability of the node being high over the total simulation clock cycles (see next section). The mapping of the RTL to power library models can use a synthesis flow relying on functional tags and operator types to partition RTL into three main logic types – sequential/standard cells, synthesizable blocks and data path blocks as shown in Fig. 4.8. These power macro-models are often augmented with a model for *glitch* power derived from simulating training input sequences on the target function block. It should be noted that the glitch power for different blocks of a microprocessor silicon was estimated to be from 0% (in the clock unit, as expected) to 20% in the data-cache unit, with average around 5% per block–when compared with total block dynamic (switching) power. However, inaccurate signal timing during logic simulation and interconnect delays and filtering effects render glitch estimation at high level fairly inaccurate.

The RTL power estimation methodology described in [286] consists of two phases: The first phase obtains zero-delay statistics of various RTL signals obtained via cycle-level RTL simulation on a test bench, including bit-level and word-level statistics for all RTL signals. The second phase estimates glitch activities by considering glitch generation and propagation.

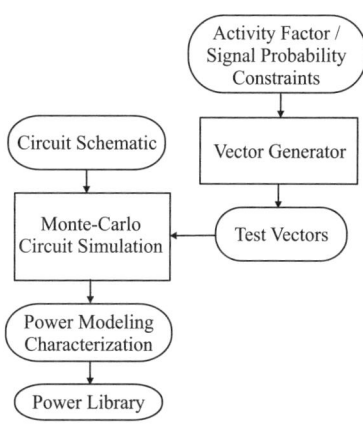

Fig. 4.7 Macro-modeling simulation flow

Fig. 4.8 Quick synthesis flow

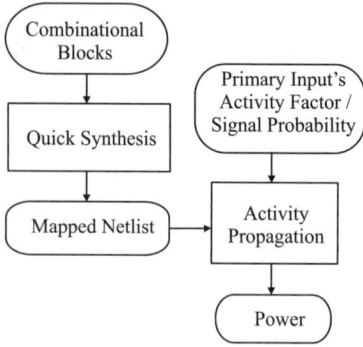

4.3 Dynamic Power Modeling and Estimation

The application target for a silicon design strongly influences the operating ranges of various circuit parameters, which in turn influence the power. Figure 4.9 shows that typical mobile and desktop processors operate at V_{cc}, higher than the minimum determined by reliability and noise margin. In this regime, power is proportional to the square of V_{cc}, a design parameter. However, some low-power applications can allow operation at the fixed V_{cc} min, the minimum reliability limit for V_{cc}.

The definitions of two properties of a signal used in finding switching statistics relevant to dynamic power estimation are given.

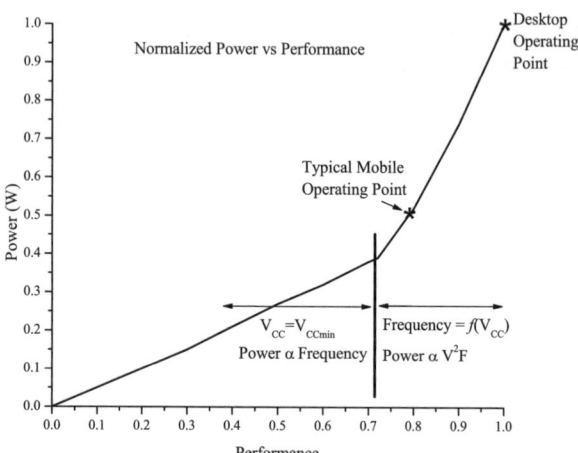

Fig. 4.9 Normalized power versus performance

- *Activity factor (AF)* is the probability of a signal (or bus) transition, measured relative to the toggle rate (i.e., transition frequency) of its reference clock. Designs that have dual-edge clocking could achieve data AFs as high as the clock's which, in turn, is defined to be one.
- *SP* is the assertion probability of a block's signal (or bus) per unit time, or per clock cycle, during the block's activation.

Statistically relevant tests could be run on a model to obtain the AF or the SP estimates in order to calculate pipe-stage power using node capacitances. AFs change as logic is performed down a pipe stage and are usually known only at the input and/or output of an architectural block. The worst-case AF is equal to the SP. Statistically computing AFs of a pipe stage's internal nodes may be a challenge, however.

4.3.1 Abstract Data Path Power Estimation

Power estimates of data path components such as muxes, bypass circuits and floating point execution units could be obtained by creating typical pipe stages for a representative number of bit slices and then rough-sizing them to compute the power. The total (dynamic) power of a pipe stage can now be approximated as

$$P_{\text{dyn, pipe stage}} = fV_{\text{cc}}^2 \left(\text{AF}_{\text{gate}} \times C_{\text{gate\_out}} \right.$$
$$\left. + \text{AF}_{\text{seq\_data}} \times C_{\text{data\_seq}} + C_{\text{clk\_seq}} \right) \times \text{AF}_{\text{Clk\_en}}. \quad (4.1)$$

Here, the summation is over all the blocks/gates in the pipe stage, AF_{gate} is the activity factor at the output of a gate (its maximum value is SP), $C_{\text{gate\_out}}$ is the total output capacitance of the gate, $\text{AF}_{\text{seq\_data}}$ is the AF of the data portion of the sequential element, $C_{\text{data\_seq}}$ is the total sequential element capacitance associated with data, $\text{AF}_{\text{Clk\_en}}$ is the activity factor of the clock enable (a gated clock) and $C_{\text{clk\_seq}}$ is the total switching capacitance of the clock portion of the sequential element.

This way of detailed power estimation, however, may not be an option in early-stage design; a quicker and more abstract estimate may be computed instead. Here, three *scaling* concepts are described that can be applied on a block from a previous design to predict power for a similar target block in the new design.

Architectural scaling: The power estimate is increased or decreased proportionately to certain architectural complexity parameters. For instance, to scale a bypass network architecturally, based on an older bypass, one could compute the architectural scaling factor for power as follows:

$$\text{Architectural scaling} = \frac{(\text{new\_read\_ports} * \text{AF}_{\text{r}} + \text{new\_write\_ports} * \text{AF}_{\text{w}})}{(\text{old\_read\_ports} * \text{AF}_{\text{r}} + \text{old\_write\_ports} * \text{AF}_{\text{w}})}$$
$$* \frac{(\text{new bus bit width})}{(\text{old bus bit width})} * \frac{(\text{new number of entries})}{(\text{old number of entries})}. \quad (4.2)$$

Pipeline scaling: A new design may stretch pipelines by reducing the number of logical gates per cycle in order to enhance the operating frequency. Often, a significant increase in sequentials occurs (which accounts for as much as half of all logic switching power) from pipeline stretching. For a constant-width logical cone such as a FIFO, the increase in count of sequential elements is a function of the new and old pipeline frequencies, and the *cycle overhead*, according to

$$N_2 = N_1 \times \left(\frac{P_1 - O_c}{P_2 - O_c} \right), \tag{4.3}$$

where N_1 is number of old sequentialsl; P_1 is the old clock period; and O_c is the clock overhead from skew, jitter and latch clock-to-out delay, etc.

Process scaling: Process scaling captures the effect of technology/process migration of the block from the old design to its new counterpart. Process scaling is performed according to the following formula:

$$\text{Process scaling} = \left(\frac{V_{cc2}}{V_{cc1}} \right)^2 \left(\frac{f_2}{f_1} \right) \times \text{PSF}, \tag{4.4}$$

where V_{cc1} is the old product's supply voltage at the package pin and V_{cc2} is the new voltage.

The scaled power is the product of these three scalings with the original (old) block's power:

Total new power of a pipeline stage = old power × architectural scaling

×pipeline scaling × process scaling. (4.5)

RTL power estimation can thus be a very useful addition to the ASIC flow with some value even for a custom flow as well.

4.3.2 Capacitance Estimation

Capacitance estimation is an important step for dynamic power modeling. However, capacitance is unfortunately not a single, static number. The effective capacitance of a node varies in time during switching as explained below. Starting from the OFF state, assume that a negative or small positive voltage is applied to the gate of an NMOS device. Then the free majority carriers at the substrate (i.e., holes) will be attracted to the Si–SiO$_2$ interface and will form an accumulation layer there. While in this region, the gate capacitance is solely determined by the gate oxide. If the gate voltage is now further increased, the free majority carriers at the substrate will be repelled from the Si–SiO$_2$ interface, and the region close to the interface will become depleted of free carriers. The gate capacitance in this region is reduced as the distance between charges is increased: effective dielectric thickness = oxide +

depletion layer. If the gate voltage is higher than the threshold voltage (V_{Th}), then minority carriers in the substrate (electrons in NMOS) will be attracted toward the Si–SiO$_2$ interface; and this region will become an 'n-*type*' region (i.e., the Si type is inverted), leading to a stop in the growth of the depletion region. The change in charge at the gate is balanced by a change in the inversion charge (i.e., the depletion charge is kept constant). This assumes the availability of minority carriers for an MOS transistor. The total gate capacitance returns back to be the thin oxide gate cap. Note that the transistor behaves as a non-linear resistor, where the saturation current I_{dsat} is proportional to $(V_{in} - V_{Th})^2$ as the transistor's current saturates at high V_{ds} values. Its effective resistance depends on the operating point. Changing V_{Th} has a high impact on I_{ds} and hence on transistor delay. This non-linear and dynamic dependence of the capacitance during operation, depicted in Fig. 4.10, makes for some of the difficulty in its accurate estimation.

4.3.3 Macro-modeling for Dynamic Power

One of the main objectives of macro-modeling is to develop a power macro-model for a module so that power dissipation can be obtained under any distribution of primary inputs. When the same module is reused, one can obtain its power simply by using a LUT. Since the power dissipation of a circuit is strongly dependent on the statistics of primary inputs, the relationship of power versus primary input probabilities and activities is a complicated surface. Once such a surface is constructed, power dissipation under any distribution of primary inputs can be easily obtained. A straightforward way is to approximate such a power surface using a large number of discrete points. The more points one chooses, the more accurate the result. However, more points directly translate to longer CPU time.

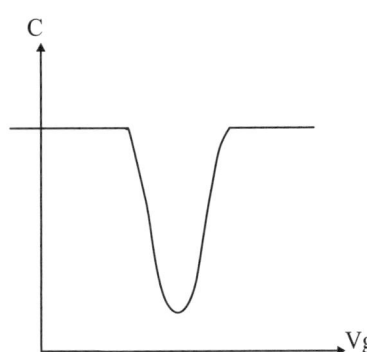

Fig. 4.10 Capacitance versus gate voltage

Power sensitivity can be used to efficiently develop a power macro-model. The power surface can be approximated by planes that are constructed by a representative point with power sensitivities. Results for power dissipation under any distribution of primary inputs demonstrated the accuracy and efficiency of this technique.

With the high demand for reliability and performance, accurate estimation of maximum instantaneous power dissipation in CMOS circuits is essential to determine the IR drop on supply lines and optimizing the power and ground routing. Unfortunately, the problem of determining the input patterns to induce maximum current and, hence, the maximum power, is NP-complete. Even for circuits with a small number of primary inputs (PIs), an exhaustive search in the input vector space requires intensive CPU time. Industry practitioners have used techniques based on automatic test generation (ATG) to efficiently generate tight lower bounds of the maximum instantaneous power for CMOS circuits with non-zero gate delays.

4.3.4 Estimation of Bounds on Average Power

Power dissipation in CMOS circuits is heavily dependent on the input signal distribution. However, because of uncertainties in the specification of input signal distribution, the average power dissipation should be specified between a maximum and a minimum possible value. Because of the problem's complexity, it is practically impossible to use traditional power estimation techniques to determine the bounds. Power sensitivity, defined as the change in average power due to changes in the specification of PIs, can be used to accurately estimate the maximum and minimum bounds for average power. Both symbolic and statistical techniques have been developed to estimate power sensitivity as a by-product of average power estimation, thereby leading to efficient implementation. For some circuits, power dissipation can be very sensitive to some PIs. A small variation in signal distribution can cause power dissipation to change drastically. Results on minimum and maximum average power show that such bounds can vary widely if the PI probabilities and activities are not specified accurately.

4.4 Leakage Modeling

Leakage power is often measured at two different microarchitecture operating points: at active state (C0–C3), where it adds to the TDP; and at sleep state (C4–C6), in which leakage is the only source of power dissipation of the core and adds to total average power and affects battery life. Frequently, cores reside in deep sleep states. A parking sequence (invoked prior to entrance to a deep sleep mode) brings the design to a known logic state called the *parking state*. The parking feature's benefits are dependent on three separate design activities: choosing the optimal parking state for leakage power, finding a parking sequence that brings the design to that state and modifying the design by implementing a parking signal that discharges the active/dynamic nodes before entering the deep sleep state.

It has been mentioned that device scaling and functionality integration trends have led to alarmingly growing total power and power densities. Dynamic power is proportional to V_{cc}^2; hence, there is a process and design trend to lower V_{cc} to the range of 0.9–1.5 V. On the other hand, the delay of a CMOS device is directly proportional to $\left(\frac{V_{cc}}{(V_{cc}-V_{Th})^2} \right)$, where V_{Th} is the threshold voltage of the transistors. Thus, devices will slow down if nothing else beyond voltage scaling is done. To compensate for this slowing effect, V_{Th} is lowered, if not proportionally, as well. However, with smaller V_{Th}, the devices are not fully turned off in their logical off-state, leading to excessive current dissipation (leakage).

Recent design approaches leverage a key observation that leakage power of a gate is strongly dependent on the logic values at its inputs. It follows that for a given circuit, one can identify an optimal state that minimizes leakage power consumption. Such techniques as *parking states* and *parking signaling* are being newly considered (see Section 4.5.1.4). An estimation technique should appropriately consider such design optimizations in the context of different active and sleep states of the processor.

4.4.1 Subthreshold and Gate-Oxide Leakage Power Modeling and Estimation

Subthreshold leakage occurs when a device is off but the source and the drain voltages differ. In contrast, gate leakage is pronounced when the gate voltage differs from the channel voltage; i.e., when the device is conducting. It is therefore called "ON-state" leakage. (However, more precisely as shown in Section 4.4.2, "backward" gate leakage happens in the OFF-state of a transistor but may be small.) What makes leakage estimation relatively inaccurate at high level is that input vectors to a combinational logic have potentially conflicting influences on subthreshold leakage vis-à-vis gate leakage. For example, simulation results for a NAND3 gate show that an input vector of 000 (110) gives the lowest subthreshold (gate) leakage, whereas an input vector of 111 is worst case for both types. An observation is that if the bottom transistor in a stack is OFF and all others in the stack are ON, it tends to give the lowest gate leakage.

A stack (series) of transistors leaking in their OFF-state may be viewed as a series of resistors. Thus, the source node of a non-bottom transistor in an NMOS stack is above ground (symmetrically, below V_{cc} for a PMOS stack) and a reverse bias between gate and source develops, which turns off the transistor harder. Stacking reduces leakage because of this increased body effect. Usually, the stacking factor is derived from previous designs. This technique gives $\approx 2 - 3x$ leakage power reduction depending on the number of devices off in the series. One needs to be careful using this technique because it can increase switching power and gate leakage significantly. In sub-micron technology, below 90 nm, gate leakage increases significantly because of lower T_{ox}. The gate leakage can be as much as 30% of total device leakage below 90 nm. Stacking is beneficial when a small number of transistors can add extra stack to a wide cone of logic or gate power supply to it.

4.4.1.1 Subthreshold Leakage

For a given technology, subthreshold or drain-to-source leakage is primarily dependent on V_{Th}, source-to-drain voltage V_{ds} and gate-to-source voltage V_{gs}. This leakage current for a transistor is often modeled as

$$I_{sub} = \mu_0 C_{ox} \frac{W_{eff}}{L_{eff}} \left(\frac{kT}{q} e^{1.8} \right)^2 \exp[\frac{q}{S_n kT} (V_{gs} - V_{Th0} - \gamma V_s + \eta V_{ds})] \left(1 - e^{\frac{qV_{ds}}{kT}} \right),$$

(4.6)

where μ_0 is the zero-bias mobility of the carrier, C_{ox} is the gate-oxide capacitance per unit area, S_n is the subthreshold swing coefficient, V_{Th0} is the zero-bias threshold voltage, γ is the linearized body effect coefficient and η is the drain-induced barrier lowering (DIBL) coefficient.

Figure 4.11 shows the $V-I$ curve for a stack of two NMOS devices with gates grounded. The subscript u(l) refers to the upper(lower) transistor. The Y-axis is normalized with respect to a stack of a single off transistor whose leakage is assumed

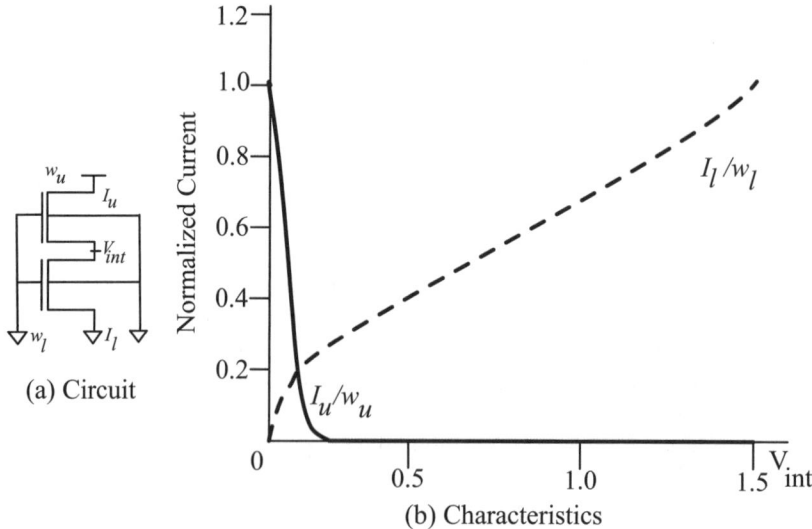

(a) Circuit

(b) Characteristics

Fig. 4.11 Stack effect on subthreshold current

to be 1. The solid and the dotted curves representing the load lines for the upper and lower devices, respectively, meet at a point representing the subthreshold leakage for the two-device stack. The low (leakage) current indicates a strong effect of stacking.

4.4.1.2 Gate-Oxide Leakage

In the recent past, while subthreshold leakage has become a dominant part of total power (especially for idle power), gate-oxide leakage and junction leakage have continued to remain subdued. However, a significant increase in gate leakage is expected because of oxide scaling (below 20 Å or equivalently, five or six atomic layers of SiO_2) when the operating voltage has not scaled proportionately. Thinner gate oxide is advanced to negate some of the performance-decreasing *short-channel* effects. Unfortunately, gate leakage increases exponentially with reduction in oxide thickness and is sensitive to voltage potential across the gate. Gate leakage through the PMOS tends to be an order of magnitude lower than the NMOS if non-nitrided SiO_2 gate insulator is used. Nitriding, which is used for raising a gate's dielectric constant to improve mobility, can lead to much higher gate leakage. For some processes such as NEC's 100 nm, gate leakage per μm is more than twice the subthreshold leakage, which is about 0.3 nA/μm.

At the same time, other innovations are being brought to bear to curb oxide leakage to tolerable levels by means of such advances as the triple-gate or fin-based field effect transistor (FinFET) and high-K dielectric [215, 240] technologies, both of which allow slower scaling of gate-oxide thickness while keeping up with the Moore's Law of transistor performance scaling. Additionally, metal gates are used to tackle gate leakage to a practical minimum. Gate leakage is proportional to total active device area and is subject to the effects of stacking factor just as subthreshold leakage is. However, the stacking factor for these two types of leakage is calculated differently.

4.4.2 Methods for Total Leakage Estimation

Described below is an analytical model to estimate both gate and subthreshold leakage, adapted from [292]: this considers both forward and backward gate leakage. Forward gate leakage occurs when the gate voltage is higher than the other nodes (drain and source) for an NMOS transistor and backward leakage is tunneling in the reverse direction when gate voltage is relatively lower than the source or drain. Gate tunneling is highly dependent on the state of the transistor; namely, the voltages at its three terminals: gate, source, and drain. For each type of CMOS transistor, [292] considers only four different prototypical states, given as a triple $< G, S, D >$ for the three terminal voltages, to be sufficiently accurate for estimation. For the NMOS device, they are $S1 =< 0, 0, 0 >$, $S2 =< V_{DD}, V_{DD} - V_{Th}, V_{DD} >$, $S3 =< 0, 0, V_{DD} >$ and $S4 =< 0, V_{DD}, 0 >$. States S1 and S2 incur no gate leakage and very small subthreshold leakage. State S3 results in backward gate leakage, which is proportional to the drain node voltage and the gate area of the device:

$$I_{\text{gate}-S3} = I_{\text{bg0}} \times \text{stackPositionFactor} \times \text{gateArea}, \tag{4.7}$$

where I_{bg0} is nominal backward gate leakage of a device of unit gate area while the drain is at V_{DD}. The stackPositionFactor is the reduction factor in leakage attributable to lower drain voltage for an NMOS by virtue of being not on the top of the device stack. These values are empirically determined – backward gate leakage I_{bg0} comes from the process technology and the other two parameters from early circuit estimates.

State S4 induces forward gate leakage when the source and drain are at logic low and the gate is at logic high for an NMOS. However, if a transistor in state S4 is "sandwiched" by OFF-state transistors in a stack, then the drain–source voltages are high enough that the gate leakage can be ignored. The nominal forward gate leakage of a device of unit gate area is

$$I_{\text{gate}-S4} = I_{\text{fg0}} \times \text{sandwichFactor} \times \text{gateArea}, \tag{4.8}$$

where I_{fg0} is the nominal forward gate leakage of a device of unit gate area; sandwichFactor is 0 when the device is sandwiched by two OFF-state devices but is 1 otherwise.

Subthreshold leakage, prominent in state S3, is modeled as

$$I_{\text{sub}-S3} = I_{\text{off0}} \times \text{stackingFactor} \times \text{deviceWidth}, \tag{4.9}$$

where I_{off0} is the subthreshold leakage of a stack of a single device with unit-width device in OFF-state; stackingFactor is an empirical value based on how many off devices exist on the stack under consideration (see Section 4.4.1); deviceWidth is the width of the transistor.

Although NMOS "states" were described, treatment of PMOS devices is analogous. Thus, the total leakage is approximated over all devices as

$$I_{\text{leak}} = I_{\text{gate}-S3} + I_{\text{gate}-S4} + I_{\text{sub}-S4}. \tag{4.10}$$

4.4.2.1 Additional Approaches

There are further reported refinements to leakage prediction as exemplified by [252], which advocates a weighted sum of devices with different subthreshold leakage as follows:

$$I_{\text{leak}-w} = \left(\frac{w_p I_p^0}{k_p}\right) \times \exp\left(\frac{\sigma_p^2}{2\lambda_p^2}\right) + \left(\frac{w_n I_n^0}{k_n}\right) \times \exp\left(\frac{\sigma_n^2}{2\lambda_n^2}\right), \tag{4.11}$$

where w_p and w_n are the total PMOS and NMOS device widths; k_p and k_n are factors determining percentage of device widths operating in their OFF state; I_p^0 and I_n^0

are the expected mean subthreshold leakage currents per unit width of PMOS and NMOS transistors in a particular chip; σ_p and σ_n are the standard deviation of channel length variation within a particular chip; and λ_p and λ_n are constants relating channel lengths of PMOS and NMOS devices to their corresponding subthreshold leakages.

Besides the foregoing sort of analytical models for leakage, the following itemizes abstractly specified steps of yet another approach taken by some researchers and practitioners:

- Perform explicit logic-level simulation at the gate level and circuit-level simulation of diffusion-connected networks (DCNs) to obtain representative leakage values to be stored in LUTs.
- Do probabilistic simulation of the states of each node modeled as a random variable, while using correlation coefficients between input patterns.
- LUTs to compute and sum up the leakage for every input combination of every gate using weighted sums as per the probability distributions of the reachable states.

4.4.2.2 Estimating for Power-Gated Circuits

Increasingly, a new circuit-level technique called "power gating" or "sleep transistors" is being used when *clock gating* is either insufficient for power saving or is ineffective. Power gating removes power completely from inactive circuits or puts them in "drowsy" (lowered voltage) mode, thereby further reducing average power, especially leakage. Typically, big arrays are power gate-enabled to support both idle-time power and sleep-time power savings for processors as shown in Fig. 4.12. A pre-characterization approach is often adopted to estimate power in these specialized circuit modes.

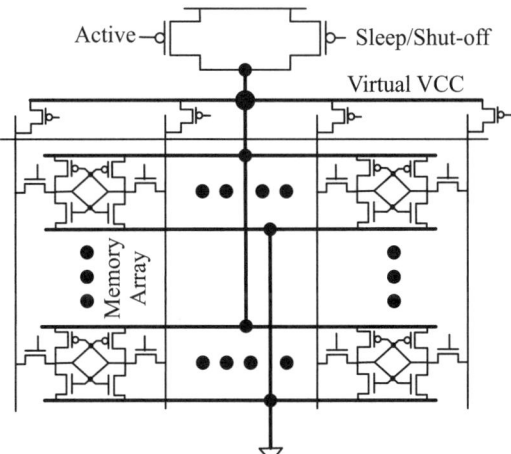

Fig. 4.12 Power gating in cache

4.5 Modeling and Analysis of Architectural Components

In an SoC, different components have different workload profiles and different power states (the C-states described earlier representing active, idle, sleep and deep sleep modes). Since dynamic power is not a strong function of variation while leakage is, one should expect to see more total component power variation for those components, such as cache banks, spending substantial time in low-computation states.

Figure 4.13 shows a typical distribution of power in various FUs of a processor. This type of data is often useful in calibrating estimations at high level in addition to providing data on which functional blocks need closer attention in the "estimation and optimization" loop. In this section, the focus is on a different method of categorization of logic blocks – not by their functionality in a microprocessor but by their general characteristics and logic type. This typification, consequently, makes the discussion useful for non-processor designs as well.

4.5.1 Design-Optimization-Aware Estimation

Absolute accuracy of power estimation at high level in early design phases is difficult to achieve without being aware of the downstream implementation techniques or without access to relevant prior design data which can be relied upon for reuse. Power gates or sleep transistors may become the part of a new design implementation methodology and ignorance about such optimizing design techniques can lead

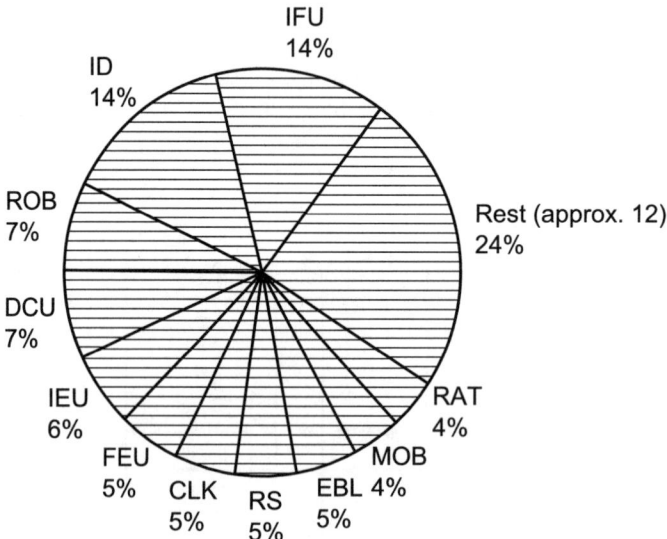

Fig. 4.13 Power distribution by units

to very inaccurate power estimation. As another example, a domino logic [72] implementation might apply sophisticated *dual-rail optimization* techniques [268] and reduce significant area or power which need to be anticipated for the block's power modeling for use in a high-level, chip-wide estimation. However, this is one of the reasons initial circuit studies and tech-readiness activities are undertaken before a major, new silicon project starts in the early "front-end design" stage.

On the other hand, clock gating can be blindly applied to essentially all structures. When clock gating is considered for a given unit of logic, several items must first be considered: what is the unit's activity factor; what is the size of the unit; is the unit larger than the gating logic itself; what is the unit power saved versus gating logic power cost? Generically applying gating to any and all microarchitectural elements leads to overly optimistic results.

4.5.1.1 Voltage Scaling, Voltage Islands, Variable V_{cc}

Because the dynamic component of power is proportional to the square of the supply voltage (V_{cc}), the reduction of V_{cc} is the most effective determinant of reduced power. The industry has thus steadily moved to lower V_{cc}. Indeed, reducing the supply voltage is most effective for low-power operation, even after considering the modifications to the system architecture required to maintain the computational throughput. However, the drive toward higher performance can sometimes outstrip the benefits of voltage scaling. An issue with voltage scaling is that to maintain performance, the threshold voltage (V_{Th}) also needs to be scaled down because circuit speed is inversely proportional to roughly ($V_{cc} - V_{Th}$). Typically, V_{cc} should be larger than $4V_{Th}$ if speed is not to suffer excessively. As the threshold voltage decreases, subthreshold leakage current increases exponentially. At present, V_{Th} is high enough such that subthreshold current is only a small portion of the total active current, although it dominates the total standby current compared to gate and well leakage components. However, with every 0.1 V reduction in V_{Th}, subthreshold current increases by 10 times. As discussed in Section 4.4 with further V_{Th} reduction, subthreshold current will become a significant portion, or even a dominant portion, of the overall chip current in the nanometer technologies severely diminishing the benefit of lower V_{cc}. In addition, design of dynamic circuits, caches, sense amplifiers, PLAs, etc. becomes difficult at higher subthreshold leakage currents. Lower V_{cc} also exacerbates noise and reliability concerns. To combat subthreshold current increase, various techniques have been developed as mentioned in the Chapter 9. *Voltage islands* and *variable V_{cc}* are variations of voltage scaling that can be used at the lower level. Voltage scaling is mainly dependent on technology and typically applied to the whole chip. Voltage island is more suitable for SoC designs, which integrates different functional modules with various performance requirements onto a single chip.

The variable voltage and voltage island techniques are complementary and can be implemented on the same block and used simultaneously. In the variable voltage technique, the supply voltage is varied based on throughput requirements. For higher throughput applications, the supply voltage is increased along with operating frequency and vice versa for the lower throughput application. Sometimes this technique is also used to control power consumption and surface temperature. On-chip sensors sense temperature or current requirements and lower the supply voltage to reduce power consumption. Many high-performance microprocessors from IBM and Intel use such techniques. Leakage power mitigation is typically achieved at the device level by applying multi-threshold voltage devices, multi-channel length devices, stacking and parking state techniques. The following section gives details on these approaches. Power gating can also be done at circuit level or module level using voltage regulator modules – an increasingly attractive technique for very low power designs. Current design techniques introduce multiple sleep states which vary in their sleep-state entry and exit times (a few hundred micro-seconds to a few tens of milliseconds) allowed for different performance goals. Thus, the estimation of necessity becomes statistical, meaning that the usage and workload mix and characteristics need to be considered more intimately.

4.5.1.2 Use of Multiple Threshold Devices

Multiple threshold voltages have been available on many, if not most, CMOS processes for a number of years. Since the standby power is so sensitive to the number of low-V_{Th} transistors, its usage, on the order of 5%–10% of total transistors, is generally limited to fixing critical timing paths, especially in the physical design stage where other design changes have very long turnaround time. For instance, if the low V_{Th} is 110 mV less than the high V_{Th}, 20% usage of the former will increase the chip standby power by nearly 50%. Since the transistor's parameters T_{ox} and L_{gate} do not change for the low-V_{Th} transistor, low-V_{Th} insertion does not affect the active power component or design size. Obvious candidate circuits are SRAMs, whose power is dominated by leakage; and a higher V_{Th} generally also improves SRAM stability (as does a longer channel). The main drawbacks of low-V_{Th} transistor are that variation due to doping is uncorrelated between the high- and low-threshold transistors and extra mask steps incur a process cost.

4.5.1.3 Long Channel Transistors for Leakage Reduction

By drawing a transistor 10 nm longer than a minimum size, the DIBL is attenuated, and the leakage can be reduced by $7x - 10x$ on a 90 nm process. With this one change, nearly 20% of the total SRAM leakage component can be alleviated at the same performance. The loss in drive current due to increased channel resistance, on the order of 10%–20%, can be made up for by increasing width. Since in a typical design, most of the transistors are in SRAM, it dominates the leakage; so long-L insertion in arrays produces maximum benefit. Compared with multiple threshold voltages, long channel insertion has similar or lower process cost, taken as size increase rather than mask cost. It allows lower process complexity, and the different channel lengths track over process variation. A potential penalty is the increase in gate capacitance and the upsized gate to meet performance.

Two other design techniques for leakage power reduction are stacking (discussed in Section 4.4) and *parking states*. In the stacking technique, two or more transistors are placed in series, which increases channel resistance of the leakage current path in the OFF state. The concept of parking states is described below.

4.5.1.4 Parking States

The main idea behind the parking states technique is to force a design into a low-leakage state when not active. The leakage power reduction depends on how effectively a design is put in the OFF state. There are three main pitfalls in this technique. The first, additional logic, is needed to generate a desirable state, which has area, switching power, as well as leakage cost. The second pitfall is that it is very specific to a design implementation; for example, a low-leakage state for a cone of logic may become a high-leakage state if implemented differently. This technique is not advisable for random logic, but careful implementation on structured data path and memory arrays can save 2–5x leakage power in the OFF state. The third pitfall is about ensuring that the design remains in the OFF state long enough to make up for the lost power in forcing the low-leakage state. It requires extensive research and simulations across all possible applications and their usage patterns.

4.5.2 Estimating Under Variation Effects

All non-quantum physical parameters of a physical system such as a chip have associated variances and tolerances in their values. However, in recent years, the importance and effect of the variations have grown, and their effect on the power characteristics of the system can be significant. Table 4.1 describes three broad categories of these variations, although researchers have only begun to address a few of these (simultaneous) effects, if any, from each category in a unified framework [214]. Die-to-die (i.e., inter-die) variations are caused by variances in processing temperatures, equipment properties, polishing, wafer placement and resist thickness. Beyond the

Table 4.1 Process variations

Process parameters	Circuit operation	Simulation tools
Channel length	Temperature	Timing analysis
Channel width	Supply voltage	RC extraction
Threshold voltage	Aging (NBTI)	Cell modeling and I–V curves
Overlap capacitance	Cross-coupling capacitance	Circuit simulations
Nesting effects	Multiple input switching	Process files

250 nm technology generation with the lithographic wavelength at 248 nm, within-die (intra-die) variation has become increasingly significant.

The process-related variations emanate from such physical effects as *random dopant fluctuation* and *line edge roughness* during manufacturing.

Because of the variations, a single deterministic estimate of power becomes less useful, and the goal of power analysis evolves toward estimating a statistical distribution. However, direct Monte Carlo simulation of a full-system, simulation-based estimation method can be prohibitively slow in early design exploration. On the other hand, spreadsheet-based methods to calculate power distribution from the given SoC characteristics, activity, and input profiles, and gate counts and process parameter distributions, are seen to be fairly inaccurate.

A middle-ground approach is to divide variation-aware modeling into three phases [61]: In phase 1, leakage power distribution for each component/IP block in an SoC is obtained. Phase 2 involves performing the conventional system simulation-based estimation of dynamic power and extracts "power-state traces" for all power states of the SoC. In the third phase, the process parameter space is sampled for all power states to perform faster Monte Carlo simulation to obtain power and store it in LUTs. Finally, trace analysis is performed to obtain the distribution of the average power consumption summing dynamic and leakage parts.

4.5.3 Estimating Power in Control and Data Path Logic

4.5.3.1 Synthesizable Arithmetic and Combinational Logic

Buyuksahin et al. [56] present a method for high-level power estimation for synthesizable CMOS combinational logic. They compute the complexity measure of a Boolean network (BN) as the product of the average fan-in of a node in the BN and the total fan-out. The average fan-in captures the computational complexity and the total fan-out captures communication complexity of an "optimized" circuit. This complexity measure is then curve-fitted to the gate counts of various representative circuits that have been synthesized/optimized using a tool such as *Synopsys Design Compiler*. Thus, the gate count of a circuit, whose Boolean network representation or Boolean function is given, can be estimated from this fitted equation (which is a power law in their case). The gate count then provides for subsequent area/capacitance estimation.

The two model parameters m and n in the estimation equation GateCount $= m \times$ (BN complexity)^n are obtained from the regression analysis of benchmark circuits and realistic delay constraints. One of the limitations of such a structural complexity measure is overestimation if a BN contains significant redundant logic. The steps of total (dynamic) power estimation are summarized below:

- Read in the Boolean equations describing the design.
- Read in the input statistics and delay constraints.
- Build a Boolean network representation of the design.

- Estimate the gate count using the technique introduced above and then scale it with the average gate (weighted by their occurrence frequency) capacitance from the given technology library.
- Run a zero-delay logic simulation of the design with random inputs to obtain the nodes' switching probabilities.
- Estimate the average activity of the circuit using the "high-level activity estimator" : $2(D_i + 2D_o)/3(p+q)$ where D_i and D_o are the *cumulative densities* [56] at the input and output nodes, respectively; p and q are the number of input and output nodes, respectively.
- Estimate the power as the sum total of the individual node power which in turn is the product of half V_{cc}^2, the estimated node switching probability and the node capacitance.

4.5.3.2 High-Level Power Simulators

Different power/energy/capacitance models for the functional or logic blocks can be integrated with a cycle-accurate performance simulator to provide more accurate estimation at the expense of some simulation cycles in contrast to pure analytical models. Although this chapter has primarily focused on analytical models, an example is given below to show how power models can be interfaced with traditional logic and performance simulators. The switching events are modeled with an embedded cycle-based logic simulator. Kim et al. [160] present power simulation techniques for data path and memory components that give high accuracy ($90 + \%$ range) without incurring too much simulation time.

There are certain pluses and minuses to the proposed levelized netlist approach of [160] to simulation: it does not have to build event queues but needs to evaluate all gates. Glitches are not simulated because timing/event is ignored. A full levelized evaluation is performed, but the convergence/fix point is faster this way.

4.5.3.3 Combinational Circuits

Several symbolic and statistical techniques have been developed [286, 56] to accurately estimate power dissipation considering simultaneous switching and temporal and spatial signal correlations. The basic idea of the symbolic method is to express the signal probability (probability of a signal being logic high) and signal activity (probability of a signal switching) of each internal node in terms of the probability and activity of primary inputs so that spatial correlation between internal nodes can be managed. Ignoring simultaneous switching can introduce an error on the order of over 21%. The basic idea of the Monte Carlo-based statistical method to estimate power dissipation is to simulate a circuit with random patterns applied to PIs. Such random patterns conform to the given probabilities and activities of PIs. The number of simulations are determined by user-specified parameters, such as confidence levels and errors that can be tolerated. The statistical technique can handle different delay models for logic gates so as to include spurious transitions in its analysis. Because of the presence of different delay paths converging through logic gates, spurious transitions (glitch) can occur and in turn increase power dissipation. Results indicate that spurious transition can account for more than 50% of power dissipation for some benchmark circuits.

4.5.3.4 Sequential Circuits

Probabilistic and statistical techniques have been implemented to estimate power dissipation in sequential circuits. Because of the feedback of input from the next state, the estimation techniques for combinational and sequential circuits are quite different. Techniques to estimate signal probability and activity work as follows. Given the state transition graph (STG) of a sequential circuit or an FSM, an extended state transition graph (ESTG) is built, and the probability of a state in the ESTG is calculated. The signal activities are then estimated from the ESTG. The exact method may require solving for a linear system of equations of size $2N + M$, where N is the number of primary inputs and M is the number of flip-flops. For large circuits with a large number of PIs and flip-flops, the exact method is not computationally feasible. Therefore, an approximate method is discussed which considers temporal correlations of PIs. The circuit is unrolled k times to calculate the probabilities and activities of internal nodes. Results indicate that this technique can have an accuracy of 90% while being several orders of magnitude faster than logic simulation. Statistical techniques for sequential circuits consider *near-closed* (NC) sets of states. A set of states is called NC if the probability of being in that set is high, given that the starting/initial state is also in that set. Techniques to determine the warm-up period and stopping criteria for Monte Carlo-based statistical simulations have been determined under the presence of NC sets. The computation time of state probability can be reduced by 50% (compared to the standard Monte Carlo techniques) by the proposed method. The relative error of the estimated individual node activity using the Monte Carlo technique with a warm-up period is within 3% of the result obtained by long run logic simulation.

4.5.4 Communication Components

4.5.4.1 Network Blocks

On-chip network processors are trending toward *systems-on-chip*, with design style becoming increasingly communication-centric to address expanding chip complexity and functionality differentiation. The use of NoCs is a solution approach to solve the problem of global physical wires, timing convergence and time-to-market issues. Major network components, such as the arbiters, links, crossbars and router buffers, may be decomposed and modeled as subcircuits in a typical high-end processor. For example, a router buffer is a simple variation of either a small-signal cache design with sense amplifiers or a large-signal multi-ported register file design.

In SoC designs, determining chip-level power based on assumptions of maximum switching allowed by the circuit is rarely a good measure of realistic power and thermal behavior. One needs to consider current and emerging *recognition-mining synthesis* or real-time analytic applications for processors and specific data or signal processing applications for embedded systems since their computation/communication requirements on one hand and the system architecture (number, type and size of cores), their interconnections and process technology on the other greatly influence the power and thermal profiles realizable on a specific design. The design space of such SoCs is huge, and an early design exploration and trade-off involving power, performance and cost necessitates fast and high-level power estimation. Consequently, one practical approach for power/thermal estimation is to first determine "utilization" of various network components on the chip based on quasi-simulation of message level traces of realistic workloads. Network congestion and arbitration models affect utilization of specific links and buffers connected to them. Then, these utilization numbers are scaled up by the respective components' average, pre-characterized switching and leakage powers. The total system power is derived by summing up these scaled values.

4.5.4.2 I/O and Clock Power

I/O circuits switch large capacitances and run at higher-than-core voltages, and their pads do not scale well with technology. An I/O pin for a processor or chipset typically drives 5 pF of capacitive load, and the switching activity for embedded processors can be high as they tend to not have sufficient on-chip L1 or L2 caches to keep storage accesses on chip. As much as 5% of total power might go into I/O pins which are loaded not only with pin capacitances but also often with PCB trace capacitances (which need to be estimated). Thus, although not paid much attention to in the past, power models for I/O are important. However, good accuracy needs good circuit-level estimation of I/O bus capacitance, sufficient details in the I/O protocol model and off-chip access frequency data. For better accuracy, counting switching activity at the I/O cycle level is argued for accuracy in [160] where the performance simulator traces I/O bus streams and feeds the power model in the relevant I/O transaction cycle.

Last, proper pre-characterization of I/O power should be done, keeping in mind that circuit implementation can greatly affect I/O power as exemplified by the following from [21] : The output buffer design was enhanced with a special dynamic on-die termination (ODT) circuit that enables disconnecting the ODT when the Pentium M processor drives the bus low, thereby reducing by half the power consumed by the I/O. Data inversion support between the processor and chip set further reduces the power dissipation because of the line termination by minimizing the number of bits driven low on the Gunning transceiver logic bus. Furthermore, to minimize the time during which the bias current flows in the input buffers of the processor side bus (PSB), a new signal was added to the interface to indicate to the processor when to operate the input buffers. The overall impact of all these optimizations is to reduce the active power of the PSB interface by a factor of 2 and the average power by a factor of 10.

High-level clock power is usually extrapolated from previous and similar designs, and it also depends on the estimation of the count of sequential elements mentioned earlier.

4.6 Register Files

Multi-ported register file type arrays are very common structures found in modern microprocessors. The architectural register files (integer, floating point, etc.), the data and instruction caches and the instruction queue are all examples of multi-ported arrays. Their large number and usually considerable size make them very important structures in terms of the power dissipation of the processor. Furthermore, their regular structure makes them ideal candidates for architectural-level power modeling.

Brooks et al. [48] introduce *Wattch*, an architectural-level power simulator based on the SimpleScalar [54] framework and various architectural-level power models for the building blocks of a processor. Their register file model is derived from Zyuban's model [379] and has the same shortcomings. Kamble et al. [146] introduce an analytical energy model for low-power caches based on the hit/miss rates and number of cache read and write accesses reported by a detailed performance simulator. Their cache energy model assumes that the capacitances of the power hungry nodes (such as bit lines and word lines) are available and concentrate on estimating the number of transitions on these nodes. This model also lacks the consideration for clock power and pipelined operation.

Cacti [322] is a well-known cache delay, area and power analysis tool. As the previous models introduced in this section, *Cacti* also lacks the consideration of clock power and pipelined operation. Furthermore, it uses very simple first-order transistor models to size the various devices.

All of these models are based on some assumptions for the structure and design style of the register files. They model the energy dissipation of a register file in terms of read- and write-access energies based on the capacitance switched at the energy dissipating nodes. Zyuban's work [379] concentrates on the energy complexity of the various register file styles and introduces an energy model for multi-ported register files. Although the model is useful for a first-order comparison of different register file styles, it is not detailed enough to be used for a more accurate analysis. The main shortcomings of the model introduced are the lack of clock power, sizing of the various devices, an inflexible timing structure and the lack of consideration for pipelined operation. Furthermore, the model is based on the assumption that the cell dimensions are wire-limited. This assumption holds only for register files with a large number of ports and breaks down for moderate or low numbers of ports. Register files in modern microprocessors are characterized by their large signal voltage swings (in contrast to cache arrays) and by their wide range of ports (in contrast to their ASIC counterparts, microprocessor register files may have anywhere from 2 to more than 15 read/write ports). The tool *Estima* described in [57] is a fast architectural-level power, area and latency modeling and estimation engine for multi-ported, pipelined register files.

4.6.1 Methodology

The relevant parameters of an architectural register file model are the number of register rows, bit width of the registers, number of read and write ports and the architectural as well as data activity factors.

Figure 4.14 shows that the dominant components of power in a typical (non-pipelined) register file are the bit lines, the word lines and the clock lines.

4.6.2 Basic Power Model

The basic power model is based on computing the energy per access (EPA) for read and write operations on a single port. This number, combined with the architectural

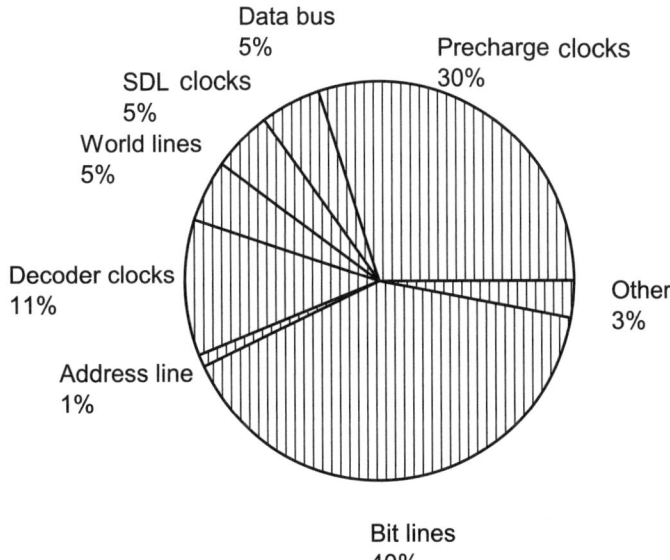

Fig. 4.14 Where in a register file is the power?

read and write activity factors and the clock frequency, gives the power consumption of the register file.

The EPA consumed at a single node, i, for charging and discharging the node can be written as

$$\text{EPA}_i = C_i \cdot V^2, \tag{4.12}$$

where C_i is the total device and interconnect capacitance at node i and V is the supply voltage. The total EPA for a read (write) access in a register file can be written as

$$\text{EPA}_{\text{rd(wr)}} = \sum_i \text{EPA}_i \cdot \#\text{nodes}_i, \tag{4.13}$$

where i enumerates all the different types of nodes that switch during a read (write) operation. EPA_i is computed by Equation (4.12), and $\#\text{nodes}_i$ is the number of nodes of type i that are switching for that particular access. This number is determined by considering the total number of such nodes in the register file and the input data statistics.

Once one has the read and the write EPA of the register file, getting the power consumption is a simple task of multiply and add:

$$P_{\text{reg file}} = (\alpha_{\text{rd}} \cdot \text{EPA}_{\text{rd}} + \alpha_{\text{wr}} \cdot \text{EPA}_{\text{wr}}) \cdot f_{\text{clk}}, \tag{4.14}$$

where α_{rd} and α_{wr} are architectural read and write activity factors, f_{clk} is the clock frequency and EPA_{rd} and EPA_{wr} can be computed using Equation (4.13).

4.6.2.1 Node Capacitance

At the heart of the power estimation algorithm lies the computation of the capacitance of a specific node. The node capacitance model lumps the capacitance of all the devices connected to a node and the interconnect capacitance in a single number for power computation purposes. This number is computed by the following:

$$C_i = l_i \cdot c_{\text{m}i} + W_{gi} \cdot c_g + W_{di} \cdot c_d, \tag{4.15}$$

where l_i is the interconnect length at node i, $c_{\text{m}i}$ is the capacitance per unit length of the metal layer that node i is on, W_{gi} and W_{di} are the total width of gate and diffusion connected devices at i and c_g and c_d are the unit gate capacitance and the unit diffusion capacitance, respectively. In this equation, $c_{\text{m}i}$, c_g and c_d are determined by the particular process technology, and lengths and widths are determined by the physical dimensions of the register file and the device sizes, respectively. While lumped capacitance is used for power computations, wires with distributed capacitance are used in device sizing and timing analysis.

As an example, Fig. 4.15 shows the model of a local read bit line from a hypothetical register file. This structure is essentially a distributed domino AND–OR–INVERT gate driving a static NAND gate. In this particular register file, there are four entries on a local bit line and hence four pull-down structures. The length of the interconnect is l_{m}; there is a pre-charge/keeper structure and a NAND gate at the end

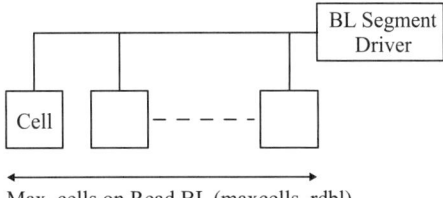

Fig. 4.15 Local read bit line

Max. cells on Read BL (maxcells_rdbl)

to combine the read value with another local bit line. Assuming the interconnect is on metal layer 2, one can get the capacitance of this node using Equation (4.15) as

$$C_{lrdbl} = l_m \cdot c_{m2} + (W_{kpinv} + W_{nand}) \cdot c_g +,$$
$$(4 \cdot W_{txrd1} + W_{txpre} + W_{txkp}) \cdot c_d. \tag{4.16}$$

4.6.3 Pipelined Register Files

As the size of the register files increases (on-chip cache sizes have increased dramatically in recent years), and the clock periods decrease, it becomes more and more difficult to read (write) from (to) a register file in a single phase or cycle of the clock. One may still try to perform reads or writes in a single cycle, but this would require stronger (i.e., larger) devices, and hence higher dynamic and leakage power. In a small register file, one may have local read bit lines driving the output data bus directly in one cycle, whereas in a larger register file, local read bit lines would usually be driving a higher level global read bit line, which in turn drives an even higher level global–global read bit line, which drives the output data bus. This will introduce new power-consuming nodes to the register file, namely, the new hierarchy of bit lines, and the latches, and latch clocks introduced at the pipe boundaries. The pipelining will also introduce latency of the reads/writes, which may effect the performance of the register file, and consequently the whole chip. Therefore, an estimate of power without an appropriate estimate of latency is not very meaningful.

The model is designed from ground up to handle multiple levels of bit-line hierarchy and pipelining and clock lines. In the next section, an algorithm used to compute the physical dimensions and the latency of a (potentially pipelined) register file is discussed.

4.6.4 Physical Dimensions and Latency

Physical dimensions play a very important role in determining the power consumption of a register file. Actually, they influence the power consumption in more than

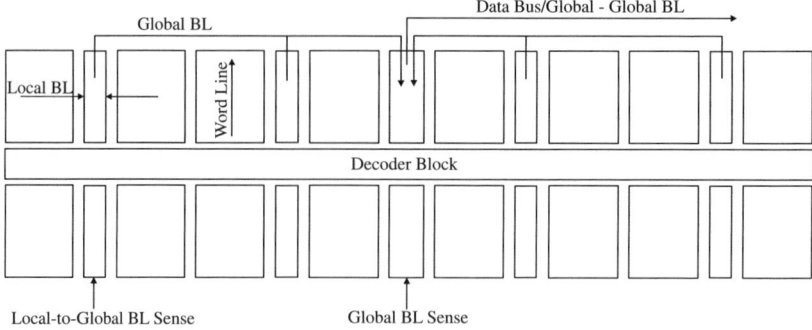

Fig. 4.16 High-level model of the register file

one way: (i) they determine the length of the wires in the register file and thus directly affect the power consumption by determining the capacitance of the nodes and (ii) they impose pipelining constraints, indirectly affecting power by introducing additional power-consuming nodes. Therefore, it is critical to have a good model to estimate the physical dimensions of the register file.

4.6.4.1 High-Level Register File Model

The high-level register file model used in *Estima* is illustrated in Fig. 4.16.

This model is of a single pipe in a pipelined configuration. A multicycle register file may have more than one of these "blocks" in either physical dimension.

4.6.4.2 Physical Dimensions of the Memory Cell

Memory cell size is the single most important factor in determining the physical dimensions of an register file. Although a single memory cell is usually small, they are replicated many times in both dimensions and hence dominate the other blocks in the register file. Two basic RAM cell types that are used in the majority of the register files are (i) the single-ended read/write type and (ii) the single-ended read/dual-ended write type. Fig. 4.17 shows the basic single-ended read/write cell, and Fig. 4.18 shows the single-ended read/dual-ended write type.

For register files with a large number of ports, it is possible to claim that the memory cell size is metal-limited and approximate the cell height and width as $mp \cdot (N_{wr} + N_{rd})$, where mp is the metal pitch in the particular dimension and N_{wr} and N_{rd} are the number of write and read ports, respectively. However, for register files with a small or moderate number of ports, this model is not completely valid. However, instead of using this approximation for the cell size, existing register files were studied for memory cell sizes, and an empirical, first-order polynomial relationship between the cell dimensions and the number of ports was derived as follows:

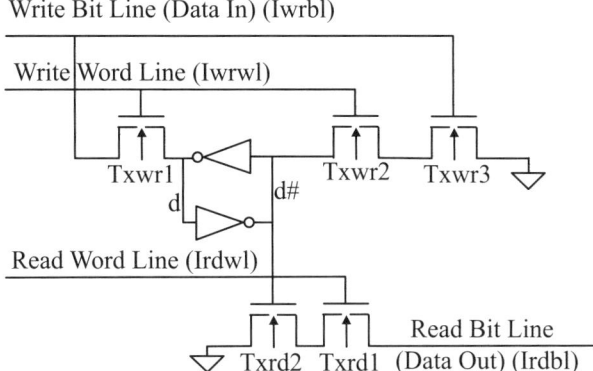

Fig. 4.17 Type I RAM cell, single-ended read/single-ended write

$$\text{height} = a_{\text{h}} + b_{\text{h}} \times \#\text{ports},\tag{4.17}$$

$$\text{width} = a_{\text{w}} + b_{\text{w}} \times \#\text{ports},\tag{4.18}$$

where a_{h}, b_{h}, a_{w} and b_{w} can be obtained from correlation analysis of existing register files and scaled for new process technologies. #ports is the total number of read and write ports of the register file. As one can see, Equation (4.18) reduces to the metal-limited form for a large number of ports.

4.6.4.3 Bit-Line Segment Driver Types

Depending on the size of the register file, one may have the read bit lines directly connected to the data-out port of the array (through a domino-to-static conversion stage) at the segment driver, or additionally the segment drivers may drive the global bit lines. An register file that has all or half of all the entries on a single bit line can

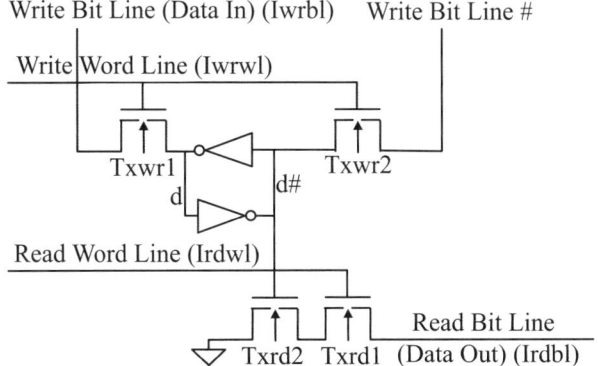

Fig. 4.18 Type II RAM cell, single-ended read/dual-ended write

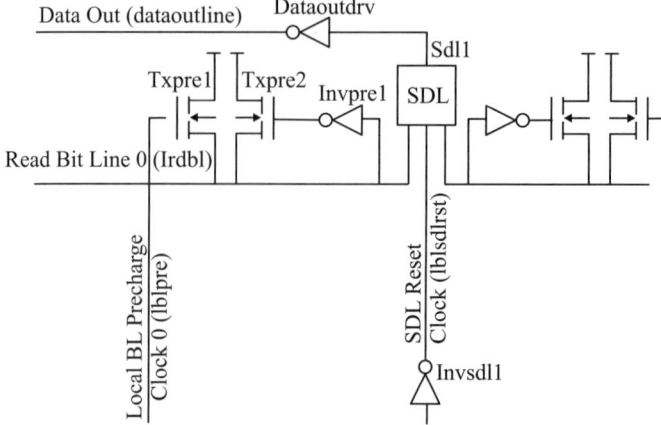

Fig. 4.19 Type I segment driver, dual pre-charge clock/SDL out

use the Type I segment driver (Fig. 4.19) to drive the data-out directly. A larger array, on the other hand, may need more than a single segment driver per bit slice and make use of a Type II segment driver (Fig. 4.20) that drives global bit lines, which in turn will drive the data-out of the array through global bit-line pre-charge stages.

If a register file uses the Type II segment drivers for the local bit lines, it will need a global bit-line pre-charge stage for both pre-charging the global bit line and driving the output as shown in Fig. 4.21.

For dual-ended write (Type II RAM cell), a write bit-line pre-charge unit is needed. Fig. 4.22 shows the circuit used for that purpose.

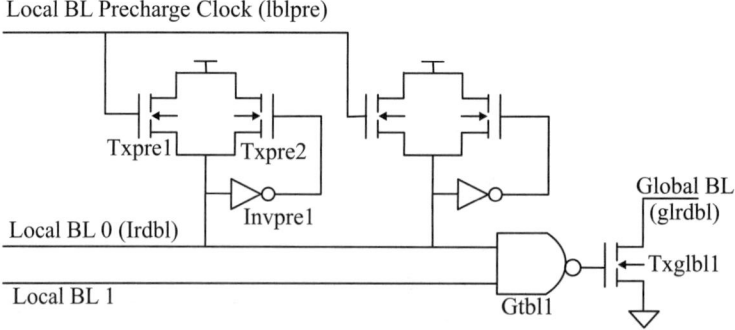

Fig. 4.20 Type I segment driver, dual pre-charge clock/SDL out

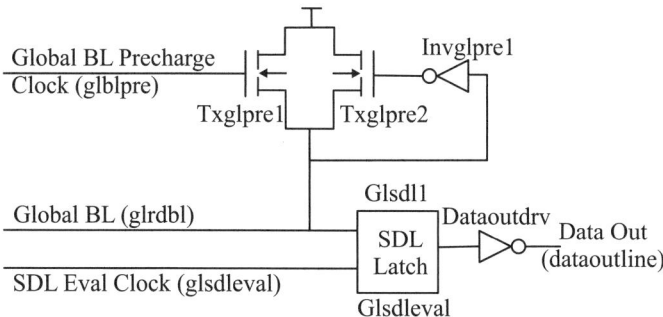

Fig. 4.21 Global bit-line pre-charge

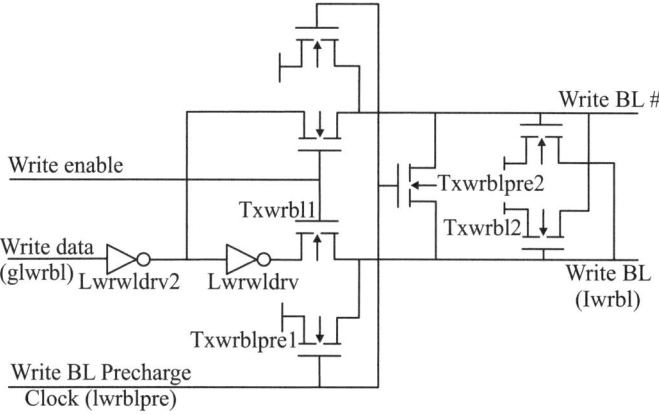

Fig. 4.22 Write bit-line pre-charge unit

4.6.5 Area, Power, Delay Models

4.6.5.1 Dimensions of the Array

The *x*-dimension of the data array is spawned by the bit lines, and the *y*-dimension is spawned by word lines. The pre-charge clocks are usually generated at the decoder stage and run parallel to word line in the *y*-dimension.

The main variable that determines the dimensions of the array structure is the size of the RAM cell. To be able to build a meaningful power model, one needs to be able to model the width and height of the RAM cell that comprises the array. Earlier studies suggest that there is a linear relationship between the number of ports of a RAM cell and its *x*- and *y*-dimensions. To verify this observation and find the corresponding relationship, we studied the cell sizes of a number of register files mentioned above. Figure 4.23 shows the relationship between the width of the RAM cell and the total number of access ports (read/write ports) in the cell. Figure 4.24 has the same data for cell height.

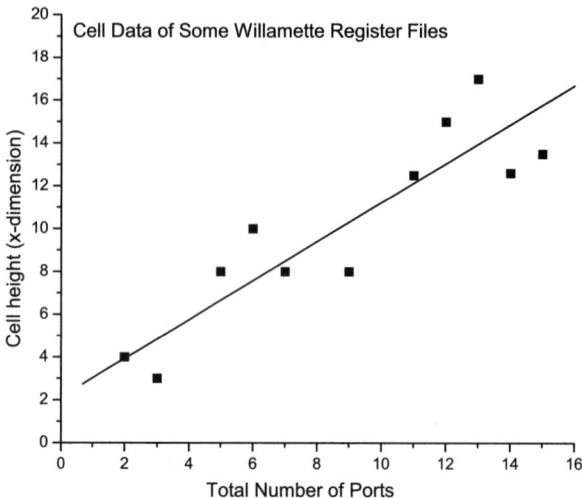

Fig. 4.23 Cell width data for existing register files

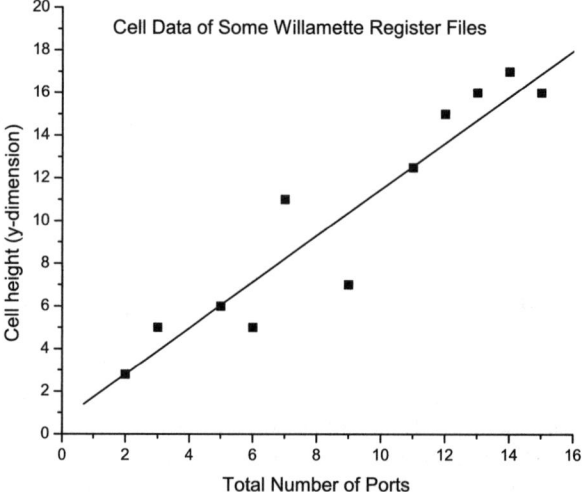

Fig. 4.24 Cell height data for existing register files

Once the dimensions of the RAM cell are found, the next important step in modeling the structure of the array is determining the tile size. To do that, one should be able to compute the maximum number of cells that can be driven by a single segment of a word line and the maximum number of cells that can drive a single local bit line.

The most straightforward way to do this would be to ask the user to supply these numbers. However, this information is not something an architect would have or care about. Therefore, one should come up with a method to determine the maximum number of cells on a word line and a bit line. One possible way to do this is to use the allowed data latency along with the RC time constant of the given process technology. Using this information, one finds the maximum length a particular line can have and the maximum number of devices that may be connected to it. Another approach would be to actually try a number of possible segmentation styles and report all the results (access time, access power and array area) of the estimation to the user to choose the best configuration.

There are a few more parameters that have to be determined to obtain the dimensions of the array, namely, the width of a bit-line segment driver, the height of a word-line repeater and the height of the decoder stage. The large number of RAM cells usually makes them the dominant determining factor for the width and height of the array and hence the length of the lines.

As mentioned above, an array is built by repeating portions vertically and horizontally. Therefore, to get the dimensions of the array, the first thing one has to determine is how many portions in each dimension are needed. From the available register file architectural parameters, these two numbers can be computed as

$$\#\text{chunk}_h = \frac{\#\text{entries}}{2} \times (\#\text{cellsonBL}) \tag{4.19}$$

$$\#\text{chunk}_v = \frac{\#\text{width}}{2} \times (\#\text{cellsonWL}) \tag{4.20}$$

These two numbers give us a fair amount of information about the structure of the array. If the number of horizontal portions (chunk_h) turns out to be 0.5 or 1, one may assume that there are no global bit lines; and the local bit lines directly drive the data-out lines via a segment driver of Type I.

If the number of vertical portions (chunk_v) is 0.5 or 1, then the word-line repeater is replaced by the decoder, and there will be no global word lines. For all other cases, the array will have a decoder block in between the upper and lower $\left(\frac{\text{chunk}_v}{2}\right)$ vertical portions.

With these numbers at hand, the dimensions of the array are given by the following:

$$\text{array}_x = 2 \times \text{tile}_x \times \text{chunk}_h + \text{chunk}_h \times \text{seg}_x, \tag{4.21}$$

$$\text{array}_y = 2 \times \text{tile}_y \times \text{chunk}_v + \text{dec}_y \text{ if } \#\text{chunk}_v \leq 1 \tag{4.22}$$

$$= 2 \times \text{tile}_y \times \text{chunk}_v + \text{chunk}_v \times \text{rep}_y + \text{dec}_y \text{ otherwise.} \tag{4.23}$$

4.6.5.2 Power Model

The switching power of a digital circuit is often given by this simplified model:

$$P_{\text{switching}} = f \times V_{cc}^2 \times \sum_{\text{nodes}} C_{\text{node}} \times \alpha_{\text{node}}, \qquad (4.24)$$

where f is the clock frequency, V_{cc} is the supply voltage, C_{node} is the total capacitance of a node and α_{node} is the activity factor of the node.

For register file type structures, often it is adequate to model the contribution of short circuit and leakage power with a simple percentage of the switching power. For dynamic power modeling, the power dissipating nodes considered are

- data output bus
- local and global bit lines
- local and global bit-line pre-charge clocks
- set-dominant latches (SDLs) at the interface of domino and static logic
- pre-charge and evaluate clocks of the word-line domino drivers.

The activity factors for these lines are determined by the architectural activity factors and the clocking scheme. The capacitance is determined by the structure of the array, dimensions and the devices connected to a given node (line).

The configuration/modeling of a register file involves determining the following:

- *Number of memory cells on a local bit line*: This determines the length of the local bit lines and the need for higher levels in bit-line hierarchy and therefore affects the power consumption of the register file. It is a design decision but cannot be assigned arbitrarily. A large number of cells on a bit line will cause the bit-line charge to leak too rapidly and result in incorrect operation.
- *Number of pull-down devices on a global bit line*: Similarly, this parameter determines the need for higher levels of bit-line hierarchy and affects power consumption. This, too, cannot be assigned arbitrarily due to leakage considerations.
- *Maximum width and height of a pipe stage*: This configuration parameter determines the need for a pipelined implementation. It affects the power indirectly by introducing more levels of bit-line hierarchy and latches in pipe boundaries. It is a function of the process technology.
- *Bit-line folding*: As seen in Fig. 4.14, most of the power in a register file is consumed in the bit-line hierarchy. Therefore, bit lines are natural targets for the power reduction techniques. One such technique that can be applied to register files with a very high aspect ratio to decrease the bit-line width is bit-line folding. A register file can be folded multiple times to correct the aspect ratio or to make the register file fit in a constrained space. Folding is a design decision that can be investigated in the design exploration phase.

The referred tool *Estima* takes in all these high-level parameters and obtains an optimum configuration satisfying these constraints. It computes the number of local bit lines in a bit slice, the number of global bit lines and the number of pipe stages in the bit-line and word-line dimensions. The last two numbers, when added, give

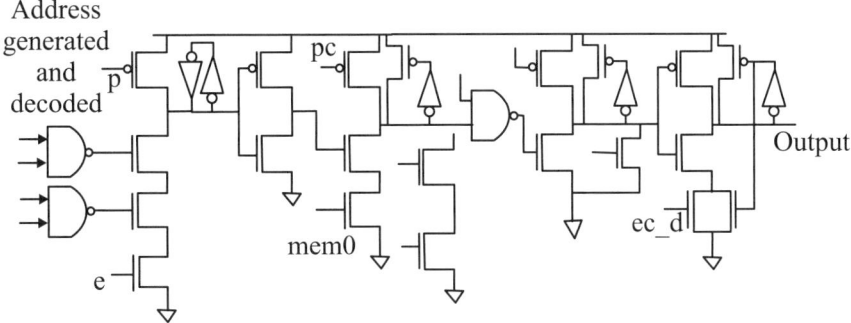

Fig. 4.25 Read path in a typical array logic structure

the latency of the register file. Combining all these numbers with the cell, segment driver and decoder sizes allows one to compute the height, width, area of the register file and, subsequently, power.

The numbers and types of nodes and the lengths of interconnect wires are derived from this data. The EPA computation performed by Equation (4.13) needs the device sizes in Equation (4.15). Figure 4.25 shows a prototypical "read timing path" in a register file. This path and other access paths (along with their time constraints, stability and reliability) such as the write and the pre-charge are optimized during circuit sizing, which determined the various device sizes.

4.6.6 Device Sizing

This section introduces the devices that are relevant to power consumption of the register file and hence need to be estimated. Figure 4.14 suggests what is important to model: one can actually get a very good estimate of the power consumption in a register file if one can estimate the power consumed by the bit lines, the drivers and various clock nodes. This means that the size of the pull-down and pre-charge devices on the bit lines (refer to Fig. 4.15 for an example) and all the clocked devices in the design should be modeled in order to get an accurate power estimate. Unfortunately, the circuit-level details of the register files are usually not available at the architectural design exploration phase. This means that, in order to be able to estimate the power of an register file, one needs an accurate device size estimation method, or historical data on similar blocks need to be mined.

One way to tackle this problem is to impose artificial timing constraints on rise and fall times of various lines and then to estimate the device sizes that will result in such timing behavior using a fast sizing tool and the simple but sufficiently thorough analytical formulas [57] described below.

In *Estima*, the device size estimation problem is addressed by using a simulation-based sizing algorithm to size various devices. At the core of the method, there is a library-independent, technology-dependent, iterative device sizer.

Finally, with the configuration information, the device sizes, and the activity factors at hand, the power consumption of the register file is computed using Equation (4.14).

Bit-line folding has significant impact on power. The power consumption of a particular register file tends to be the lowest when the layout aspect ratio is close to one. This is understood by observing that when the register file is skewed in the bit-line direction (has more registers than the number of bits in a register), the bit lines are long, the devices are larger to meet the timing requirements and therefore they consume most of the power. As the register file is folded, the bit lines become shorter, bit-line devices get smaller and the word lines and the clock lines get longer. This reduces the power consumption while shifting it from bit lines to word lines and clock lines. If one keeps folding the register file, the word lines and the clock lines become too long, and the power consumption goes back up.

4.7 Cache Arrays

Array structures such as the instruction and data caches (and respective tag arrays), issue queues, load/store buffers, register files, branch target buffers, translation lookaside buffers (TLBs), reservation stations and virtual channel buffers (of any on-chip routers) occupy a significant die area of modern processors. In high-end SoCs, memory may contribute as much as 50% of the system power. Correspondingly, the leakage power in these structures accounts for a large portion of the total die leakage and thus often determines the average power. Hence, power estimation in the early stages of the design and subsequent optimization are of paramount importance. However, system designers normally do not have the ability to perform early estimation of such leakage power. While there exists some research on leakage estimation based on pre-characterizing the array blocks at the RTL, such an approach is practical only for already existing designs and not applicable to new designs in a different technology that might require early estimates of power dissipation prior to RTL definition. Structure-aware power estimation of regular arrays is important for accuracy. Significant delay and power savings are achievable in a segmented cache design as depicted in Fig. 4.26. Moreover, multiple banking of a cache may be needed for floor plan reasons and to enable bank-specific power gating – i.e., a bank that is predicted to be inactive for a long enough time may be put into a "drowsy" or even "powered-off" state. A typical banking scheme is illustrated in Fig. 4.27. Next, the estimation of dynamic power and leakage in caches is presented.

4.7.1 CACTI Dynamic Power Model for Caches

CACTI [322] is a well-known tool that also provides a flow for estimating array area and timing that includes device size estimation. Because the CACTI model tracks

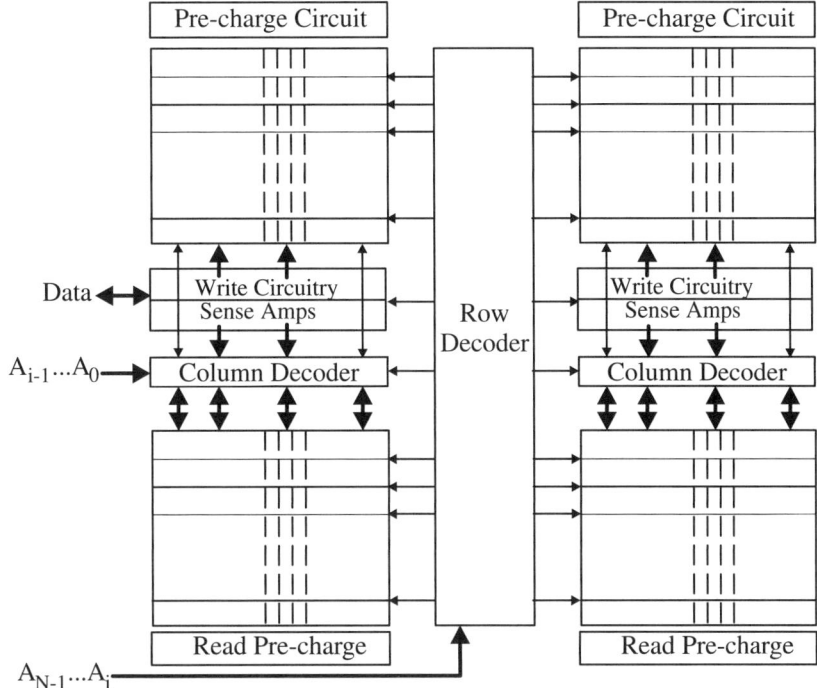

Fig. 4.26 Cache segmentation structure

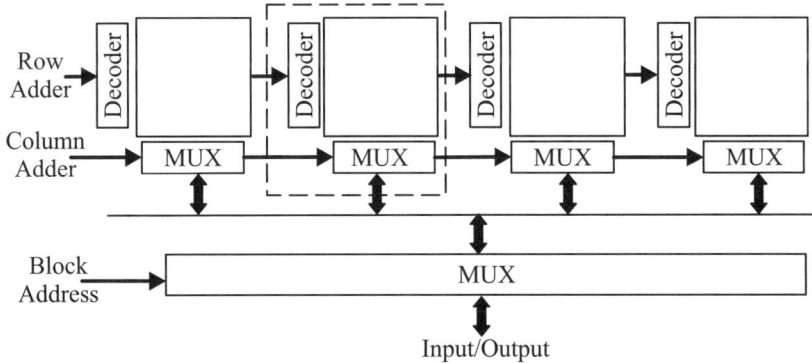

Fig. 4.27 Banked cache structure for smaller bit/word lines and bank-specific power gating

the physical capacitance of each stage of the cache model, the energy consumption, Equation (4.29), is used to calculate the power consumed at each stage. Additionally, one needs to factor in the switching activity and the number of such devices in the cache (as CACTI models the activity along one particular path in a cache).

As an example, consider the power consumption modeled for the decoder on the data path of a set associative cache. CACTI models the decoder, such as in Fig. 4.28,

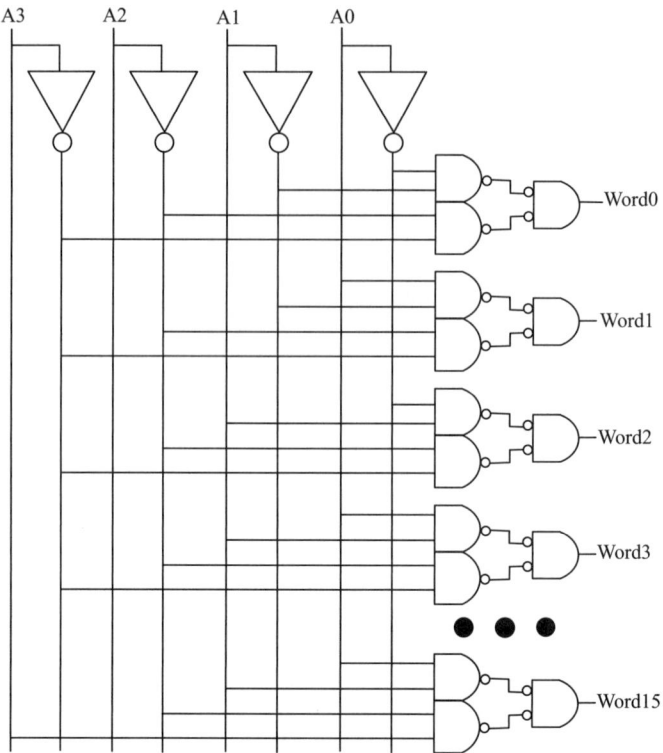

Fig. 4.28 Typical cache decoder circuitry in a cache core

as a composition of three stages: the inverter that drives the probe address bit, the NAND gate that generates the 1-of-8 code and the NOR gate that combines the 1-of-8 codes and drives the word-line driver.

For the first stage, it is estimated that a quarter of the total address bits $\left(= \log_2 \frac{C}{B \times A \times N_{dbl} \times N_{spd}}\right)$ will require the inverter to undergo a 0 to 1 transition (i.e., half of the address bits will be *0*s and half of these were *1*s before). However, both the true and complement forms of the address bits are needed. So the energy consumption of the first stage can be represented as

$$E_{DD1} = C_{stage1} \times V_{cc}^2 \times 0.25 \times \log_2 \frac{C}{B \times A \times N_{dbl} \times N_{spd}} \times 2. \qquad (4.25)$$

The next stage is composed of $\left(\frac{1}{3}\log_2 \frac{C}{B \times A \times N_{dbl} \times N_{spd}} \times 2\right)$ blocks in each subarray. Each $N_{3 \rightarrow 8}$ block is composed of eight NAND gates. One observes that half of these gates will undergo energy-consuming switching. Since there are $N_{dbl} \times N_{dwl}$ decoders, the energy consumption is thus given by

$$E_{\text{DD2}} = C_{\text{stage2}} \times V_{\text{cc}}^2 \times N_{\text{dbl}} \times N_{\text{dwl}} \times 4\frac{1}{3} \log_2 \frac{C}{B \times A \times N_{\text{dbl}} \times N_{\text{spd}}} \times 2. \quad (4.26)$$

Finally, the last stage is composed of a NOR gate that will drive a single word line. Only one of the NOR gates in the decoder will be selected, which implies

$$E_{\text{DD3}} = C_{\text{stage3}} \times V_{\text{cc}}^2. \quad (4.27)$$

Note that in general, bit lines of a small signal array undergo a smaller swing than the full V_{cc}, and hence the corresponding energy equations have the following form:

$$E = C \times V_{\text{cc}} \times \delta V_{\text{cc}}, \quad (4.28)$$

where δV_{cc} is the voltage swing experienced by switching bit lines.

Analogously, power is represented as a sum over all nodes n:

$$E = \frac{1}{2} \sum_n f_n C_n \times \delta V_{\text{cc}} \times V_{\text{cc}}. \quad (4.29)$$

4.7.2 Leakage Modeling for Arrays

An approach to analytical estimation models for high-level leakage of array structures without pre-characterized data is given in [195]. There, they develop models parameterized in terms of the organizational parameters of the array (number of rows, columns, read multiplexer size and write multiplexer size) and the operations (read, write, pre-charge and idle) on the array, considering transistor leakage at the level of the major sub-blocks of the array. Figure 4.29 shows structural components of a typical array. It is primarily composed of the following main sub-blocks: address decoder, memory core, read column logic, write column logic, read control and write control logic. Arrays typically support read and write operations. However, the read and write operations may happen on alternate clock phases. The bit lines (read or write) are pre-charged in the phase they are inactive (not reading or writing, respectively). If no operation is being performed in a clock cycle, all the word lines remain deactivated (logic Low); the bit lines stay pre-charged to V_{cc}, and the operation phase is termed an idle phase. Since different transistors assume different states of conduction in different operation phases, leakage is modeled in the following simplified way. Transistor leakage current I_{leak} is directly proportional to width w of the transistor but is an exponential function of the threshold voltage V_{Th} and temperature T. However, leakage current reduces drastically for stacked transistors as discussed in introducing the "stacking factor". Thus for a given technology and temperature, I_{leak} can be estimated if the stacking factor and the transistor sizes are approximated.

The methodology to estimate transistor sizes is similar to that discussed under the dynamic power estimation for register files and the CACTI models earlier. The leakage current can be estimated as follows:

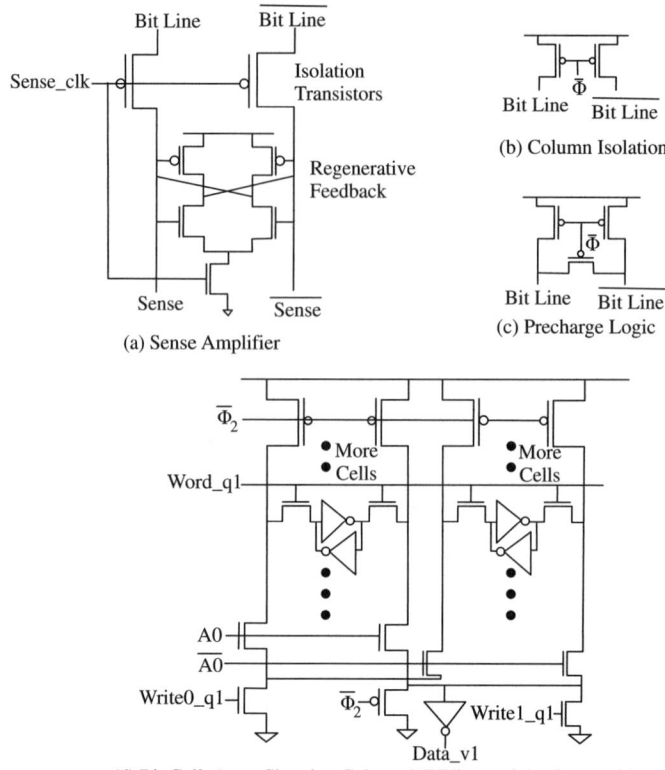

Fig. 4.29 Component of a typical array structure

$$I_{\text{leak}} = I_{\text{off}} \left(\frac{W_{\text{tot}}}{X_{\text{s}}} \right) X_t X_n, \tag{4.30}$$

where I_{off} is leakage/micron expected from silicon, X_t is a temperature factor, X_{s} is the stacking factor and X_n is a noise factor.

The memory cell in Fig. 4.30 can be analyzed for its subthreshold leakage as follows under different phases of operation. When the cell is idle, the word lines are asserted low whereas the bit lines are pre-charged high. The case described here assumes that the cell holds a high value at Q; the opposite case of a low value is analogous. With Q at 1, transistors P1, N2, N3 and N4 are off. However, N4 has both its drain and source at the same logic level, namely high. Thus, only the other three are leaking, giving

$$I_{\text{idle}_{\text{memcell}}} = \text{WP1} \times I_{\text{lp}} + (\text{WN2} + \text{WN3}) \times I_{\text{ln}}, \tag{4.31}$$

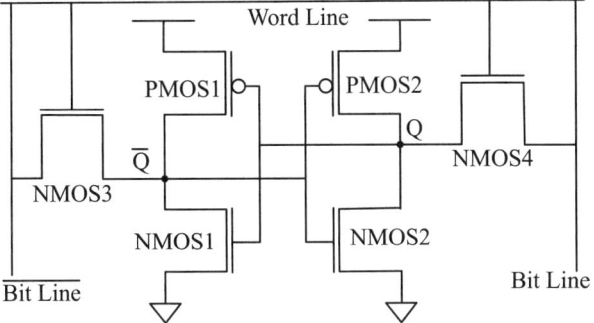

Fig. 4.30 Typical six-transistor cache SRAM cell

where I_{ln} and I_{lp} denote leakage of a unit-width n and p device, respectively. WN and WP represent device widths used in a typical memory cell for the technology in question. To obtain total memory core leakage, one multiplies this estimated value with the total number of memory cells in the array $(N_{rows} * N_{cols})$. During the read phase, exactly one word of the array has its word line asserted. In that case (again assuming that a "1" is stored; i.e., $Q = 1$), transistor N3 is turned on and thus not leaking. Consequently, leakage current of the memory cell core is as follows:

$$I_{read_{memcore}} = N_{cols} \times N_{rows} \times (WP1 \times I_{lp} + WN2 \times I_{ln} + WN3 \times I_{ln})$$
$$-N_{cols} \times WN3 \times I_{ln}. \tag{4.32}$$

The write-phase leakage can be modeled as the same as read-phase leakage. Precharge phase leakage in the memory core is approximated as the same during idle.

Column read and write circuitry: Each column contains bit-line conditioning/precharge and isolation transistors, sense amplifier and column MUX. A typical write circuitry is presented in Fig. 4.31.

4.8 Validation and Accuracy

Some of the validation methods used in the industry and the issues of accuracy arising in high-level estimation, with register file power estimation as an example, are briefly discussed below.

4.8.1 Model Validation: Arrays as an Example

For validation of the model of array structures, a switch level simulation flow was used. The validation process has two levels. The first level involves comparing the extracted capacitance values with the estimated capacitance values from the line

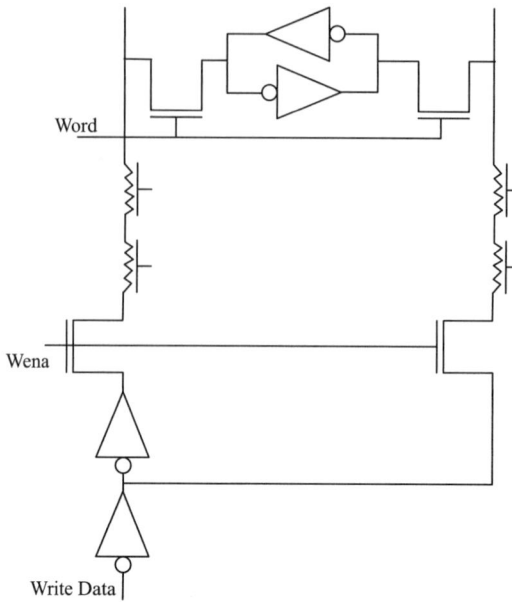

Fig. 4.31 Typical write circuitry in the cache core

models. Once this is done, the actual power model should compare the power numbers generated by the switch level simulator.

4.8.1.1 Capacitance Validation

For all the "lines" in the array such as local and global bit lines, word lines for both read and write logic, the line models compute the metal length (*length*), the load devices at the end of the line (*load*), the distributed loads connected to the line (*dist*), the number of such distributed loads (*segments*), whether the line is driven by a driver (*driven*) and the number of such lines (*num*). For the diverse set of register files that were studied, the capacitance values obtained from the line models (with actual device sizes) were compared with the values extracted from the layout. The accuracy of the capacitance numbers was within 15% for most of the lines. That means that one can use these line models in the power model with confidence.

4.8.1.2 Power Validation

To do the power validation, detailed simulators such as *PowerMill* should be run with appropriate traces to exercise the register file and generate activity factors for the nodes. Then, one uses the activity factors along with the extracted capacitance values to compute the power numbers to be compared against. Correlating models with test-chip silicon measurements and tuning of the models may also be undertaken during this phase.

4.8.2 Simulator Accuracy

It can be argued that simulator accuracy for a processor is not necessarily as dependent upon microarchitectural representation, such as the number of pipe stages, as it is on correctly capturing the behavior of the simulated architecture. For example, if a test workload has three loads and five adds, then whether the add takes place in one or four cycles has a far smaller effect on simulator accuracy than matching the throughput and time base of the modeled architecture.

Some previous architectural power modeling works have stated that comparing low-level power capacitance values is the most precise means of validating a power simulator. This could not be farther from the truth; in fact, accurately representing the underlying architecture is the first-order accuracy and validation constraint. It is easily seen that $P = \frac{E}{T}$. If the underlying representation of the architecture, its instruction throughput and time base are not accurately captured, then it is implicit that the power values derived therein must be erroneous. To sharpen this point: even if an implemented structure were to have the same circuit characteristics (i.e., capacitance) and power for a given published architecture, if the events concerning that structure are different from the actual architecture over the same time period (or if the time bases are different), then the underlying architecture is not being captured and the power profiles for that structure are not the same. In general, event throughput/ rates are the implicit definitions of the underlying architecture's pipelines, structure and balance.

This example serves to capture an oft-missed point in architectural research as a whole and especially so in the power realm: get the big things right first. Several existing power modeling tools are built upon the SimpleScalar simulator [48, 322]. While this simulator has proven invaluable in academic architectural research, further enhancements or replacements are necessary to accurately capture and represent modern microarchitectures in order to provide meaningful and useful results.

4.8.3 Power Model Accuracy

Architectural power modeling by and large is dependent upon activity counters and the power associated with that activity. This lends to two fundamental areas for error: first in the derivation of event and counter placement within the simulator and second in the determination of the associated power value. Of these two items, event activity counters is more significant. The power value itself may be derived with varying accuracy, without affecting the overall legitimacy of the results. To illustrate this, consider three clusters with $C1 = X$ Watts, $C2 = Y$ Watts and $C3 = Z$ Watts.

Regardless of the literal values of $X - Z$, if, say, the relative powers of $X < Y < Z$ are preserved, then implementing an architectural-level change which lowers Y to $Y < X$ would still be valid. Furthermore, it can be argued that implementation methodology A may be as valid and effective for cluster $C1$ as method B. But, given two different designers, the resulting cluster design and corresponding power may be different.

If the event monitoring is broken, however, it would be impractical to expect the results to be representative of cluster power, even if the power value itself is exact. Again, this recalls the criticality of the underlying simulator capturing the modeled architecture as a basis. Further, although the modeled architecture has an event (E) for every memory access, the simulator only captures one fourth of these events due to incorrect accounting (e.g., placement); thus, the results are inconclusive. The interested researcher may feel that from the resultant data event E is not worth looking further into for power savings when E could in fact be a prime target. Conversely, if event E is over-counted, interested researchers may focus their efforts incorrectly. Compounding this, results claiming that by modifying the implementation, algorithm or design of the cluster handling event E might be erroneously published, with overly optimistic power savings reported.

4.9 Effect of Temperature on Power

Strong dependence of leakage on temperature requires the estimation process to be iterative. Most of the techniques of power estimation described below assume a certain chip-wide uniform temperature. However, the temperature gradients, some times as high as $70°$ C differential, are created when real applications run, due to localities of computation/communication energy and uneven thermal resistances. These temperatures then influence leakage which in turn affects densities of power dissipation in different regions, and the cycle of cause and effect continues until a stable (fixed point) is reached. Figure 4.32 illustrates the dependence of leakage power on temperature. As a result, thermal estimation is fast becoming an integral part of power estimation. A typical procedure for generating a converged thermal map is as follows:

1. Pre-characterize and obtain leakage per transistor width and type, from simulation or measurement (transistor types such as PMOS/NMOS, low/high V_{Th}, extended/normal channel length).
2. Collect transistor type and width information for all blocks. For each block, sum the leakage over total transistor widths proportionally, considering their type. Finally, scale the block's leakage by the block's estimated average stacking factor.
3. Generate a leakage power estimate per block according to transistor width/type information, calculated as a function of local temperature and voltage.
4. Generate the power map based on the leakage and the dynamic power estimates along with the floor plan location for each block.
5. Then, simulate thermal dissipation based on the power map and the die and package thermal resistances.

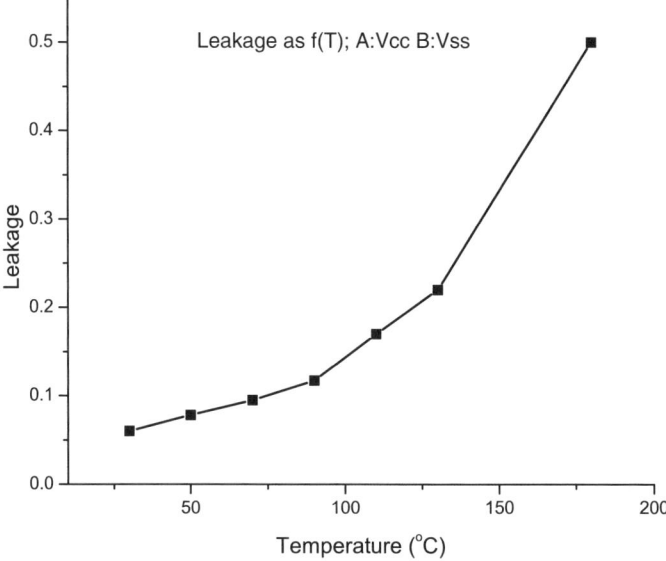

Fig. 4.32 Leakage versus temperature

6. If the newly computed maximum or minimum temperature in the thermal map has changed by more than some threshold (sometimes a threshold on change in power is used), repeat from Step 3.

Burn-in power estimation assumes negligible dynamic power but models significant leakage power because of elevated temperature and normally higher voltage. On the other hand, under normal system operation, dynamic power is substantial and dominates leakage. Thermal convergence would normally take 5–10 iterations. As much as 5° C or more has been observed in hot spots between the first and the last iterations. The hot spots and the profiles of power/temperature can dramatically vary between test and normal use conditions. The estimation of hot spots is needed in selecting appropriate spots to embed thermal sensors on die for passive monitoring or active control.

4.10 Summary and Conclusions

Various techniques for power estimation at abstract RTL were presented, and their role in early design exploration was discussed. Why and how estimation is integral to optimization and design for reliability and how high-level estimation enables power versus performance trade-off at the most effective level of abstraction were also addressed. Estimation techniques for specific, important architectural blocks such as register files and caches and synthesizable control logic are discussed. The effect of variation and the interplay of power with temperature are discussed. The importance of performing technology readiness that predicts characteristics of major design structures and use of any significant circuit and process techniques in power optimization are discussed such that high-level power modeling and estimation can anticipate these trends to be successful and sufficiently accurate.

In trying to make a traditional academic simulator more accurate, some suggest correlating the academic simulator directly to silicon measurement. But this has a deficiency: a single-chip-wide power value would be used for correlation in such an approach. Consequently, this approach loses the level of granularity that can be accomplished via correlating a high-level power simulator to a design team's internal microarchitecture simulator. This chapter is concluded by mentioning some of the desirable properties of any high-level power simulator that better correlates to a design team's detailed microarchitecture simulator able to model faithfully the micro-operations and the events streaming through the machine. Thus, a good power simulator for a typical Intel P4-like IA-32 processor should include such microarchitectural features as (i) the front-end prefetch and micro-instruction translation engine, which translates macro-instructions into the component micro-ops; (2) the trace cache and control; (3) the out-of-order logic that performs allocation, reorder, register renaming and scheduling; (4) the execution engine, which includes integer, floating point and multimedia units; (5) the memory subsystem, which provides memory access logic and buffers; and (6) the system interface unit defining the buses, queues, arbiters and contention logic.

The ability to accurately model the structure and balance of an architecture including the (architectural) throughput of the machine being modeled is important for building an accurate, yet efficient high-level power model. However, the onus is on the researcher to ensure that the components of interest for his or her particular study are simulated and assigned power for the particular type and level of research are being performed.

Chapter 5
Power Reduction Fundamentals

5.1 Introduction

Low-power circuit design is a three-dimensional problem involving area, performance and power trade-offs [275, 239, 309, 66]. The trend of decreasing device size and increasing chip densities involving several hundred millions of transistors per chip has resulted in tremendous increase in design complexity [324, 341, 340, 47, 179, 348]. Additionally, this trend of reducing the feature size along with increasing the clock frequency has made reliability a big challenge for the designers, mainly because of high on-chip electric fields [89, 326]. Designing chips of such complexity using traditional *capture and simulate* methodologies is time consuming and difficult. The industry has started looking at the development cycle to reduce design time and to gain a competitive edge. Power dissipation occurs in various forms, such as dynamic, subthreshold leakage, gate leakage, etc., and there is need to reduce each of these [211, 235, 213, 270, 373, 241, 341]. In addition, it is essential to study the dissipation profile of CMOS circuits and formulate the optimization problem accordingly for more effective power optimization [268, 115, 116, 291, 218, 226, 232, 213, 254]. High-level synthesis of digital circuits has become necessary due to several inherent advantages such as reduction of design time, exploration of different design styles and meeting design constraints and requirements [288, 287, 149].

5.2 Power Dissipation or Consumption Profile of CMOS Circuits

In low-power design for battery-driven portable applications, the reduction of peak power, peak power differential, cycle difference power, average power and energy is equally important. These are different forms of total power dissipation of a CMOS circuit. Each one of these power dissipation measures affects different attributes of a CMOS circuit and needs to be optimized for meaningful low-power design. Various

S.P. Mohanty et al., *Low-Power High-Level Synthesis for Nanoscale CMOS Circuits*,
DOI: 10.1007/978-0-387-76474-0_5, © Springer Science+Business Media, LLC 2008

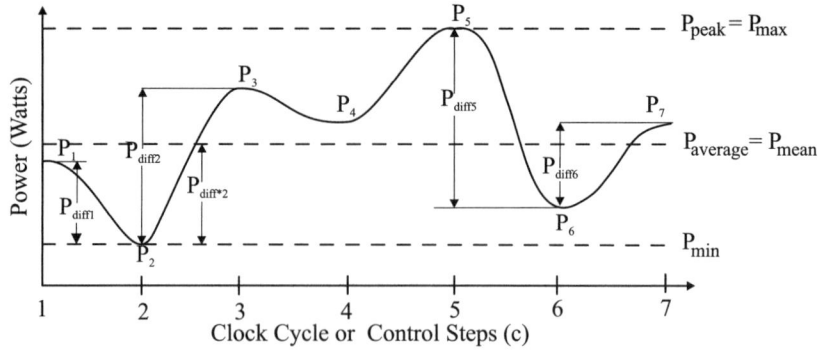

Fig. 5.1 Power dissipation or consumption profile of CMOS circuits

interrelated terms that can describe the power dissipation profile of a CMOS circuit are explained in Fig. 5.1 [115, 116, 291, 218, 226, 232, 202, 213, 254, 316, 319, 317, 227, 220, 228, 222, 210].

1. *Total energy consumption* (E): This is the total power dissipated by a circuit over the period of operation. This is a true measure of energy consumption for which a consumer pays the electricity bill.
2. *Energy per cycle* (E_c): This is the energy consumption of a circuit in a clock cycle or control step c.
3. *Power per cycle* (P_c): This is the power dissipation of a circuit in a clock cycle or control step c. This is defined as $\left(\frac{E_c}{T_c}\right)$ with T_c being the clock cycle width. When expressed as power at any instance of circuit operation, then this is the instantaneous power dissipation of a circuit.
4. *Peak power* $\left(P_{peak} \text{ or } P_{max}\right)$: This is the maximum power dissipation by the circuit over a period of time, i.e., $\text{maximum}(P_c)_{\forall c}$. This has direct and indirect impact on the power dissipation profile, such as average power, transient power, etc., while being manifested as peak power dissipation. This is also known as maximum instantaneous power or maximum sustained power of a circuit.
5. *Minimum power* (P_{min}): This is the minimum power dissipation by the circuit over a period of time, i.e., $\text{minimum}(P_c)_{\forall c}$. This has direct and indirect impact on the power dissipation profile, such as average power and transient power.
6. *Average power* $(P_{average})$ or *mean power dissipation* (P_{mean}): This is the average power dissipation by the circuit over a period of time. This can be defined in two different ways. In the first approach, the average power is the total energy dissipated divided by the total time, i.e., $\left(\frac{E}{N_c \times T_c}\right)$, where N_c is the number of clock cycles. In the second way, the average power dissipation can be quantified as the mean of power per cycle (P_c), i.e., $\text{mean}(P_c)_{\forall c}$ or $\frac{1}{N_c}\Sigma_c(P_c)$.
7. *Cycle difference power* $(PD_c \text{ or } P_{diff_c})$ or *cycle differential power*: The cycle differential power, a measure of cycle power fluctuation or transient power, is

modeled either as the absolute deviation from the mean cycle power (P_{mean}) or as the cycle-to-cycle power gradient. Thus, mathematically it is expressed as $PD_c = (|P_{mean} - P_c|)$ or $PD_c = (|P_{c+1} - P_c|)$.

8. *Peak power differential* (PD_{peak} *or* PD_{max}): There are two different alternatives to measure peak power differential. It can be defined as a difference between maximum power and minimum power dissipation, i.e., $P_{max} - P_{min}$. The other way to measure the same quantity is the maximum of cycle difference power, i.e., $\text{maximum}(PD_c)_{\forall c}$. This is a measure of transience or fluctuation in power dissipation profile.

9. *Power delay product (PDP)*: The power delay product of a circuit is defined as the product of the average power consumption and critical path delay, i.e., $PDP = P_{average} \times T_{cp}$, where T_{cp} is the critical path delay. The unit of PDP is the same as that of energy. PDP is considered when optimizing for both power and performance.

10. *Energy delay product (EDP)*: The energy delay product of a circuit is defined as the product of the total energy consumption and critical path delay, i.e., $EDP = E \times T_{cp}$, where T_{cp} is the critical path delay. EDP has to be optimized for optimization of both energy and performance. This is one of the difficult metrics to optimize.

5.3 Why Low-Power Design?

Low-power design of VLSI circuits and systems is very much essential due to several reasons, as presented in Fig. 5.2 [39, 41, 67, 373, 210, 289, 309, 193, 220, 66, 326, 341, 340, 37, 203]. Of course, what source of power dissipation to be reduced and what type of power profile metric to be considered may vary depending on the applications.

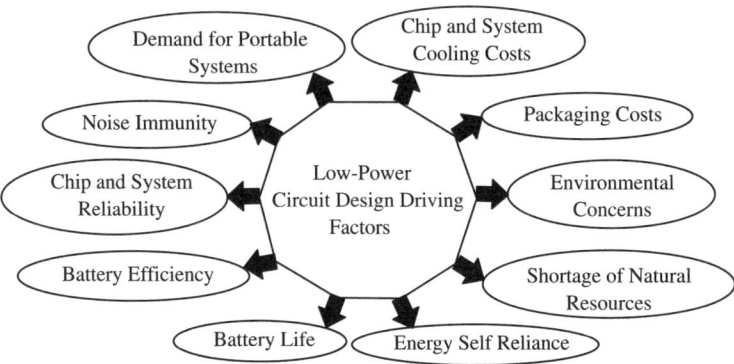

Fig. 5.2 Driving factors for low-power circuit design

The need for low-power synthesis is driven by several interrelated factors:

1. *Battery life*: Battery life has been one of the major motivating factors of low-power design. Battery life is determined by its ampere-hour (Amp-Hr) rating. The lower the energy consumption, the longer the battery life for the same Amp-Hr rating [203, 37, 204, 254, 346, 301, 140].

2. *Battery efficiency*: Besides battery life, another factor that affects battery life indirectly is the battery efficiency. Portable electronic appliances like mobile phones and notebooks use different types of batteries, such as lithium polymer (Li-Poly), lithium ion (Li ion), nickel-metalhydride (NiMh), nickel–cadmium (NiCd), etc., as source of energy [301, 296, 104, 36, 306, 295, 140]. In the battery, energy is stored in the form of chemical energy which is converted to electric energy whenever needed by the appliance. At the same time, when the battery is being recharged, electrical energy is being converted to chemical energy. The electro-chemical conversion efficiency is affected by the power dissipation profile of the systems and chips.

3. *Chip and system reliability*: The reliability of a chip or system is the chance that it works correctly when needed. In other words, the probability that a system or chip works correctly at any instance of time can be represented mathematically as a number in the range $(0, 1)$ [359]. If the power consumption is higher, the temperature in the circuit is increased. Thermal effects have multi-fold impact on the circuit and system performance and reliability [45]. The active current of the circuit decreases and leakage of the circuit increases with the increase in temperature, thus reducing the driving capability [365]. In other words, there is a shift in the I–V characteristics of a CMOS device. The increase in temperature has another direct impact on the reliability of the system and the chip, as for every $10°C$ rise in operating temperature, the failure rate of the components [289, 331, 129, 258] roughly doubles. Thermal effects may also lead to phenomena such as electromigration and hot-electron effects, as well as dielectric break down, causing circuit degradation [371, 309, 189]. The effect is quantified in the mean time to failure (MTF), which states $MTF \propto J^{-2}$, where J is the average current density.

4. *Noise immunity*: Noise in a CMOS circuit is any activity that causes the deviation of a node voltage from its nominal value [92]. Noise in a circuit originates from various sources, a degrades the circuit performance, and affects noise margins; thus the circuit requires sufficient noise immunity as a design constraint. Excessive power consumption can lead to high current flow and higher IR drop (i.e., resistive voltage drop) which affects circuit performance. Similarly, supply bounce and ground bounce can be caused due to large $\left(L\frac{di}{dt}\right)$ and cross-talk $\left(L_m\frac{di}{dt}\right)$ arising from high current changes [220, 210]. Here L and L_m are self and mutual inductance, respectively.

5. *Demand for portable systems*: The emergence of portable devices and mobile Internet devices such as laptop/palmtop computers, mobile phones, personal digital assistants (PDAs) and multimedia-rich wireless devices is one of the major driving factors for low-power design [289, 193, 210, 142]. In portable systems,

battery life and efficiency is one of the major issues. In addition, form factors and lack of room for installing sophisticated cooling systems further strengthens the need for low-power design.

6. *Chip and system cooling costs*: Chip and system cooling costs are particularly important for high-performance systems as they push the limits of cooling systems [289, 189, 153, 45]. Increase in power dissipation of a chip and system will need more powerful cooling systems, such as radiators and fans, thus increasing the cooling cost. At the same time, the fans themselves may need energy for their operation which will have impact on the energy bill.

7. *Packaging costs*: Proper packaging of chips is very important since they not only have to protect the active portion of the chip but also have to transfer the heat from them to keep them cooled. Thus, packaging (see Fig. 4.1) does add to the cost of the chip. The cost of the packaging is dependent on the operating temperature of the chip, which is affected by power consumption. For example, ceramic package can be replaced by plastic package to reduce cost if the operating temperature (power dissipation) can be reduced [289, 365, 258].

8. *Environmental concerns*: The higher the power dissipation in a circuit, the higher the heat pumped into the rooms in which a chip and the corresponding system are stationed and thus, the higher the impact on the environment. Therefore, the electricity consumption will be increased, the impact on room air conditioning will be increased and the impact on the environment will be high [289, 210].

9. *Shortage of natural resources*: If the energy dissipation of a chip or a system is higher, then they intrinsically need more energy which is created from natural resources like coal. Similarly, the associated cooling systems also need energy and consume natural resources. If a chip is installed in an automobile or aircraft, etc., it consumes gas or petrol, hence it consumes natural resources. All of these add to environmental pollution and shortage of natural resources [19, 1].

10. *Energy self-reliance*: CMOS chips are everywhere, namely household appliances (washing machine, refrigerators, etc.), automobiles, aircrafts, office appliances. The higher the energy consumption by the chips, the higher the power consumption by each of these appliances. The higher the energy needs of a nation, the greater its dependence on other nations with more natural resources. Thus, from an energy self-reliance point of view, the power/energy consumption needs to be minimized [19, 1].

5.4 Why Energy or Average Power Reduction?

Energy or average power has been the target of optimization for most of the low-power design works, as it has profound impact on circuit performance. Irrespective of applications of whether it is a desktop, server, laptop or portable device, energy or average power reduction is an important attribute of the chips or systems. Energy and average power reduction is essential for the following reasons [67, 341, 275, 239, 210, 309, 63]:

- To increase battery lifetime
- To enhance noise margin
- To reduce cooling and energy costs
- To reduce energy costs
- To reduce use of natural resources
- To increase system reliability

The battery lifetime is determined by the Amp-Hr rating of the battery. If the average power (and/or energy) consumption is high, then battery lifetime may be reduced because of high ampere consumption. This factor is important for portable applications. The reduction of average power is essential to enhance noise margins (to decrease functional failures). The cost of packaging and cooling is determined by average current flow and hence the average power and energy. The high energy consumption of computer systems leads to environmental concerns due to the need for more power generation. If the average power is large, the operating temperature of the chip increases, which may lead to failures.

5.5 Why Peak Power Minimization?

The peak power is the maximum power consumption of the integrated circuit (IC) at any instance, which is also known as the maximum instantaneous power or maximum sustained power of a circuit. Peak power reduction is more important for desktop applications than portable applications. The literature is rich in efforts to reduce total energy consumption and average power consumption of CMOS circuits. At the same time, the reduction of peak power consumption is essential for the following reasons [201, 73, 316, 202, 291, 218, 317, 225, 227, 375, 228, 222]:

- To maintain supply voltage levels
- To increase reliability
- Smaller heat sinks
- Cheaper packaging

If the current flow is large, then the IR drop of the interconnects becomes large which can reduce the supply voltage levels at different parts of an IC. This can lead to malfunctioning of the circuit due to insufficient voltage levels. On the other hand, if a device is over-designed with higher supply voltage and/or general power-grids allowing for these voltage drops, then there is a significant increase in overall power and metal-area cost. High current flow can reduce reliability because of hot-electron effects and high current density. The hot-electron effects – the injection and trapping of "hot" carriers into the SiO_2 gate-oxide layer – may lead to runaway current failures and electrostatic discharge failures. High current density can also cause electromigration, which is the undesirable displacement of metal atoms in conductors caused by momentum transfer between conducting electrons and diffusing atoms, leading to interconnect failure. It is observed that the mean time to failure (MTF) of a CMOS circuit is inversely proportional to the current density (or power density).

If the current (power) dissipation is large, then the heat generated out of the system is high. The system then needs bigger sink and costlier heat dissipation mechanisms in order to maintain the operating temperature of the ICs in its tolerance limit. If high switching activity happens at the same time then it can have impact on the $\left(\frac{di}{dt}\right)$ value.

5.6 Why Transient Power Minimization?

Transience or fluctuation in power is yet another important attribute of the power dissipation profile of a CMOS circuit. The peak power, the cycle difference power and the peak power differential drive the transient characteristics of a CMOS circuit. Transient power needs to be reduced for the following reasons [289, 291, 218, 226, 232, 220, 213, 228]:

- To reduce power supply noise
- To reduce cross-talk and electromagnetic noise
- To increase battery efficiency
- To increase reliability

Power fluctuation leads to larger $\left(\frac{di}{dt}\right)$ causing power supply noise or bounce, because of the self-inductance of power supply lines. Cross-talk is the noise voltage induced in a (victim) signal line due to the switching in another (attacker) signal line. The voltage induced by the mutual inductance is expressed as $\left(L\frac{di}{dt}\right)$ and that induced by the mutual capacitance as $\left(C\frac{dv}{dt}\right)$. If the power fluctuation is high, then large $\left(\frac{di}{dt}\right)$ and $\left(\frac{dv}{dt}\right)$ can introduce significant noise in the signal lines. As the power fluctuation increases, it reduces the electrochemical conversion and hence there is decrease in battery life. High current peaks (power fluctuation) in short time spans can cause high heat dissipation in a localized area of the silicon die which may lead to permanent failure of the integrated circuit.

5.7 Why Leakage Power Minimization?

In nanoscale CMOS circuits, besides dynamic power consumption, energy is dissipated in terms of various forms of leakage, such as gate-oxide leakage, subthreshold leakage, junction tunneling leakage. The leakage power dissipation occurs both in active (ON) and passive (OFF) states of a device. Hence, leakage reduction has become more critical for nano-CMOS transistors and circuits [155, 193, 309, 373, 241, 159, 308, 358]. Leakage power dissipation will be additive to instantaneous as well as average power and hence will have impact on most of the power profile metrics. However, it is of the highest significance in devices designed for applications where the system goes to standby mode very often such as mobile phones and PDAs.

5.8 Power Reduction Mechanisms at Different Levels of Abstraction

Power reduction in general can be achieved at various levels of design abstraction, such as system, architectural (behavioral, algorithmic, high level) and logic and transistor level. At each level of design abstraction researchers have proposed different techniques for reduction of various sources of power dissipation as summarized in Fig. 5.3 [193, 136, 276, 289, 309, 373, 221].

Dynamic power management (DPM) techniques, dynamic voltage (or frequency) scaling (DVS) and clock gating are popular system-level methods [35, 325]. Similarly, multiple voltage (multi-V_{DD}) techniques have been explored as attractive methods for behavioral- and RTL-level dynamic power minimization [320, 199, 221]. Moreover, multiple threshold (multi-V_{Th}) options have been proposed for the reduction of subthreshold current [155, 106, 261, 213]. Recently, a dual-T_{ox} method is proposed as a transistor-level method for tunneling current reduction [339, 328, 235, 234]. As evident from the figure, the higher is the design abstraction, the higher is the energy/power reduction and the lower is the number of iterations needed to perform analysis and optimization [289, 210].

5.9 Why Power Optimization During High-Level or Behavioral Synthesis?

The behavioral level is not as highly abstracted as the system level nor as detailed as the logic gate or transistor level. Hence, at behavioral level there is a balanced degree of freedom to explore power reduction mechanisms and it can help in investigating

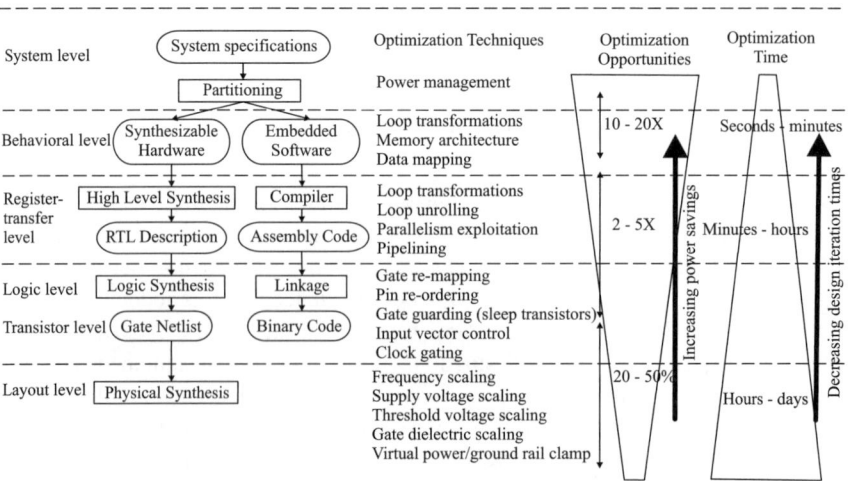

Fig. 5.3 Reduction mechanisms at different levels of design abstractions

lower power design alternatives prior to circuit layout in actual silicon. Moreover, correct design decisions at early phases of circuit abstraction (like high-level) will ensure that design errors are not propagated to lower levels, which may be costly to rectify [289, 276, 210]. The power analysis and optimization approaches can be done with minimal time and with slight accuracy trade-offs, which can tremendously reduce the design cycle time, as presented in Fig. 5.3.

5.10 Methods for Power Reduction in High-Level Synthesis

Based on the research available in existing low-power high-level synthesis literature [193, 289, 309, 373, 221, 320, 199, 221, 155, 106, 213, 339, 328, 235, 234, 210], the techniques can be technology independent and technology dependent. Technology-independent methods include transformation, operator shutdown and increased parallelism and technology-dependent methods use V_{DD}, V_{Th} or dielectric (K or T_{ox}) scaling. Parameter "scaling" implies changing values to reduce power, e.g., scaling V_{DD} implies reduction in its value but scaling V_{Th}, K or T_{ox} implies an increase in their values.

- *Transformation*: The basic approach is to scan the design space by utilizing various flow graph transformations with high-level power estimation techniques and transform data flow graphs into less-power-consuming data flow graphs.
- *Operator shutdown*: Massive switching in large components, such as adders, multipliers and registers, consumes a large amount of power. By disabling the clock signal the internal nodes remain at static voltage levels and do not consume power. Shutdown or isolation techniques rely on disabling idle circuit blocks during some proper cycles [194]. Guarded evaluation, precomputation and gated clock essentially use this approach.
- *Increased parallelism*: Slower operations can be used on non-time critical paths, while parallelism can be increased to compensate for slower components. The parallel option consumes less power and has a shorter total delay. However, extra area might be needed to achieve parallelism.
- *Lower supply voltages*: In a CMOS circuit, power consumption decreases quadratically with voltage while the speed reduction is linear. When intensive computation is not needed, the supply voltage is lowered and consequently can save power consumption. The lowering of supply is implemented either as a variable voltage or as mixed voltage approach.
- *Lower operating frequency*: In a CMOS circuit, power consumption decreases linearly with operating frequency. When intensive computation is not needed, the frequency can be lowered and consequently can save power consumption. The lowering of frequency is implemented either as a variable frequency or as multi-domain clocking approach.
- *Threshold voltage scaling*: This approach is used to reduce subthreshold leakage reduction. This approach is implemented using multiple threshold voltage or variable threshold approach.

- *Dielectric scaling*: This approach is recently used to reduce gate-oxide leakage. In this method either the gate-oxide thickness is scaled or alternative dielectrics are used.

5.11 Frequency and/or Voltage Scaling for Dynamic Power Reduction

Voltage and/or frequency scaling has been used as a technique for energy or average power reduction. It is argued that they need to be varied in a coordinated manner to get better results in terms of power, energy or performance. In this kind of design technique, special circuitry like voltage converters or dynamic clocking unit is needed. The design-related issues of having multiple supply voltages in a processor are discussed. Design of level converters and dynamic frequency clocking units are also presented.

5.11.1 What Is Voltage or Frequency Scaling?

5.11.1.1 Voltage Scaling

This type of controlled scaling or variation is implemented in two different ways: multiple supply voltages [223, 224, 221, 144, 135, 352, 310] and variable supply voltages [118, 277, 52, 53, 278, 130, 184, 191, 192]. The two approaches are distinguished in Fig. 5.4. In a multiple supply voltage approach, different portions of a chip are provided with different supply voltages as shown in Fig. 5.4(a). A special circuit called voltage-level converter is used to interface the two voltage islands. In a variable supply voltage scheme, shown in Fig. 5.4(b), the supply voltage of the overall system is varied on the fly based on certain criteria. In this scheme, an overhead in the form of a voltage–frequency controller is needed. The multiple supply voltage technique is a static approach in which decisions are taken during the design

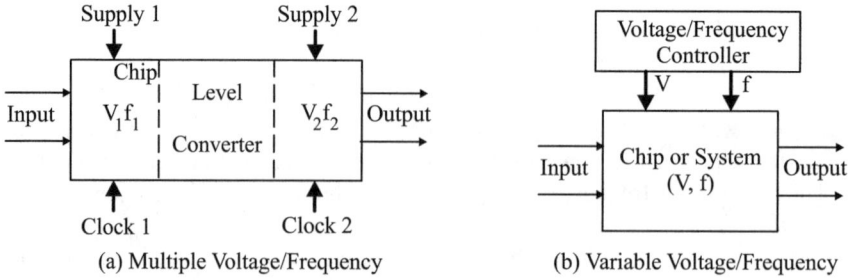

(a) Multiple Voltage/Frequency   (b) Variable Voltage/Frequency

Fig. 5.4 Multiple voltage–frequency versus variable voltage–frequency

time and once a decision is made it cannot be changed. On the other hand, the variable voltage technique is a dynamic approach that may need some type of servo mechanism or help of an operating system for its realization.

5.11.1.2 Frequency Scaling

The frequency scaling approach is realized in different ways such as variable frequency [169, 50, 49, 221, 157, 23, 190, 257, 184, 131] and multiple frequency [263, 223, 224]. As with voltage scaling, these two approaches fall under the broad categories of spatial and temporal techniques. In the multiple frequency scheme, various portions of a circuit are provided with different clocks whereas in the variable frequency technique the overall core or system frequency is varied on the fly. The variable frequency approach is a dynamic mechanism and the multiple frequency technique is a static mechanism. The variable frequencies are determined in various ways as well. They can be determined off-line using a compiler-assisted approach or for a chip during high-level synthesis. However, for varying voltage, on the fly use of an intelligent voltage–frequency controller or the use of an operating system is necessary. In a slightly different context, similar concepts have been presented as variable latency components [290] or telescopic units [38, 40] in which units complete execution in a variable number of clock cycles depending on the input data.

5.11.1.3 Voltage and Frequency Variation in a Coordinated Manner

Due to the interdependency of voltage and frequency, the optimal power and performance of a circuit or a system can be achieved if they are applied in a coordinated manner. This aspect is explained in Fig. 5.5. The clock frequency needs to be adjusted according to the supply voltage to guarantee correct operation of the circuit. In Fig. 5.5, three regions can be identified as normal operation, non-functional

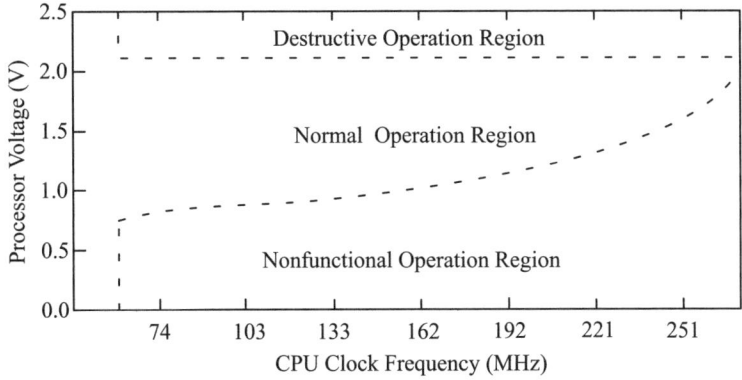

Fig. 5.5 Variable voltage processor operation: voltage versus frequency [280]

Table 5.1 How much we save by varying voltage (V_{DD}) and/or frequency (f)

Cases	Voltage (V_{DD})	Frequency (f)	Power (P_{dyn})	Energy (E_{dyn})
1	V_{DD}	f_{max}	P_{dyn}	E_{dyn}
2	$\frac{1}{2}V_{DD}$	f_{max}	$\frac{1}{4}P_{dyn}$	$\frac{1}{4}E_{dyn}$
3	$\frac{1}{2}V_{DD}$	$\frac{1}{2}f_{max}$	$\frac{1}{8}P_{dyn}$	$\frac{1}{4}E_{dyn}$
4	V_{DD}	$\frac{1}{2}f_{max}$	$\frac{1}{2}P_{dyn}$	E_{dyn}

operation and destructive operation. Table 5.1 examines the reduction of power and energy while simultaneously varying the voltage and frequency. Let the maximum frequency of a circuit for a supply voltage of V_{DD} be f_{max}. As in case 2, when both voltage and frequency are changed simultaneously, both power and energy reduce. Only reducing frequency reduces the power but not the energy, as in case 4. Reducing voltage reduces both power and energy as shown in case 1, but this may be a non-functional mode.

The variable voltage processor has special instructions for controlling power. The supply voltage and clock frequency can be changed at any time by the instructions in the application programs or operating systems. Examples of such processors are Transmeta Crusoe, Itsy, Intel StrongARM, etc. The four approaches to manage variable voltage processor are as follows [280, 278]: (1) hardware based (no information), (2) interval based (load information only), (3) integrated schedulers (all operating system statistics) and (4) application specific (complete knowledge). In this section, the scheduling algorithms proposed for variable voltage core-based systems under the assumption that the operating system has a voltage scheduler (as in case 3) are discussed. Instruction scheduling for a variable voltage processor which assigns voltage and frequency at compiler level is also discussed. The variable scheduling scheme may be either static (off-line) or dynamic (online), but the instruction scheduling schemes are off-line. Variable voltage or instruction scheduling schemes can be either preemptive or non-preemptive. It may be noted that variable voltage processors are also referred to as variable frequency processors. In the context of high-level synthesis the operating frequencies of a chip are determined during the synthesis process; these frequencies are stored as indices in the controller that supplies them to the clocking unit for frequency generation [221, 210].

5.11.2 Why Frequency and/or Voltage Scaling?

To analyze the effect of voltage and frequency on power and/or energy reduction, let us consider the following equations for a CMOS circuit [229, 219, 221]:

- Energy dissipation per operation due to capacitance switching is

$$E_{dyn} = C_{eff}V_{DD}^2, \qquad (5.1)$$

where C_{eff} is the effective switched capacitance and V_{DD} is the supply voltage.

- The corresponding dynamic power dissipation for the operation is

$$P_{\text{dyn}} = C_{\text{eff}} V_{\text{DD}}^2 f, \tag{5.2}$$

where f is the operating frequency.
- Further, the critical delay (T_{delay}) in a device that determines the maximum frequency (f_{max}) is

$$T_{\text{delay}} = \gamma_{\text{tech}} \frac{V_{\text{DD}}}{(V_{\text{DD}} - V_{\text{Th}})^\alpha}, \tag{5.3}$$

where V_{Th} is the threshold voltage, α is a technology-dependent factor and γ_{tech} is a constant.

From the above three equations, the following can be deduced [229, 219, 221]:

- By reducing only V_{DD}, both energy and power can be saved at the cost of performance (speed/time).
- Slowing down the CPU by reducing only f will save power but not energy.
- However, by scaling frequency and voltage in a coordinated manner, both energy and power can be saved while maintaining performance.

The third factor above forms the major motivation for voltage and frequency scaling. The objective is to generate a data path schedule that attempts energy and power reduction without degrading the performance by using multiple voltages and dynamic frequency clocking in a coordinated manner. Moreover, simultaneous voltage and frequency reduction opens opportunities for power reduction in three areas. The power and energy reduction due to combined use of multiple supply voltages, dynamic frequency clocking and multicycling are investigated.

5.11.3 Energy or Average Power Reduction Using Voltage or Frequency Scaling

The power and energy consumption for the data flow graph in several possible modes of data path operation are analyzed: (i) single supply voltage and single frequency (SVSF), (ii) single supply voltage and multicycling (SVMC), (iii) single supply voltage and dynamic frequency clocking (SVDFC), (iv) multiple supply voltages and single frequency (MVSF), (v) multiple supply voltages and multicycling (MVMC) and (vi) multiple supply voltages and dynamic frequency clocking (MVDFC) [229, 219, 221, 210]. While SVSF is a baseline case, the other cases like SVMC, SVDFC and MVSF are not practically used, and hence will be dropped from further discussion. In addition to this, pipelined operation needs to be considered. To understand how multiple supply voltage, variable frequency and multicycling can be helpful in energy or power reduction, let us consider the small data flow graph shown in Fig. 5.6(a). Let t_a and t_m be the delays of the adder and the multiplier, respectively, at the maximum supply voltage V. The DFG is scheduled to three control steps.

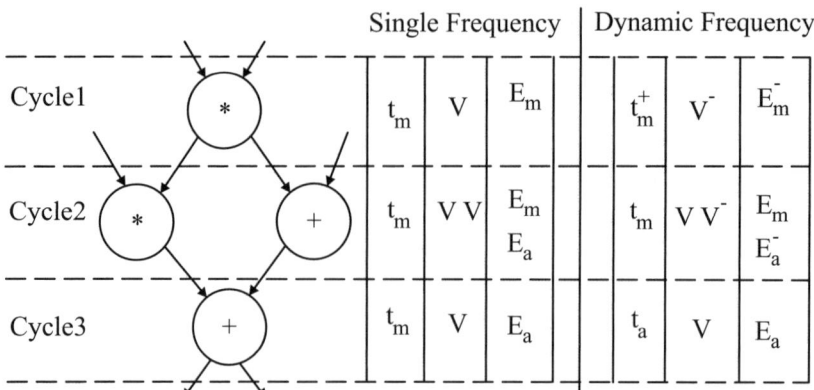

(a) Data Flow Graph: Variable Frequency Vs Single Frequency

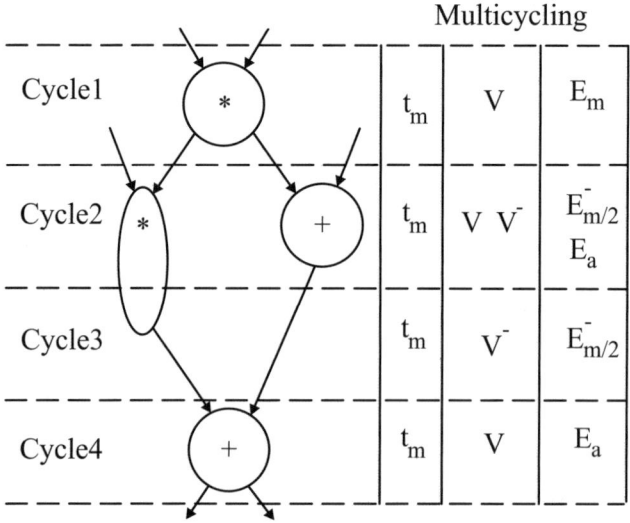

(b) Data Flow Graph: Multicycling → Performance Degradation

Fig. 5.6 Data flow graph in three modes of operation

Single supply voltage and single frequency (SVSF): Each cycle has its clock cycle width determined by the slowest operator delay t_m. The total energy consumption is given by $E_{sf} = 2E_m + 2E_a$ and the total delay is $T_{sf} = 3t_m$. In this case, the peak power consumption is given by $P_{peak,sf} = \left(\frac{E_m + E_a}{t_m} \right)$.

Multiple supply voltages and dynamic frequency (MVDFC): Suppose E_m^- and E_a^- are some energy values less than E_m and E_a, respectively, and let t_m^+ be the delay of the multiplier at the lower voltage V^-. In the data flow graph shown in Fig. 5.6(a), we assume that the clock cycle width for the third cycle is t_a which is smaller than t_m. This allows us to increase the clock width of some other cycles from t_m to some t_m^+ without violating the time constraints (or without time penalty).

In this case, the total delay $T_{dfc} = t_m^+ + t_m + t_a$ and the energy consumption is given by $E_{dfc} = E_m + E_a + E_m^- + E_a^-$. Since, $T_{dfc} \approx T_{sf}$ and $E_{dfc} < E_{sf}$, energy reduction is achieved without degrading performance. However, the energy overhead of level converters has to be considered for this case. The peak power consumption is given by, $P_{peak,dfc} = \left(\frac{E_m + E_a^-}{t_m} \right)$.

Multiple supply voltages and multicycling (MVMC): In this mode of operation, the functional units are operated at multiple supply voltages. The functional units operating at low voltage are made to run in more than one consecutive control steps. Let us assume that the multiplier takes two control steps when it is operated at a lower supply voltage. The example data flow graph for the multicycling case is shown in Fig. 5.6(b) with the assumption that the clock cycle width is determined by the slowest operator delay t_m. In this case, the total energy consumption $E_{mc} = E_m + E_m^- + 2E_a$ and total delay $T_{mc} = 4t_m$. Since $T_{mc} > T_{sf}$ and $E_{mc} < E_{sf}$, energy reduction is obtained with a degradation in performance of the circuit. For the multicycling case, level converters are the only overhead. The peak power consumption of the DFG will be determined by the multiplication operation in control step 1: $P_{peak,mc} = \left(\frac{E_m}{t_m} \right)$. This is based on the observation that the power consumption of a multiplier is much higher than that of an adder. It may be noted that the above-mentioned performance degradation may not always happen. Moreover if the clock cycle width is determined to be t_a then the scenario will change, in particular, for peak power.

5.11.4 Peak Power Reduction Using Voltage and Frequency Scaling

To have a clear understanding of the scheduling for energy minimization and peak power minimization, let us refer to the data flow graph (DFG) in Fig. 5.7 [225, 227, 222]. The figure shows two different possible schedules of the same DFG using a multiple supply voltage scheme. Since in both cases there are two multipliers

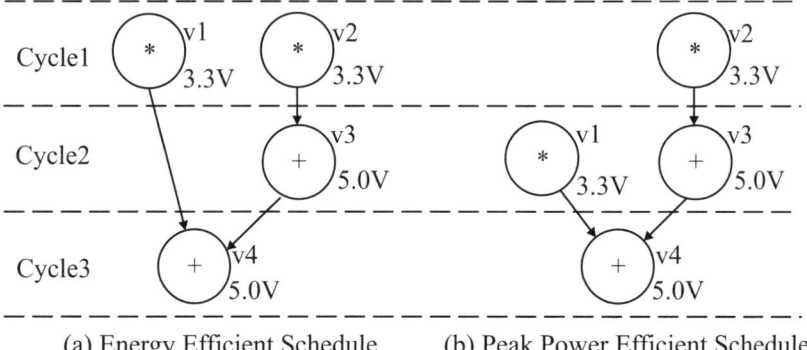

(a) Energy Efficient Schedule   (b) Peak Power Efficient Schedule

Fig. 5.7 Energy versus peak power efficient schedule

operating at 3.3 V and two adders operating at 5.0 V, the energy and average power consumption of both scheduled DFGs is the same. However, the peak power consumption of the schedule in Fig. 5.7(a) is more than that of Fig. 5.7(b), as in the previous DFG there are two energy-hungry operators (two multipliers) scheduled in one control step. The objective of the peak power reduction approach is to generate peak power efficient schedules similar to the one in Fig. 5.7(b).

5.11.5 Issues in Multiple Supply Voltage-Based Design

A designer needs to take into consideration several design issues when a multiple voltage design is targeted for fabrication. The effects of multiple voltage operation on IC layout and power supply requirements need to be considered [210, 70, 143, 145, 311, 320, 199, 198, 231, 51, 221, 309, 193]. Multiple voltages in a design may affect integrity and packaging of the circuit in the following ways:

- If the multiple supplies are generated off-chip, additional power and ground pins will be required.
- It may be necessary to partition the chip into separate regions, where all modules in a region operate at the same voltage.
- Some kind of isolation will be required between the regions operated at different voltages.
- There may be some limit on the voltage difference that can be tolerated between the regions.
- Protection against latch-up may be needed at the logic interfaces between regions of different voltages.
- New design rules for routing may be needed to deal with signals at one voltage passing through a region at another voltage.
- A choice between generating the voltage on-chip or off-chip has to be made depending on the application.
- The clocking scheme needs to be modified.

5.11.6 Voltage-Level Converter Design

It is well known that whenever one resource has to drive an input of another resource operating at a different voltage, level conversion is needed. Thus, the level converter or level shifter is a very essential component for multiple supply voltage designs [172, 279, 283, 137, 171, 374]. This results in overhead in the form of area and power for multiple supply voltage designs as compared to single supply voltage designs.

The process of converting a low-voltage signal to a higher voltage level is called level-up voltage conversion. For level-up conversion, a cross-coupled level converter (CCLC) is employed as shown in Fig. 5.8(a). In this circuit, there are two

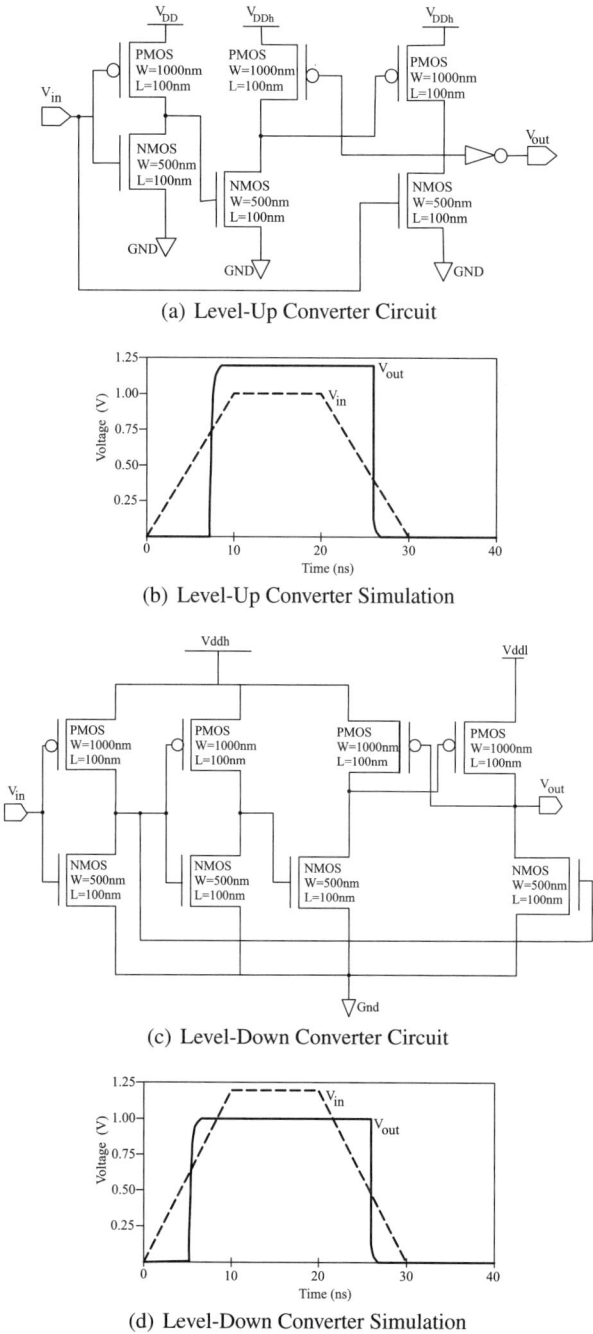

(a) Level-Up Converter Circuit

(b) Level-Up Converter Simulation

(c) Level-Down Converter Circuit

(d) Level-Down Converter Simulation

Fig. 5.8 Voltage-level converter circuit design and simulation waveform

cross-coupled PMOS transistors that form the circuit load. The cross-coupled PMOS transistors act as a differential pair [137, 233]. Thus, when the output at one side gets pulled low, the opposite PMOS transistor will be turned on. The output on that side will be pulled high. Below the PMOS load, there are two NMOS transistors that are controlled by the input signal V_{in}. The NMOS transistors operate with a reduced overdrive $V_{DD1} - V_{Th}$, compared to the PMOS devices. They have to be made large to be able to overpower the positive feedback [285]. The constant output voltage indicates that the level converter can step up the voltage to produce a constant supply voltage as shown in Fig. 5.8(b).

The process of converting a higher voltage signal to a lower voltage signal is called level-down voltage conversion. In this design, a differential input level converter is employed. It has a differential input which enables a stable operation for low voltage and high speed use [147, 233, 279]. The differential input also offers immunity against power supply bouncing. The circuit diagram is shown in Fig. 5.8(c). As evident from Fig. 5.8(b), the constant output voltage indicates that the level converter can step down the voltage to produce a constant supply voltage.

5.11.7 Dynamic Frequency Clocking Unit Design

5.11.7.1 Concept

In dynamic frequency clocking, the functional units can be operated at different frequencies depending on the computations occurring within the data path during a given clock cycle. The strategy is to operate high-energy units such as multipliers at lower frequencies such that they can be operated at lower voltages to reduce energy consumption and the low-energy units, such as adders, at higher frequencies, to compensate for speed. In this clocking scheme, all the units are clocked by a single clock line which switches at run time. Figure 5.9(a) shows the single frequency and dynamic frequency diagrams [229, 219, 221, 210].

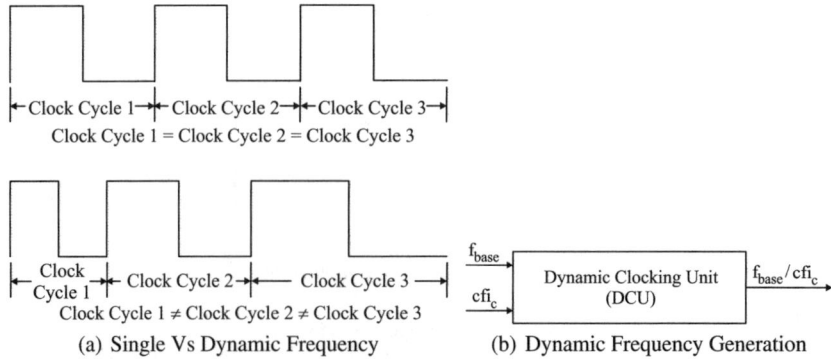

(a) Single Vs Dynamic Frequency   (b) Dynamic Frequency Generation

Fig. 5.9 Dynamic frequency generation using dynamic clocking unit

The dynamic clocking unit (DCU), which generates the required clock frequencies, uses a clock divider strategy to generate frequencies which are submultiples of the base frequency. Base frequency (f_{base}) is the maximum frequency (or multiple of maximum) of any functional unit (FU) at the maximum supply voltage. A parameter called *cycle frequency index for control step c* (cfi_c) is loaded as an input to the DCU which comes from the controller. The scheme for dynamic frequency generation is shown in Fig. 5.9(b) [229, 219, 221, 210]. Loading a value of cfi_c into the counters provide a divided output clock of frequency $\left(\frac{f_{base}}{cfi_c}\right)$.

5.11.7.2 DCU Design – 1

The block diagram of a dynamic clocking unit (DCU) design is shown in Fig. 5.10 [210, 300]. The DFC scheme is more suitable for data flow intensive applications (such as DSP and image processing). The DCU is a series of cascaded clock divider stages whose inputs are controlled by the pass logic blocks. The output of one clock divider is presented at the input of the next stage when the pass logic is enabled. The pass logic block is controlled by a set of signals generated by the enable encoder. Based on the instruction class, the appropriate pass logic blocks are activated by the enable encoder. The master clock is accordingly divided by the clock divider circuit to generate the resultant output clock.

5.11.7.3 DCU Design – 2

A dynamic programmable clock divider (DPCD) for use in conjunction with FPGA clock managers is presented in Fig. 5.11 [50, 49, 210]. Clock division by ordinary clock dividers can lead to glitches or distortions of the output clock. Distortions at the output clock can result in metastability and latching errors. The DPCD is capable of performing dynamic frequency division without undesired effects at the output. Division of the input clock is performed by creating a loop of D-flip-flops {A-D} driven by the input clock and feeding the signal back into the loop through an inverter {D} to create the necessary clock inversion. To expand the length of

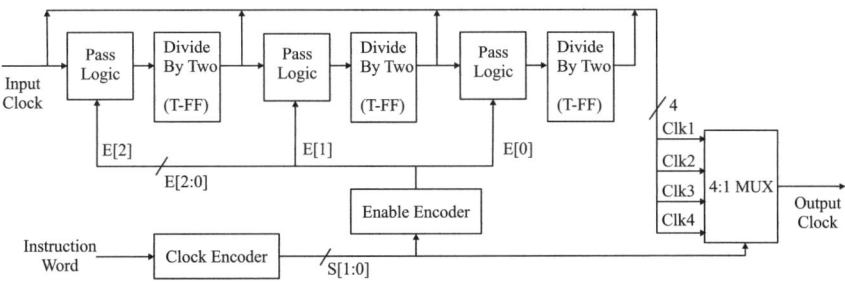

Fig. 5.10 Dynamic clocking unit design – 1

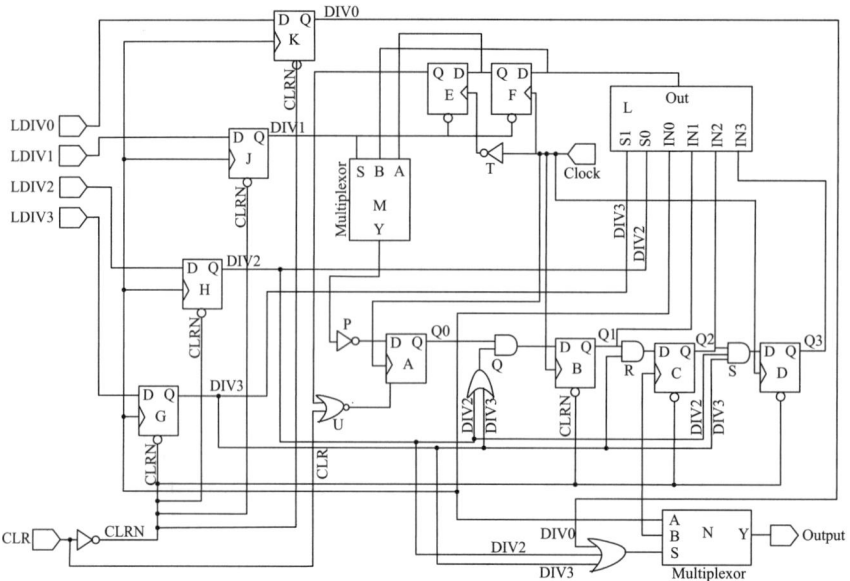

Fig. 5.11 Dynamic clocking unit design – 2

the output clock, the number of D-flip-flops in the loop is increased by multiplexor {L}. In order to perform an odd division, flip-flops {E, F} extend the loop, by half a period, with an asynchronous clear of flip-flop {A} on the falling edge of the input clock. For the divider output, multiplexer {N} chooses between the original input clock, for a division of one, and the output of {A}. To prevent output glitches, D-flip-flops {G, H, J, K} latch the new program value on the rising edge of the output from {A}. Combinational logic {Q, R, S} help to prevent glitches, but also prevent transient patterns from being captured and fed back, thus causing irregular oscillation in the circuit.

5.12 V_{Th} Scaling for Subthreshold Leakage Reduction

5.12.1 The Concept

Subthreshold leakage current of a CMOS device can be expressed as [309, 365]

$$I_{sub} = \gamma_{tech} \ \exp\left(\frac{V_{gs} - V_{Th}}{S/\ln(10)}\right)\left(1 - \exp\left(-\frac{V_{ds}}{V_{therm}}\right)\right), \qquad (5.4)$$

where γ_{tech} is a technology-dependent factor, V_{Th} is the threshold voltage, S is the subthreshold swing and V_{therm} is the thermal voltage. The subthreshold swing of a device can be calculated as follows [309, 365]:

$$S = V_{therm} \ln(10) \left(1 + \frac{C_d}{C_{ox}} \right), \tag{5.5}$$

where C_d is the depletion region capacitance and C_{ox} is the gate-oxide capacitance. The thermal voltage is calculated as follows:

$$V_{therm} = \left(\frac{k\,T}{q} \right), \tag{5.6}$$

where k is the Boltzmann constant, T is the absolute temperature and q is the elementary charge. From the above equations it is deduced that subthreshold current can be reduced by decreasing V_{gs} and/or by increasing S and V_{Th}. Threshold voltage scaling for subthreshold leakage reduction has been realized in various ways, such as multiple threshold CMOS (MTCMOS), variable threshold CMOS (VTCMOS) and dynamic threshold CMOS (DTCMOS) [247, 343, 188, 29, 261, 241, 106, 107, 155, 154, 213, 108, 373, 247, 178].

To understand the threshold variation or scaling consider the following equation:

$$V_{Th} = V_{Th0} + \Delta V_{Th}, \tag{5.7}$$

$$= V_{Th0} + \gamma_{body} \left(\sqrt{\psi_S - V_{bs}} - \sqrt{\psi_S} \right), \tag{5.8}$$

$$= V_{fb} + 2\phi_F + \left(\frac{1}{C_{ox}} \right) \sqrt{2q\varepsilon_{Si}N_{sub}\,(2\phi_F + V_{bs})}, \tag{5.9}$$

$$= V_{fb} + 2\phi_F + \left(\frac{T_{ox}}{\varepsilon_{ox}} \right) \sqrt{2q\varepsilon_{Si}N_{sub}\,(2\phi_F + V_{bs})}, \tag{5.10}$$

where V_{Th0} is the zero body bias threshold voltage, V_{bs} is the body bias, γ_{body} is the body effect coefficient and ψ_S is the surface potential. The Fermi level ϕ_F is calculated as $\left[V_{therm} \ln \left(\frac{N_{sub}}{n_i} \right) \right]$, and assuming strong inversion the surface potential ψ_S is $2\phi_F$. Here, N_{sub} is the substrate doping concentrations per cm^3 and n_i is the intrinsic concentration per cm^3. Thus, it can be concluded that the threshold voltage V_{Th} of a device can be changed in the following ways:

- By varying the body bias voltage V_{bs}
- By changing the substrate doping concentration
- By changing the oxide thickness

5.12.2 Multiple Threshold CMOS (MTCMOS) Technology

In multiple threshold CMOS (MTCMOS) technique, two high-V_{Th} transistors are put in series with a low-V_{Th} logic block to cut off the path from power supply rail to ground rail during the idle or standby period to reduce the leakage as shown in Fig. 5.12(a). Even though two high-V_{Th} transistors are shown in the figure, one of them is sufficient to control the leakage. This type of technique uses doping-level variation during processing to achieve different thresholds.

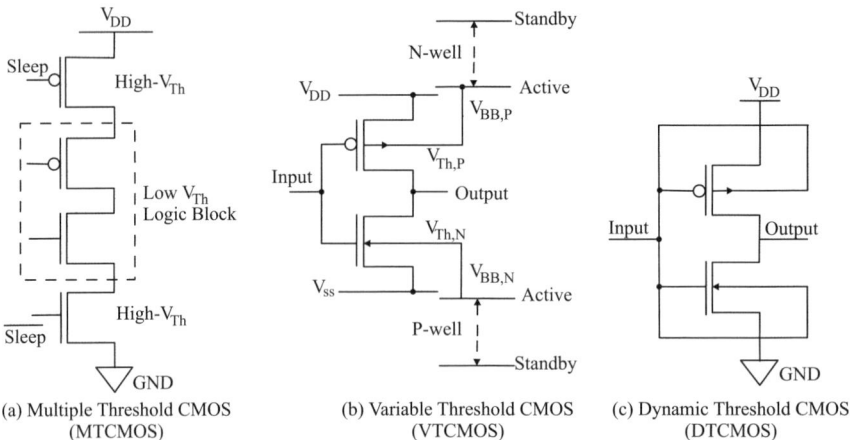

Fig. 5.12 Multiple threshold voltage (MTCMOS), variable threshold voltage (VTCMOS), and dynamic threshold voltage (DTCMOS) CMOS techniques for subthreshold leakage reduction

5.12.3 Variable Threshold CMOS (VTCMOS) Technology

As shown in Fig. 5.12(b), the body (substrate) biasing technique is used to obtain variable threshold CMOS (VTCMOS) design. In this design, nearly zero body bias is applied in active mode and reverse body bias is applied in standby mode to reduce the subthreshold leakage current. This technique relies on variation of body bias voltage V_{bs}.

5.12.4 Dynamic Threshold CMOS (DTCMOS) Technology

DTCMOS is a technique presented in Fig. 5.12(c) that varies V_{Th} as per the operating state of a circuit, i.e., high V_{Th} for standby mode and low V_{Th} for active mode to contain the leakage. This technique also relies on variation of body bias voltage V_{bs}.

5.12.5 Leakage Control Transistor (LECTOR) Technique

In a new technique for subthreshold reduction called leakage control transistor (LECTOR) two leakage control transistors are introduced within the logic block [119]. This is demonstrated in Fig. 5.13. For each leakage control transistor (LCT), the gate terminal of one is controlled by the source of the other. Thus, one of the LCTs is always near its cut-off voltage for any input combination consequently increasing the resistance of the path from supply to ground and decreasing the subthreshold leakage currents.

Fig. 5.13 Leakage control
transistor (LECTOR) tech-
nique for subthreshold leakage reduction

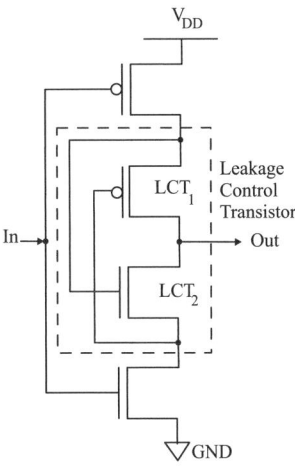

5.12.6 The Issues

The issues of this kind of technology is the increase in fabrication cost when
multi/dual-V_{Th} is used. It may need special design rules at the boundary of the different V_{Th} devices [193]. It may also require special processing to vary the doping
concentration and may need larger number of masks.

5.13 T_{ox}, K or L Scaling for Gate-Oxide Leakage Reduction

5.13.1 The Concept

For nanoscale CMOS technology, gate-oxide leakage has become a major challenge
for design engineers. Gate-oxide leakage arises due to tunneling of carriers through
the gate dielectric. The tunneling mechanism between the substrate and gate can
be either Fowler–Nordheim (FN) tunneling or direct tunneling; both differ in the
form of potential barrier. For sub-65 nm technology direct tunneling is predominant.
Tunneling current dissipation in a CMOS with a supply voltage V_{DD} and effective
gate-oxide thickness T_{ox} is given by the following expression [159, 62, 217, 223]:

$$I_{gate} = \alpha W_{gate} \left(\frac{V_{DD}}{T_{ox}} \right)^2 \exp \left(-\beta \frac{T_{ox}}{V_{DD}} \right). \tag{5.11}$$

Here, α and β are experimentally derived factors. Based on this equation, we have
two options for the reduction of gate leakage: reduction of supply voltage and/or
increase of the gate-oxide thickness. The popular option of scaling down of supply
voltage [220, 221] continues to play its role in the reduction of dynamic power as

well as leakage power, but it is not sufficient to control the exponential growth of gate leakage. Increase in the gate-oxide thickness leads to an increase in propagation delay and area. Thus, the use of multiple gate-oxide thicknesses can serve as a leakage current, performance and area trade-off.

5.13.2 Multiple Oxide Thickness CMOS (MOXCMOS) Technology

The use of multiple oxide thickness CMOS (MOXCMOS) technology has been proposed for reduction of gate-oxide leakage [193, 276, 328, 329, 235, 234, 211, 217, 241, 29, 338, 337, 133, 196, 217, 240, 213] as in Fig. 5.14(a). In this technique L is scaled along with T_{ox} for effective gate leakage reduction.

5.13.3 Multiple Dielectric (κ) (MKCMOS) Technology

Recently, the "dual dielectric" approach is being introduced for gate leakage reduction, where SiO_2 is selectively replaced with high-K materials such as SiON, Si_3N_4, etc. [234, 240]. The use of multiple dielectric (κ) technology (MKCMOS) has been proposed as an option for gate leakage reduction while maintaining the performance [234, 240]. This is presented in Fig. 5.14 (b).

5.13.4 The Issues

MOXCMOS and MKCMOS will have several issues [240, 337, 234, 235]. There are process integration problems due to the use of dual thicknesses and dual dielectrics during fabrication processes. Issues such as chemical reactivity with existing or new

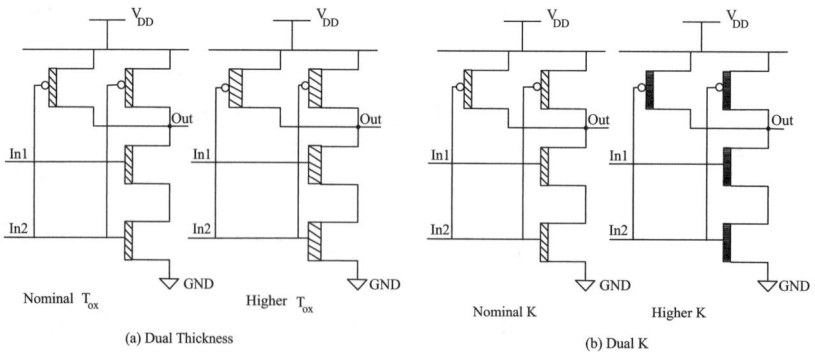

Fig. 5.14 (**a**) A NAND logic gate for the dual thickness approach that uses devices with either higher T_{ox} or lower T_{ox}. (**b**) A NAND logic gate for dual dielectric approach that uses devices with either higher K or lower K

gate dielectric materials to ensure optimal CMOS compatibility need to be resolved. These technologies may also require the use of more masks for the lithographic process of circuit fabrication. However, it is believed that this additional cost would be compensated by the reduction of energy or power costs. However, the research on these materials by the material science and engineering as well as electrical engineering communities is in full swing. It is expected to see new process technologies in the future addressing these issues.

5.14 Transformation Techniques for Power Reduction

Operation transformations such as operation reduction and operation substitution are used to reduce dynamic power [65, 64, 373, 289, 309, 109]. These transformations essentially reduce dynamic power or energy through the reduction of switching capacitance. However, they may have impact on the circuit performance as they often result in increase in critical path delay.

5.14.1 Operation Reduction

In the operation reduction technique, switching capacitance is reduced by reducing the number of operations in the DFG [64, 289, 309]. The smaller the number of operations in the DFG, the smaller the capacitance of the corresponding system, and hence there is reduction in dynamic power or energy. To understand this technique, consider the implementation of a polynomial, $y = x^2 + ax + b$. This is implementable through the DFG of Fig. 5.15(a). The expression can be rewritten using mathematical transformations as follows: $y = x^2 + ax + b = (a + x)x + b$. The transformed expression can then be implemented as shown in the DFG of Fig. 5.15(b).

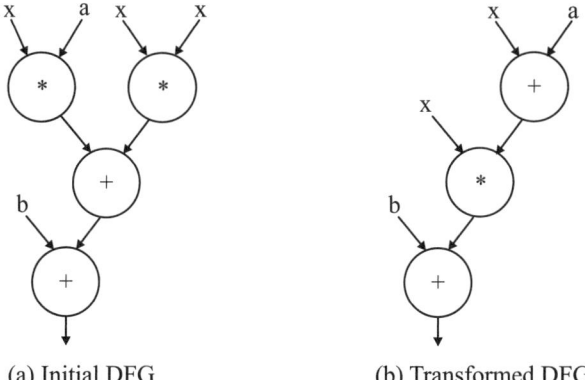

(a) Initial DFG   (b) Transformed DFG

Fig. 5.15 Operation reduction for switching capacitance reduction to reduce dynamic power or energy

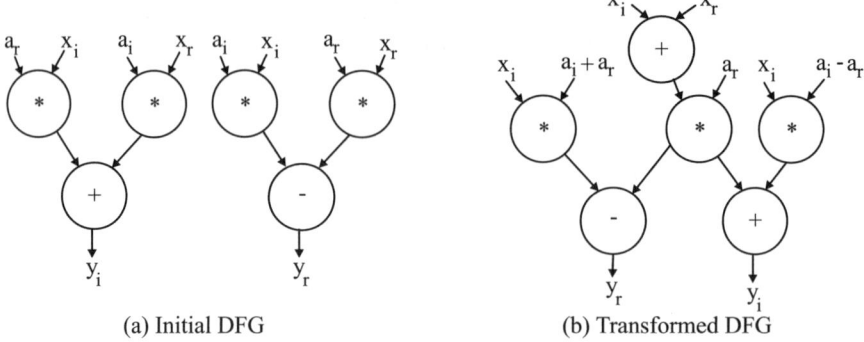

(a) Initial DFG   (b) Transformed DFG

Fig. 5.16 Operation substitution for switching capacitance reduction to reduce dynamic power or energy

5.14.2 Operation Substitution

In the operation substitution technique, switching capacitance is reduced by substituting high-energy operations with low-energy operations in the DFG [64, 373, 309]. For example, a multiplication operation that needs a multiplier is more high energy consuming than an addition operation that needs an adder, as a multiplier consumes more energy than an adder. The smaller the number of multiplication operations in the DFG, the smaller the capacitance of the corresponding system, and hence there is reduction in dynamic power or energy. As an example, let us consider the DFGs in Fig. 5.16. In the DFGs, a_i and a_r are constants and x_i, x_r, y_i and y_r are variables. In the DFG of Fig. 5.16(b) one multiplier of DFG of Fig. 5.16(a) is substituted by an adder, but it needs one more clock cycle. This operation substitution is achieved by using *strength reduction* compiler optimizations in which a function of some systematically changing variable is calculated more efficiently by using previous values of the function [64].

5.15 Increased Parallelism and Pipelining with Architecture-Driven Voltage Scaling for Power Reduction

In order to achieve power and performance trade-off, techniques that use voltage scaling along with pipelining and parallelism are proposed [373, 309, 66]. While parallelism is spatial parallelism that uses higher number of resources to reduce delay, pipelining is temporal parallelism that uses additional registers. To understand how the two alternatives reduce power dissipation, consider the reference or baseline data path in Fig. 5.17. Let us assume that V_{base}, f_{base} and C_{base} are supply voltage, operating frequency and total effective capacitance being switched per clock cycle,

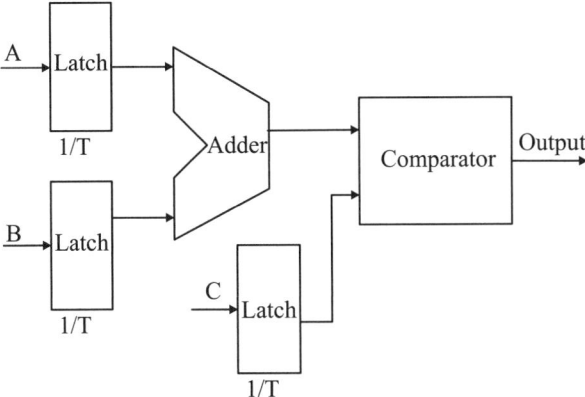

Fig. 5.17 Example data path – the baseline

respectively, for the reference or baseline data path. Then the power for the baseline
data path is given by

$$P_{\text{base}} = C_{\text{base}} \, V_{\text{base}}^2 \, f_{\text{base}}. \tag{5.12}$$

5.15.1 Parallelism with Voltage Scaling

Let us assume that the voltage is scaled by a factor α_V to make the data path run at
half of the baseline operating frequency. As shown in Fig. 5.18, an approach to maintain throughput while reducing the supply voltage is to utilize a parallel architecture
[373, 309, 66]. In this architecture, two identical data paths are used, allowing each
unit to work at half the original rate while maintaining the original throughput. Unfortunately, there is also a slight increase in the total "effective" capacitance by a
factor of β_C introduced due to the extra routing. The power dissipation of the parallel data path is then expressed as

$$P_{\text{para}} = C_{\text{para}} \, V_{\text{para}}^2 \, f_{\text{para}} \tag{5.13}$$
$$= (\beta_C C_{\text{base}}) \, (\alpha_V V)_{\text{base}}^2 \, (0.5 f_{\text{base}}), \; 0 < \alpha_V < 1 \text{ and } \beta_C \approx 2 \tag{5.14}$$
$$= 0.5 \alpha_V^2 \beta_C \times C_{\text{base}} \, V_{\text{base}}^2 \, f_{\text{base}}, \text{ where } 0 < 0.5 \alpha_V^2 \beta_C < 1 \tag{5.15}$$
$$= 0.5 \alpha_V^2 \beta_C \times P_{\text{base}}. \tag{5.16}$$

As evident from the above equation, since $0.5 \alpha_V^2 \beta_C < 1$, $P_{\text{para}} < P_{\text{base}}$, and hence
there is reduction in the power dissipation.

5.15.2 Pipelining with Voltage Scaling

Another alternative in the power and performance trade-off is voltage scaling combined with pipelining, as shown in Fig. 5.19 [373, 309, 66]. The pipelining approach

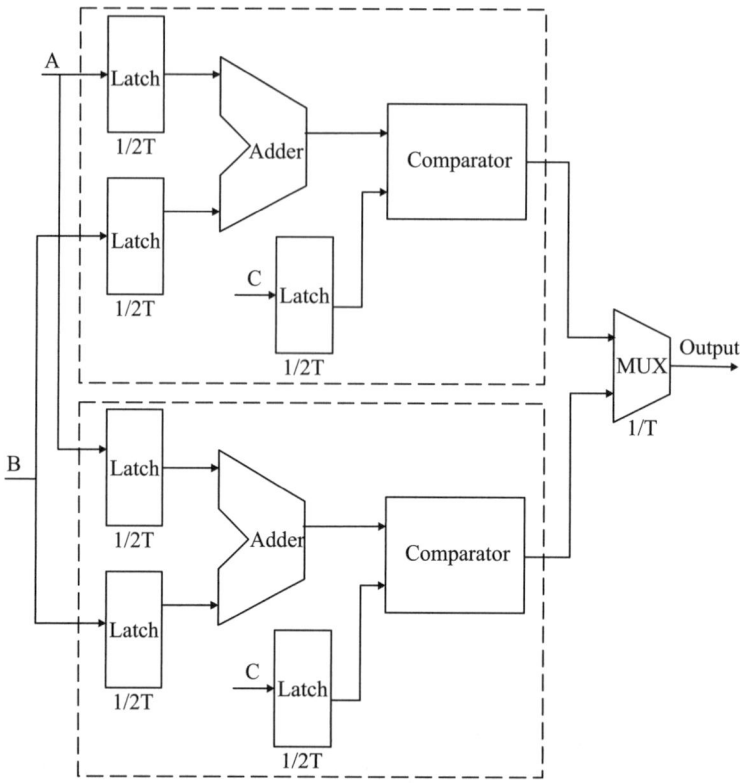

Fig. 5.18 Example data path – the parallel alternative

has advantages over parallelism in that it is area efficient. The critical path delay of the pipelined data path is the maximum of the adder and comparator delay, i.e., maximum $\left(T_{\text{delay,adder}}, T_{\text{delay,comparator}}\right)$. Thus, the adder and comparator can be operated at slower speed. Assuming the two delays are equal, let the supply voltage be scaled by a factor α_V to increase the delay by two while maintaining the same throughput. The addition of extra latches increases the effective capacitance by a factor γ_C. The power dissipation of the pipelined data path is then expressed as

$$P_{\text{pipe}} = C_{\text{pipe}} \, V_{\text{pipe}}^2 \, f_{\text{pipe}} \tag{5.17}$$

$$= (\gamma_C C_{\text{base}}) \, (\alpha_V V)_{\text{base}}^2 \, (f_{\text{base}}), \ 0 < \alpha_V < 1 \text{ and } \gamma_C \approx 1 \tag{5.18}$$

$$= \alpha_V^2 \gamma_C \times C_{\text{base}} \, V_{\text{base}}^2 \, f_{\text{base}}, \text{where} 0 < \alpha_V^2 \gamma_C < 1 \tag{5.19}$$

$$= \alpha_V^2 \gamma_C \times P_{\text{base}}. \tag{5.20}$$

As evident from the above equation, since $\alpha_V^2 \gamma_C < 1$, $P_{\text{pipe}} < P_{\text{base}}$, and hence there is reduction in the power dissipation.

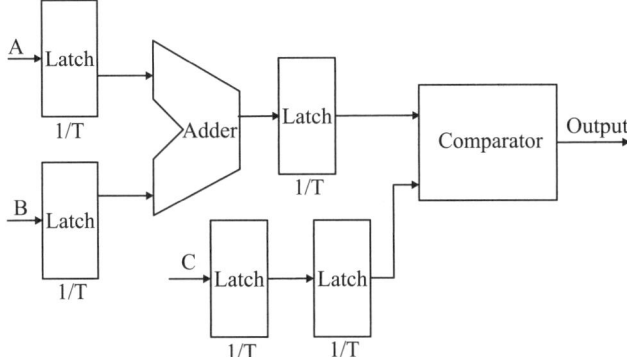

Fig. 5.19 Example data path – the pipelining alternative

5.16 Guarded Evaluation to Reduce Power

Guarded evaluation is a shutdown or operand isolation technique that dynamically disables particular computing units at times when the results they produce are not used by other parts of a circuit or system [289, 194, 347]. In this technique latches are inserted at all the inputs of an embedded logic block to isolate and hold values as shown in Fig. 5.20. An important circuit element needed in this technique is the circuit to detect idle conditions of the embedded block. In order to reduce power dissipation during the idle state of the logic block, the latches at the inputs are disabled and they retain values of the previous cycle. For successful realization of this technique two conditions need to be met:

• The circuit should compute when the enable signal of latch is "0".
• Latches are disabled in time which is early enough to cut off input transitions.

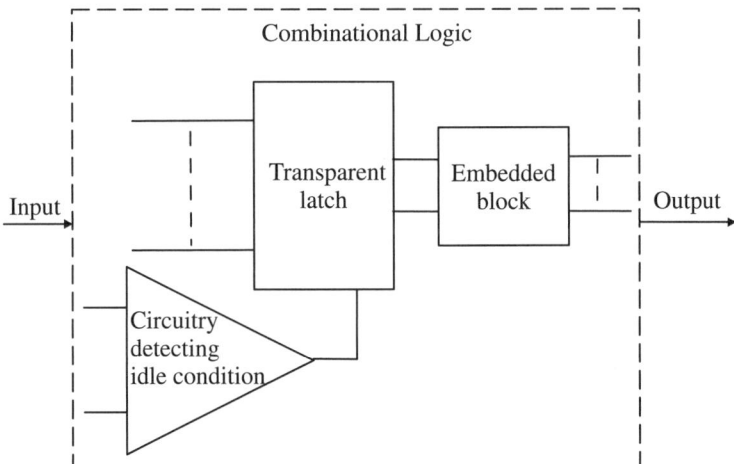

Fig. 5.20 Guarded evaluation using operand Isolation

5.17 Precomputation-Based Power Reduction

As the name signifies, the precomputation-based technique calculates logic values of the circuit one clock cycle in advance [373, 27]. The precalculated logic values are then used in the following control step to reduce the switching activities at the internal nodes of a circuit. While the technique is found to reduce significant amount of power for circuits, it suffers from large area overhead.

Figure 5.21 shows one of the precomputation architectures proposed in [27]. In the architecture, the inputs to the logic block have been partitioned into two sets $(x_1, ..., x_i)$ and $(x_j, ..., x_n)$, corresponding to the input registers R_1 and R_2, respectively. The output bits of the combination logic block are fed to the output register R_3. The inputs of the first register R_1 are also supplied to two logic blocks f_1 and f_2. The functions f_1 and f_2 are predictor functions such that

$$f_1 = 1 \Rightarrow \text{Output} = 1, \tag{5.21}$$
$$f_2 = 1 \Rightarrow \text{Output} = 0. \tag{5.22}$$

The functions f_1 and f_2 never evaluate to logic 1 simultaneously. Therefore, if either of the functions evaluates to 1 during a clock cycle c then the enable line of register R_2 is turned off and hence the outputs of register R_2 do not change during clock cycle $c + 1$. However, the output evaluates to the output of the register R_1. It may be noted that only a subset of input values to the combinational logic block changes, thus resulting in the reduction of the switching activities at the internal nodes of the combinational logic block. For better effectiveness of precomputation, the following conditions are desirable:

- The subset of inputs fed to register R_2 should be large.
- Complexity of logic blocks f_1 and f_2 should be small.
- Signal probability of $(f_1 + f_2)$ which is $P(f_1) + P(f_2) - P(f_1)P(f_2)$ has to be large.

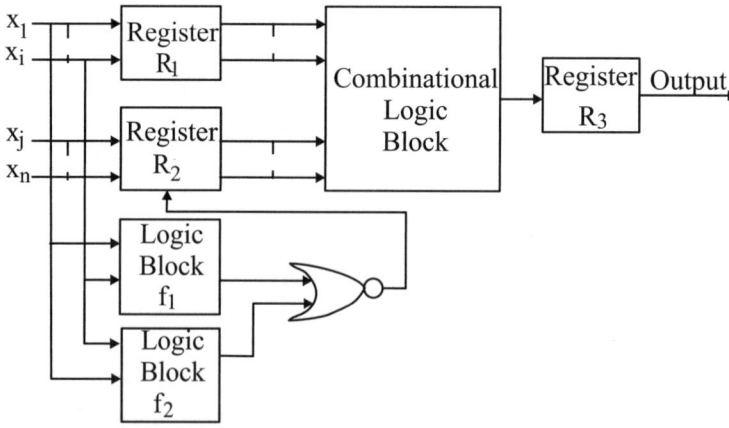

Fig. 5.21 Precomputation-based power reduction

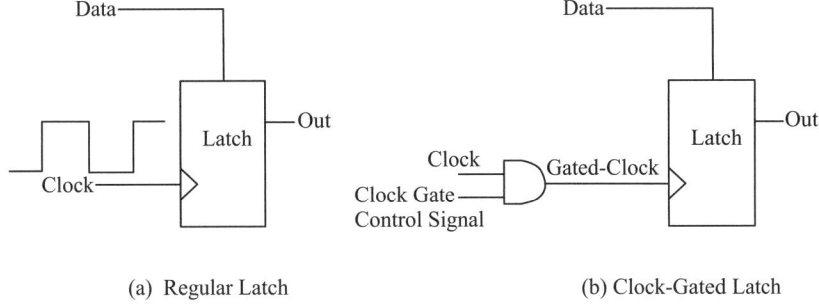

Fig. 5.22 Clock gating for power reduction

5.18 Clock Gating to Reduce Clock Power Dissipation

Clock power is a major component of synchronous circuits and systems [373, 289]. In a typical circuit, the individual subcircuits are not used all the time and with equal performance importance thus providing an opportunity for power reduction. Clock gating is a shutdown technique to reduce clock power. This technique disables the clock to a subcircuit whenever it is not used, thus avoiding unnecessary charging and discharging of unused subcircuits. A clock-gated latch is shown in Fig. 5.22 [373]. As the name signifies the clock signal is ANDed with a control signal called "clock-gate control" that enables or disables it.

5.19 Interconnect Power Minimization

In nanoscale circuits, the power dissipation and the delay of interconnects with respect to the active devices [376, 281, 170] cannot be ignored . Power dissipation of on-chip interconnect is necessary and research is in full swing at all levels of design abstraction. During high-level synthesis that can target either bus-based or multiplexer-based on-chip interconnection, interconnect power delay optimization has been considered in many recent research works [376, 281, 170]. The power estimation in interconnects is typically based on an empirical formula called Rent's Rule:

$$N_T = N_{Avg} \times N_B{}^p, \qquad (5.23)$$

where N_T is the number of terminals, N_B is the number of blocks within a chip, N_{Avg} is the average number of terminals for a block and p is the Rent's exponent or constant. In high-level synthesis different binding and scheduling algorithms generate circuits with different p and N_{Avg} [376, 170].

5.20 Summary and Conclusions

Low-power design is important for various reasons. A prominent one is the significant demand of portable appliances for which the battery life is an important factor. Power optimization can be performed at different levels of design abstraction. Different methods have been proposed at each particular level of design abstraction for power reduction. Power optimization during high-level synthesis can significantly reduce power at full-chip or system level, and the optimization process is much faster, as needed by early design exploration. Low-power high-level synthesis techniques, such as multiple V_{DD}, multiple V_{Th}, multiple T_{ox} have been used. At the same time technology-independent techniques such as pipelining, parallelization, precomputation, shutting-down, etc., are used for high-level power optimization. Both leakage and dynamic power reductions are important for nanoscale circuits and were investigated here. While active device power is important and has received major attention, modeling and reduction of interconnect power dissipation, especially for communication-centric designs, needs further attention.

Chapter 6
Energy or Average Power Reduction

6.1 Introduction

A CMOS circuit can be operated in different modes: single supply voltage and single frequency; multiple supply voltages and single frequency; and multiple supply voltages and dynamic frequency. Traditionally, CMOS circuits are operated in the single supply voltage and single-frequency mode where the clock width (thus, frequency) is dictated by the slowest operator delay and each FU is operated at the same voltage level. In multiple supply voltages and single-frequency mode, different FUs are operated at different voltage levels to reduce energy consumption [144, 70, 186, 221]. In this case, energy consumption of the level converters must be taken into account. More recently, MVDFC mode of operation is being explored as a possible strategy for low-power, high-performance operation. In DFC, the clock frequency is varied on the fly based on the FU active during that cycle. In this scheme, all the FUs are clocked by a single clock line that switches at run time. This scheme, in particular, is suitable for data-intensive or computationally intensive DSP applications. Dynamic frequency scaling has been explored at the central processing unit (CPU) and system levels for power optimization. In this chapter, data path scheduling algorithms that use MVDFC in a coordinated manner in order to reduce energy and the energy delay product (EDP) are discussed. The strategy is to schedule high-energy units, such as the multipliers at lower frequencies so that they can be operated at lower voltages to reduce energy consumption, and low-energy units, such as adders, at higher frequencies to compensate for speed. First, an ILP-based data path scheduling algorithm which incorporates MVDFC or MVMC for EDP reduction is presented [210, 230]. Then, heuristic-based time-constrained scheduling (TC-DFC) and the resource-constrained scheduling (RC-DFC) algorithms are elaborated for energy reduction using the MVDFC approach [210, 219, 221, 229]. Subsequently, several existing scheduling algorithms are summarized that use either voltage reduction or switching activity reduction for energy minimization.

S.P. Mohanty et al., *Low-Power High-Level Synthesis for Nanoscale CMOS Circuits*,   163
DOI: 10.1007/978-0-387-76474-0_6, © Springer Science+Business Media, LLC 2008

6.2 Target Architecture and Data Path Specifications for Multiple Voltage

The target architecture model assumed in the design of scheduling schemes is shown in Fig. 6.1. All FUs have one register each and one multiplexer. Each FU feeds into a single register. The register and the multiplexer operate at the same voltage level as that of the FUs. Level converters are used when a low-voltage FU is driving a high-voltage FU [144, 320, 210, 219, 221]. A controller decides which FUs are active in each control step and those that are not active are disabled using the multiplexers. The controller has a storage unit to store the parameters cfi_c obtained from the scheduling. The cycle frequency $f_c = \left(\frac{f_{\text{base}}}{\text{cfi}_c} \right)$ is generated dynamically, and a FU operating at one of the supply voltages is activated [210, 219, 221, 229].

The data path is specified as a sequencing DFG [210, 219, 221, 209]. Each vertex of the DFG represents an operation and each edge represents a data flow or dependency. The DFG does not support hierarchical entities. The conditional statements are handled using comparison operation. Since the dynamic frequency clocking scheme is useful only in the case of signal-processing applications, we assume that the above does not exist in the directed acyclic DFG representation of data paths. Each vertex has attributes that specify the operation type such as addition, subtraction, multiplication or null operations (NOPs).

The delay of a control step is dependent on the delays of the FU and the multiplexer and register pair. Let d_{REG} be the delay of the register, d_{MUX} be the delay of the multiplexer, d_{FU} be the delay of the FU and d_{Conv} be the delay of the level converter. The worst-case operational delay of a control step can be written as

$$d_c = d_{\text{REG}} + d_{\text{MUX}} + d_{\text{FU}} + d_{\text{Conv}}. \tag{6.1}$$

The register delays include the set-up and propagation delays. The delay of control step d_c is the delay of the slowest FU in the control step c. Using the above delay model, the worst-case delays of the library components are estimated.

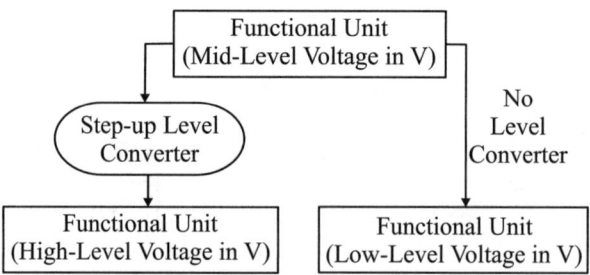

Fig. 6.1 Level converters needed for stepping up signal

For a given base frequency (f_{base}), maximum frequencies of each FU are scaled down to operating frequencies (f_c). These parameters are determined as follows:

$$f_{\text{base}} = \left\lceil \frac{\lfloor 1/d_c^{\min} \rfloor}{2^{L_f}} \right\rceil 2^{L_f}, \tag{6.2}$$

$$cfi_c = \left\lceil \frac{\lfloor d_c/d_c^{\min} \rfloor}{2^n} \right\rceil 2^n \text{ and} \tag{6.3}$$

$$f_c = \left(\frac{f_{\text{base}}}{cfi_c} \right), \tag{6.4}$$

where $d_c^{\min}$ is the minimum of the control step delays and L_f is the number of allowable frequencies. The value of n is chosen in such a way that cfi_c is the closest value greater than or equal to $\left\lceil \frac{d_c}{d_c^{\min}} \right\rceil$. The value of cfi_c is bound by the product of the total number of resource types and number of voltage levels. For example, for three frequency levels, the possible frequencies are $\text{ALU}_{\text{High}}(cfi_c = 1)$, $\text{ALU}_{\text{Med}}(cfi_c = 2)$, $\text{ALU}_{\text{Low}}(cfi_c = 4)$, $\text{MULT}_{\text{High}}(cfi_c = 2)$, $\text{MULT}_{\text{Med}}(cfi_c = 4)$ and MULT_{Low} ($cfi_c = 8$). If the base frequency fed to the DCU is 36 MHz, then the frequencies generated are 28, 9 and 4.5 MHz. The clock frequency for a given control step is the minimum of the operating frequencies of all FUs active in that step.

6.3 ILP-Based Scheduling for EDP Reduction

6.3.1 Introduction

The architecture for DFC-based systems consists of a data path, a controller and a DCU. The data path consists of FUs with registers and multiplexers. The controller decides which FUs are active in each control step, and those not active are disabled with a multiplexer. The DCU generates the required clock frequency, usually after a clock divider strategy with submultiples of the base frequency [210, 219, 221, 229]. The base frequency is the maximum frequency (or a multiple of the maximum frequency) of any FU at maximum supply voltage. The controller has storage units to store a parameter called "clock frequency index" for each control step derived during the data path scheduling. This clock frequency index parameter serves as the clock-dividing factor for the DCU. The cycle frequency is generated dynamically, and the FUs with the appropriate supply voltages are activated. The main overhead in this scheme are level converters, the DCU and some additional storage in the control unit. When a value of cfi_c is loaded into the DCU, it provides a divided output clock frequency $\left(\frac{f_{\text{base}}}{cfi_c} \right)$.

6.3.2 EDP Modeling of a DFG

Assuming that the data path is represented as a sequencing DFG the notations given in Table 6.1 will be used for developing the following energy and EDP for a data path. The energy consumption in any cycle c is the energy consumption of all the resources active in c, given as

$$E_c = \sum_{i=1}^{R_c} \alpha_{i,c} C_{i,c} V_{i,c}^2. \qquad (6.5)$$

The voltage-level converters are considered as resources operating in the control step in which it needs to step up the signal. The total energy consumption of the whole DFG (or data path) is the sum of the energy consumption for all cycles as given below:

$$E_{\text{MVDFC}} = \sum_{c=1}^{N} E_c = \sum_{c=1}^{N} \sum_{i=1}^{R_c} \alpha_{i,c} C_{i,c} V_{i,c}^2. \qquad (6.6)$$

The DCU is responsible for generating the dynamic clock and is considered as a resource operating in all the control steps. The energy consumption of the DCU is to be added along with the energy dissipation corresponding to a schedule but need not be considered for minimization.

The critical path delay of the DFG is given by the summation of the inverse of the clock frequencies,

Table 6.1 Notations used in the description

N	: Total number of control steps in the DFG
N_v	: Total number of vertices (operations) in the DFG excluding source and sink
v_i	: Any vertex (operation) such that $1 \leq i \leq N_v$
c	: Any control step or clock cycle in DFG
R_c	: Number of resources active in step c
f_c	: Cycle frequency for control step c
$\alpha_{i,c}$	: Switching at resource i used by operation o_i operating in step c
$C_{i,c}$	: Load capacitance of resource i used by operation o_i operating in control step c
$V_{i,c}$	: Operating voltage of resource i used by operation o_i operating in control step c
E_c	: Energy consumption of all FUs active in cycle c
EDP_c	: EDP of all FUs active in cycle c
T_{cp}	: Critical path delay of the DFG
E	: Total energy consumption of the DFG
EDP	: Total EDP of the DFG
SVSF	: Subscript used for single supply voltage and single-frequency operation
MVDFC	: Subscript used for multiple supply voltage and dynamic frequency operation
MVMC	: Subscript used for multiple supply voltage and multicycling operation
f_{clk}	: Operating clock frequency for single frequency or multicycling operations
V	: Operating voltage for single voltage and single-frequency operations

$$T_{\mathrm{cpMVDFC}} = \sum_{c=1}^{N} f_c. \tag{6.7}$$

The total EDP can be calculated as the product of the total energy consumption and the critical path delay as shown in the following equation:

$$\mathrm{EDP}_{\mathrm{MVDFC}} = E_{\mathrm{MVDFC}} * T_{\mathrm{cpMVDFC}} = \left(\sum_{c=1}^{N} \sum_{i=1}^{R_c} \alpha_{i,c} C_{i,c} V_{i,c}^2 \right) * \sum_{c=1}^{N} f_c. \tag{6.8}$$

This is the objective function for minimization during high-level synthesis to reduce energy while maintaining performance requirements.

The objective function can be minimized by using voltage- and frequency-scaling technology. Since it involves the product of the two variables and is a non-linear function, ILP cannot be used directly for its minimization. Hence, instead of finding the energy consumption for each cycle c as in Equation (6.5), the EDP must be derived for each cycle. The EDP for each cycle can be expressed as

$$\mathrm{EDP}_c = \frac{E_c}{f_c} = \frac{\sum_{i=1}^{R_c} \alpha_{i,c} C_{i,c} V_{i,c}^2}{f_c}. \tag{6.9}$$

The total EDP of the DFG is the sum of the above EDP_c for all control steps which is given as follows:

$$\mathrm{EDP}_{\mathrm{MVDFC}} = \sum_{c=1}^{N} \mathrm{EDP}_c = \sum_{c=1}^{N} \frac{\sum_{i=1}^{R_c} \alpha_{i,c} C_{i,c} V_{i,c}^2}{f_c} = \sum_{c=1}^{N} \sum_{i=1}^{R_c} \frac{\alpha_{i,c} C_{i,c} V_{i,c}^2}{f_c}. \tag{6.10}$$

For MVMC operation, f_c is the same for all control steps and will be denoted as f_{clk}. Following the same steps as above, the total energy of the DFG for MVMC operation is modeled by the following equation:

$$E_{\mathrm{MVMC}} = \sum_{c=1}^{N} E_c = \sum_{c=1}^{N} \sum_{i=1}^{R_c} \alpha_{i,c} C_{i,c} V_{i,c}^2. \tag{6.11}$$

It may be noted that even though the above expression is the same as the MVDFC case, the quantitative values will be different because of different values of parameters, like $\alpha_{i,c}$, $C_{i,c}$ and $V_{i,c}^2$. Following the same steps as above, the total EDP of the DFG for MVMC operation is modeled by the following equation:

$$\mathrm{EDP}_{\mathrm{MVMC}} = \sum_{c=1}^{N} \mathrm{EDP}_c = \sum_{c=1}^{N} \frac{\sum_{i=1}^{R_c} \alpha_{i,c} C_{i,c} V_{i,c}^2}{f_{\mathrm{clk}}} = \sum_{c=1}^{N} \sum_{i=1}^{R_c} \frac{\alpha_{i,c} C_{i,c} V_{i,c}^2}{f_{\mathrm{clk}}}. \tag{6.12}$$

For SVSF mode of operation, $V_{i,c}$ and f_c are the same for any clock cycle (c) and any operation (i). Hence, the energy and EDP can be evaluated using the following expressions, respectively:

$$E_{\text{SVSF}} = \sum_{c=1}^{N} E_c = \left(\sum_{c=1}^{N} \sum_{i=1}^{R_c} \alpha_{i,c} C_{i,c} \right) \times V^2, \qquad (6.13)$$

$$\text{EDP}_{\text{SVSF}} = \sum_{c=1}^{N} \text{EDP}_c = \left(\sum_{c=1}^{N} \sum_{i=1}^{R_c} \alpha_{i,c} C_{i,c} \right) \times \left(\frac{V^2}{f_{\text{clk}}} \right). \qquad (6.14)$$

6.3.3 ILP Formulations for EDPs

In this section, the ILP formulations to minimize the EDP of a data path circuit are discussed [210, 230]. The formulations for MVDFC-based circuits are followed by MVMC-based circuits. In order to formulate an ILP-based model for the objective function and the scheduling scheme for the DFG, the notations given in Table 6.2 are required.

6.3.3.1 ILP Formulations: MVDFC

The ILP formulation for the objective function given in Equation (6.10) for multiple supply voltages and DFC is derived.

Objective Function: The objective function minimizes the total EDP of the entire DFG. Using the decision variable x_{i,c,L_V,L_f}, the objective function is written as

Table 6.2 Notations used in ILP formulations

L_V	: Number of voltage levels for multiple supply voltage operation
L_f	: Number of frequency levels for dynamic frequency operation
F_{k,L_V}	: FU of type k operating at voltage level L_V
M_{k,L_V}	: Maximum number of resources of type k operating at voltage level L_V
$C_S[i]$	: ASAP time stamp for the vertex v_i
$C_L[i]$	: ALAP time stamp for the vertex v_i
$\text{EDP}(i, L_V, L_f)$	: EDP of a resource used by vertex v_i operating at voltage level L_V and frequency level L_f
x_{i,c,L_V,L_f}	: Decision variable which takes the value of 1 if vertex v_i is scheduled in control step c using the FU F_{k,L_V}
$y_{i,L_V,l,m}$	: Decision variable which takes the value of 1 if v_i is using the resource $F_{k,v}$ and scheduled in control steps $l \rightarrow m$
L_{i,L_V}	: Latency in terms of number of clock cycles for vertex v_i using resource operating at voltage level L_V

$$Minimize : \text{EDP}_{\text{MVDFC}}, \tag{6.15}$$

$$Minimize : \sum_c \sum_i \sum_c \sum_{L_V} \sum_{L_f} x_{i,c,L_V,L_f} \times \text{EDP}(i,L_V,L_f). \tag{6.16}$$

Uniqueness Constraints: These constraints ensure that each vertex v_i is scheduled to a unique control step within the mobility range $(C_S[i], C_L[i])$ with a particular supply voltage and operating frequency represented as: $\forall i,\ 1 \le i \le N$,

$$\sum_c \sum_{L_V} \sum_{L_f} x_{i,c,L_V,L_f} = 1. \tag{6.17}$$

Precedence Constraints: These constraints guarantee that for an vertex v_i, all its predecessors are scheduled in earlier control steps and its successors are scheduled in later control steps. These are modeled as: $\forall i, j, v_i \in \text{Pred}_{v_j}$,

$$\sum_{L_V} \sum_{L_f} \sum_{d=C_S[i]}^{C_L[i]} d \times x_{i,d,L_V,L_f} - \sum_{L_V} \sum_{L_f} \sum_{e=C_S[j]}^{C_L[j]} e \times x_{j,e,L_V,L_f} \le -1. \tag{6.18}$$

Resource Constraints: The resource constraints make sure that no control step contains more than FU_{k,L_V} operations of type k operating at voltage level L_V. These can be enforced as $\forall c,\ 1 \le c \le N$ and $\forall L_V$,

$$\sum_i \sum_{L_f} x_{i,c,L_V,L_f} \le M_{k,L_V}. \tag{6.19}$$

Frequency Constraints : This set ensures that if an FU is operating at a higher voltage level then it can be scheduled in a lower frequency control step; whereas, if an FU is operating at a lower voltage level, then it cannot be scheduled during a higher frequency control step. These constraints are written as $\forall i,\ 1 \le i \le N, \forall c,\ 1 \le c \le N$, if $L_f < L_V$, then $x_{i,c,L_V,L_f} = 0$.

6.3.3.2 ILP Formulations: MVMC

The ILP formulation for the objective function for MVMC operation mode is now presented.

Objective Function: The objective is to minimize the EDP of the whole DFG over all control steps using MVMC.

$$Minimize : \text{EDP}_{\text{MVMC}}, \tag{6.20}$$

$$Minimize : \sum_l \sum_i \sum_{L_V} y_{i,L_V,l,(l+L_{i,L_V}-1)} \times \text{EDP}(i,L_V,L_{f_{\text{clk}}}). \tag{6.21}$$

Uniqueness Constraints: These constraints ensure that each vertex v_i is scheduled in the appropriate control step within the mobility range $(C_S[i], C_L[i])$ and is assigned the specific supply voltage. An operation may be executed with more than one clock cycle, depending on the supply voltage. These constraints are represented as $\forall i$, $1 \leq i \leq N_v$,

$$\sum_{L_V} \sum_{l=C_S[i]}^{C_S[i]+C_L[i]+1-L_{i,L_V}} y_{i,L_V,l,(l+L_{i,L_V}-1)} = 1. \tag{6.22}$$

When an operation is scheduled at the highest voltage, then it is scheduled in one unique control step; whereas when it is to be operated at lower voltages it needs more than one clock cycle for completion. Thus, for lower voltages, mobility is restricted.

Precedence Constraints: These constraints guarantee that for a vertex v_i, all its predecessors are scheduled in earlier control steps and its successors are scheduled in later control steps. These constraints should also take care of the multicycling operations. These are modeled as $\forall i, j, v_i \in \mathrm{Pred}_{v_j}$,

$$\sum_{L_V} \sum_{l=C_S[i]}^{C_L[i]} (l + L_{i,L_V} - 1) \times y_{i,L_V,l,(l+L_{i,L_V}-1)}$$

$$- \sum_{L_V} \sum_{l=C_S[j]}^{C_L[j]} l \times y_{j,L_V,l,(l+L_{j,L_V}-1)} \leq -1. \tag{6.23}$$

Resource Constraints: These constraints ensure that each control step contains no more than $F_{k,v}$ operations of type k operating at voltage level L_V. This can be enforced as $\forall L_V$ and $\forall l, 1 \leq l \leq N$,

$$\sum_i \sum_l y_{i,L_V,l,(l+L_{i,L_V}-1)} \leq M_{k,L_V}. \tag{6.24}$$

6.3.4 ILP-Based Data Path Scheduling Algorithm

In this section, the solution for the ILP formulations obtained in the previous section is discussed [210, 230]. The same target architecture and the same characterized data path components used in [210, 230] are assumed. The ILP-based scheduler that attempts to minimize the EDP is outlined in Algorithm 14. The first step is to determine the ASAP and ALAP time stamps of each operation. The ASAP time stamp is the start time and ALAP time stamp is the end time of each operation. These two times provide the mobility of an operation, and the operation must be scheduled in this mobility range. Then the scheduler finds the ILP formulations based on the models described in the previous section. The scheduler determines the cycle frequencies in step 6 that contribute the smallest frequencies of all operations

---

Algorithm 14 ILP-Based Scheduling for EDP Minimization

1: Determine the ASAP and ALAP schedules of the unscheduled DFG.
2: Determine the mobility graph of each node.
3: Construct the ILP formulations for the DFG.
4: Solve the ILP formulations using LP-Solve.
5: Find the scheduled DFG.
6: Determine the cycle frequencies.
7: Find the energy and EDP estimates of the scheduled DFG.

---

scheduled in a particular cycle. Finally, the EDP and the energy consumption of the whole DFG are estimated.

6.3.4.1 For MVDFC

The solution for the ILP formulation in the MVDFC case is illustrated with the help of the DFG shown in Fig. 6.2. The ASAP schedule is shown in Fig. 6.2(a), and the ALAP schedule is shown in Fig. 6.2(b). From the ASAP and ALAP schedules, the mobility graph as in Fig. 6.2(c) is obtained. Using this mobility graph, the ILP formulation shown in Algorithm 15 is obtained for the resource constraint to RC2, three multipliers at 2.4 V, one ALU at 2.4 V and one ALU operating at 3.3 V. In Algorithm 15, the following additional notations are used, Mmult1: number of multipliers at voltage level 1, Mmult2: number of multipliers at voltage level 2, Malu1: number of ALUs at voltage level 1 and Malu2: number of ALUs at voltage level 2. The formulation is solved using LP-Solve and based on the results obtained, the scheduled DFG shown is Fig. 6.2(d) is obtained.

6.3.4.2 For MVMC

The solution for the ILP formulation in the MVMC case is illustrated using the example DFG shown in Fig. 6.3. The ASAP schedule is shown in Fig. 6.3(a) and the ALAP schedule is shown in Fig. 6.3(b). From the ASAP and ALAP schedules, we obtain the mobility graph shown in Fig. 6.3(c). It may be noted that this mobility graph is different from that shown in Fig. 6.2(c). In the MVMC case, the mobility graph considers the multicycle operations. We assume two operating voltage levels; and when a multiplier is operated at the lower voltage level, it takes two clock cycles for completing the operation. For the characterized cells used in our experiment, the operating clock frequency, f_{clk}, is 9 MHz. Using this mobility graph, the ILP formulation shown in Algorithm 15 is obtained for the resource constraint (RC2), three multipliers at 2.4 V, one ALU at 2.4 V and one ALU operating at 3.3 V. In Algorithm 16, the notations, such as, Mmult1, Mmult2, Malu1 and Malu2 are the same as those used in the case of the MVDFC. The formulation is solved using LP-Solve; and based on the results, the scheduled DFG shown in Fig. 6.3(d) is obtained.

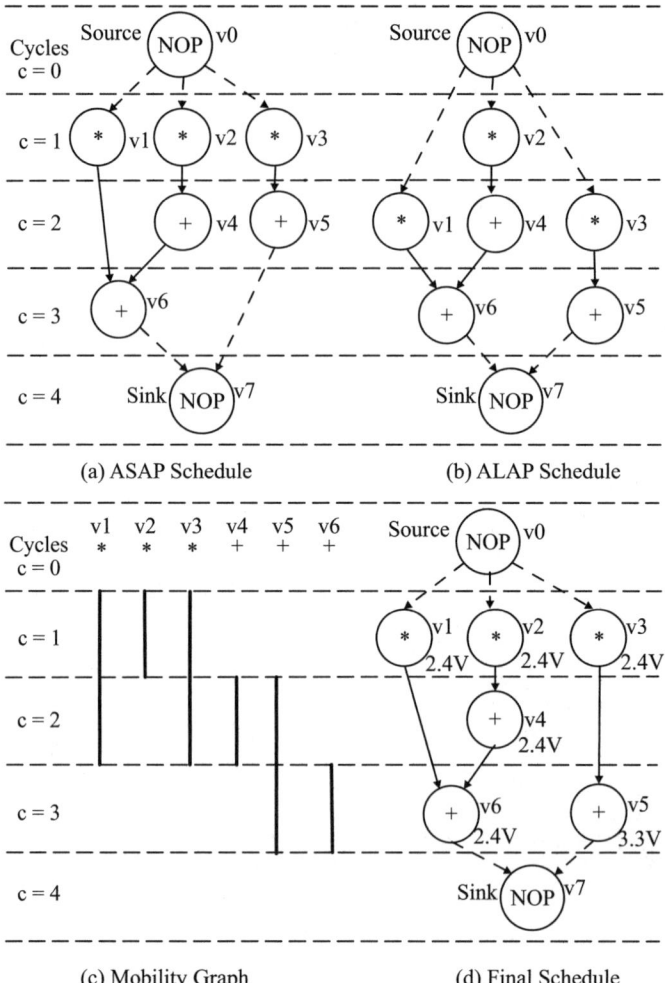

(a) ASAP Schedule   (b) ALAP Schedule

(c) Mobility Graph   (d) Final Schedule

Fig. 6.2 Example DFG for MVDFC

6.3.5 Experimental Results

The ILP scheduler is tested with selected benchmark circuits, such as (1) Example circuit, (2) FIR filter, (3) IIR filter, (4) HAL differential equation solver and (5) Auto Regressive filter (ARF). The FUs (FUs) assumed are ALUs and MULTs. The data path cells and their characterization are considered from [210, 230]. The data path scheduling algorithms were tested using the different sets of resource constraints. Figure 6.4 shows the results for the various benchmarks averaged over different resource constraints. The energy estimation includes the energy consumption of the overheads. The results reported are based on the assumption of two supply voltages and switching activity of 0.5. The energy reduction and EDP reduction are shown.

Algorithm 15 ILP Formulation for Example DFG for MVDFC

1: /* Objective Function */
2: min: 106.6 x1111 + 213.2 x1112 + 56.4 x1121 + 112.8 x1122 + 106.6 x1211 + 213.2 x1212 +
   56.4 x1221 + 112.8 x1222 + 106.6 x2111 + 213.2 x2112 + 56.4 x2121 + 112.8 x2122 + 106.6
   x3111 + 213.2 x3112 + 56.4 x3121 + 112.8 x3122 + 106.6 x3211 + 213.2 x3212 + 56.4 x3221
   + 112.8 x3222 + 2.8 x4211 + 5.5 x4212 + 1.5 x4221 + 2.9 x4222 + 2.8 x5211 + 5.5 x5212 +
   1.5 x5221 + 2.9 x5222 + 2.8 x5311 + 5.5 x5312 + 1.5 x5321 + 2.9 x5322 + 2.8 x6311 + 5.5
   x6312 + 1.5 x6321 + 2.9 x6322;
3: /* Uniqueness Constraints */
4: x1111 + x1112 + x1121 + x1122 + x1211 + x1212 + x1221 + x1222 = 1;
5: x2111 + x2112 + x2121 + x2122 = 1;
6: x3111 + x3112 + x3121 + x3122 + x3211 + x3212 + x3221 + x3222= 1;
7: x4211 + x4212 + x4221 + x4222 = 1;
8: x5211 + x5212 + x5221 + x5222 + x5311 + x5312 + x5321 + x5322 = 1;
9: x6311 + x6312 + x6321 + x6322 = 1;
10: /* Precedence Constraints */
11: 3 x6311 + 3 x6312 + 3 x6321 + 3 x6322 − 2 x1211 − 2 x1212 − 2 x1221 − 2 x1222 − x1111
    − x1112 − x1121 − x1122 ≥ 1;
12: 2 x4211 + 2 x4212 + 2 x4221 + 2 x4222 − x2111 − x2112 − x2121 − x2122 ≥ 1;
13: 3 x6311 + 3 x6312 + 3 x6321 + 3 x6322 − x4211 − x4212 − x4221 − x4222 ≥ 1;
14: 3 x5311 + 3 x5312 + 3 x5321 + 3 x5322 + 2 x5211 + 2 x5212 + 2 x5221 + 2 x5222 − 2 x3211
    − 2 x3212 − 2 x3221 − 2 x3222 − x3111 − x3112 − x3121 − x3122 ≥ 1;
15: /* Resource Constraints */
16: x1111 + x2111 + x3111 + x1112 + x2112 + x3112 ≤ 0; /* mult1 */
17: x1121 + x2121 + x3121 + x1122 + x2122 + x3122 ≤ 3; /* mult2 */
18: x1211 + x3211 + x1212 + x3212 ≤ 0; /* mult1 */
19: x1221 + x3221 + x1222 + x3222 ≤ 3; /* mult2 */
20: x4211 + x5211 + x4212 + x5212 ≤ 1; /* alu1 */
21: x4221 + x5221 + x4222 + x5222 ≤ 1; /* alu2 */
22: x5311 + x6311 + x5312 + x6312 ≤ 1; /* alu1 */
23: x5321 + x6321 + x5322 + x6322 ≤ 1; /* alu2 */
24: /* Frequency Constraints */
25: x1121 = 0; x1221 = 0; x2121 = 0; x3121 = 0; x3221 = 0; x4221 = 0; x5221 = 0; x5321 = 0;
    x6321 = 0;
26: /* Zero-One Type Cast */
27: INT x1111, x1112, x1121, x1122, x1211, x1212, x1221, x1222, x2111, x2112, x2121, x2122,
    x3111, x3112, x3121, x3122, x3211, x3212, x3221, x3222, x4211, x4212, x4221, x4222,
    x5211, x5212, x5221, x5222, x5311, x5312, x5321, x5322, x6311, x6312, x6321, x6322;

It is observed that both the energy and the EDP are reduced considerably for both
MVDFC and MVMC schemes. The MVDFC scheme results in better savings than
those of the MVMC scheme for most of the cases, except the FIR benchmark. The
energy savings of both the MVDFC and MVMC schemes are the same for most
cases except for few resource constraints. The savings would have been the same
for both schemes on using energy as objective function, as the energy savings is due
to the voltage reduction and not due to the dynamic frequency clocking or multicycling. However, use of energy as objective function would have increased the EDP,
thus reducing performance. A comparative perspective with other existing works is
shown in Table 6.3 that proves the efficacy of the presented approach.

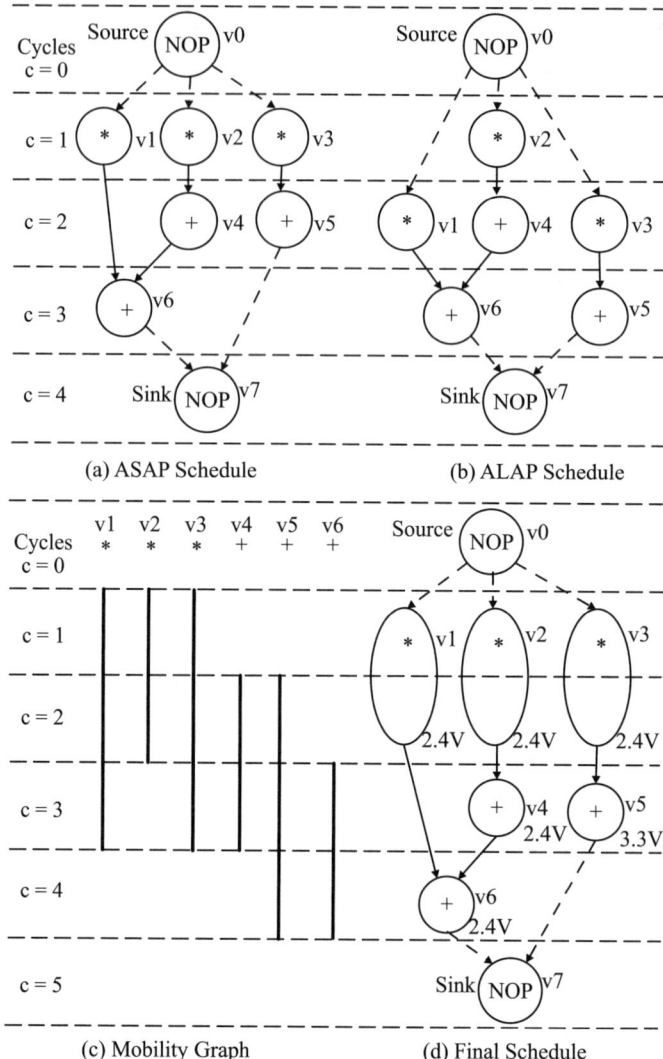

Fig. 6.3 Example DFG (for RC2) (MVMC)

6.3.6 Conclusions

This section introduced an ILP-based resource-constrained data path scheduling algorithm using both multiple supply voltages and DFC. It is observed that using two supply voltage levels, an average energy reduction of 46% and an average EDP reduction of 21% are obtained with MVDFC; whereas, for the MVMC scheme an average energy reduction of 43% and average EDP reduction of 16% are obtained. If there is a proportionate number of multipliers and ALUs in the critical path such that net performance degradation due to the low-frequency operation of multipliers can be overcome by high-frequency operation of ALUs, then the reduction is sig-

---

Algorithm 16 ILP Formulation for Example DFG for MVMC

1: /* Objective Function */
2: min: 106.6 x1111 + 106.6 x1122 + 106.6 x1133 + 56.4 x1212 + 56.4 x1223 + 106.6 x2111 + 106.6 x2122 + 56.4 x2212 + 106.6 x3111 + 106.6 x3122 + 106.6 x3133 + 56.4 x3212 + 56.4 x3223 + 2.8 x4122 + 2.8 x4133 + 1.5 x4222 + 1.5 x4233 + 2.8 x5122 + 2.8 x5133 + 2.8 x5144 + 1.5 x5222 + 1.5 x5233 + 1.5 x5244 + 2.8 x6133 + 2.8 x6144 + 1.5 x6233 + 1.5 x6244;
3: /*Uniqueness Constraints*/
4: x1111 + x1122 + x1133 + x1212 + x1223 = 1;
5: x2111 + x2122 + x2212 = 1;
6: x3111 + x3122 + x3133 + x3212 + x3223 = 1;
7: x4122 + x4133 + x4222 + x4233 = 1;
8: x5122 + x5133 + x5144 + x5222 + x5233 + x5244 = 1;
9: x6133 + x6144 + x6233 + x6244 = 1;
10: /* Resource Constraints */
11: x1111 + x2111 + x3111 $\leq$ 0; /* Mmult1 */
12: x1212 + x2212 + x3212 $\leq$ 3; /* Mmult2 */
13: x1122 + x2122 + x3122 $\leq$ 0; /* Mmult1 */
14: x1212 + x1223 + x2212 + x3212 + x3223 $\leq$ 3; /* Mmult2 */
15: x1133 + x3133 $\leq$ 0; /* Mmult1 */
16: x1223 + x3223 $\leq$ 3; /* Mmult2 */
17: x4122 + x5122 $\leq$ 1; /* Malu1 */
18: x4222 + x5222 $\leq$ 1; /* Malu2 */
19: x4133 + x5133 + x6133 $\leq$ 1; /* Malu1 */
20: x4233 + x5233 + x6233 $\leq$ 1; /* Malu2 */
21: x5144 + x6144 $\leq$ 1; /* Malu1 */
22: x5244 + x6244 $\leq$ 1; /* Malu2 */
23: /* Precedence Constraints */
24: 4 x6144 + 4 x6244 + 3 x6133 + 3 x6233 - 3 x1133 - 3 x1223 - 2 x1122 - 2 x1212 - x1111 $\geq$ 1;
25: 4 x6144 + 4 x6244 + 3 x6133 + 3 x6233 - 3 x4133 - 3 x4233 - 2 x4122 - 2 x4222 $\geq$ 1;
26: 3 x4133 + 3 x4233 + 2 x4122 + 2 x4222 - 2 x2122 - 2 x2212 - x2111 $\geq$ 1;
27: 4 x5144 + 4 x5244 + 3 x5133 + 3 x5233 + 2 x5122 + 2 x5222 - 3 x3133 - 3 x3223 - 2 x3122 - 2 x3212 - x3111 $\geq$ 1;
28: /* Integer Constraints */
29: INT x1111, x1122, x1133, x1212, x1223, x2111, x2122, x2212, x3111, x3122, x3133, x3212, x3223, x4122, x4133, x4222, x4233, x5122, x5133, x5144, x5222, x5233, x5244, x6133, x6144, x6233, x6244;

---

nificant. Such a scheduler incorporated into a low-power data path synthesis tool will greatly benefit low-power processor design, especially for computer-intensive applications.

Table 6.3 Savings for various scheduling schemes

Benchmark circuits	Percentage average energy savings						
	This work		Shiue [320]	Sarrafzadeh [311]	Johnson [144]	Chang [70]	Mohanty [219]
	MVDFC	MVMC					
FIR	47	44	–	23	53	–	46
IIR	44	40	–	–	36	–	–
HAL	47	43	24	–	–	36	40
ARF	46	46	12	18	39	29	39

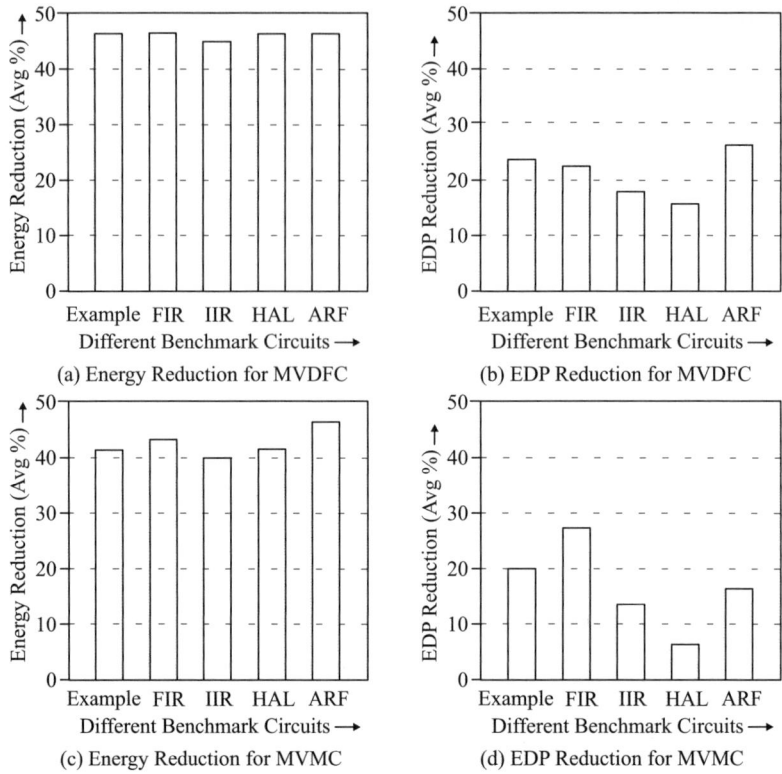

Fig. 6.4 Reduction for different benchmarks expressed as average percentage

6.4 Heuristic-Based Scheduling Algorithm for Energy Minimization

6.4.1 Introduction

The ILP-based optimization approach presented in the previous section is an optimal solution for a given resource constraint. However, as evident from the ILP formulations, their complexity grows exponentially as the number of vertices or operation of a given DFG increases. The solutions consume significant time to converge to a solution. To overcome these shortcomings, a heuristic-based solution is presented in this section. The algorithm is presented for MVDFC scenario. Two heuristic-based scheduling algorithms are presented, of which one is a time-constrained approach (TC-DFC) and the other is a resource-constrained approach (RC-DFC) [210, 219, 221].

6.4.2 Time-Constrained Scheduling: TC-DFC

The data path is represented in the form of a DFG constructed as a sequencing graph. Figure 6.5 shows such a graph for the HAL benchmark. The inputs to the algorithm are an unscheduled DFG (UDFG), the scaled-down operating frequencies and the execution time constraint T_{con} for the whole schedule. To obtain higher energy savings and at the same time maintain performance, the multipliers are to be operated at as low a frequency as possible and the adders at as high a frequency as possible. This objective can be achieved if adders/subtractors are not operated along with multipliers in the same duty cycle. When they are to be operated during the same cycle to meet the time constraint, energy savings will come from the multipliers only. Initially, TC-DFC generates a schedule such that the low-frequency operators are scheduled at earlier steps and the high-frequency operators are scheduled at later steps. Later on, the TC-DFC modifies the schedule by moving operations from one step to another with the objective of meeting the time constraint. It then finds the appropriate clock-cycle width and assigns appropriate voltage.

6.4.2.1 Algorithm Flow

Algorithm 17 shows the flow of the TC-DFC scheduling algorithm. In step 1, an ASAP schedule for the DFG is determined. In step 2, the scheduler creates a priority

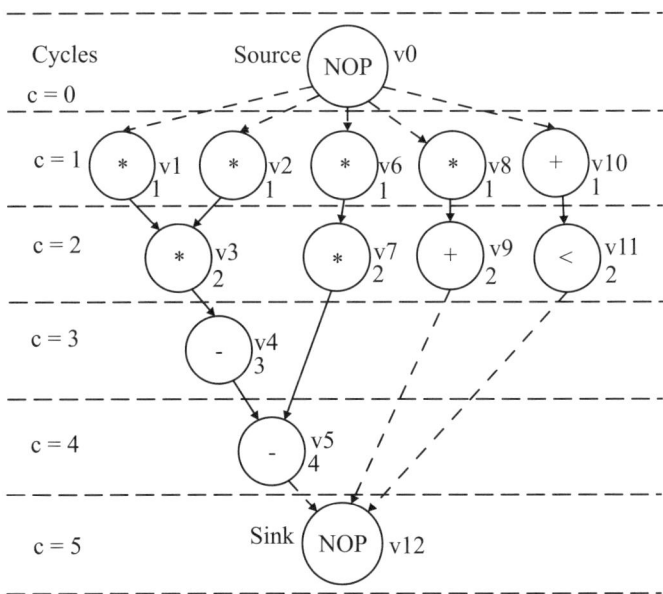

Fig. 6.5 HAL differential equation solver (with ASAP labels)

Algorithm 17 TC-DFC Scheduling Algorithm Flow

1: Find an ASAP schedule for the sequencing UDFG.
2: Create a priority list of vertices using the ASAP schedule in step 1.
3: Assign control steps to the operations such that the higher priority vertex is scheduled at an earlier time stamp, precedence is satisfied, and the multiplications and ALU operations are not scheduled in the same cycle.
4: Find the cycles having only ALU operations, those with only multiplications and those with both ALU operations and multiplications (mixed) for the currently obtained schedule.
5: Create a priority list of clock cycles such that cycles with only ALU operations get higher priority than the cycles with only multiplications or those with mixed operations (cycles with only multiplications get higher priority than the cycles with mixed operations).
6: Initialize cycle frequency to the minimum operating frequency.
7: If time constraint is not satisfied, the highest priority cycle is assigned the next higher frequency and repeat the step for the next higher priority cycle if necessary.
8: If any cycle has multiplier operating at highest frequency, then eliminate the cycle having minimum number of ALU operations, adjust the schedule and go to step 4.
9: Do voltage assignment and determine energy details.
10: Find the cycle frequency index for each cycle.

list of the vertices such that all multiplications (i.e., low-frequency operators) are grouped with higher priority than the ALU operations (i.e., high-frequency operators, such as additions, subtractions, comparisons, etc.). Among the multiplication operations higher priority is given to the operations with smaller ASAP time stamp, and the same is done for the group of ALU operations. In step 3, the vertices are time-stamped such that no multiplication and ALU operations are scheduled to function concurrently. In addition, it is ensured that operation precedence is satisfied and higher priority vertices are scheduled at earlier time stamps. In step 4, for the current schedule, the cycles are categorized as cycles having only ALU operations, only multiplication and both ALU operations and multiplications (mixed operations). In step 5, a priority list of clock cycles is created such that cycles with only ALU operations get higher priority than cycles with only multiplications or mixed operations. The cycles with only multiplications get higher priority than the cycles with mixed operations. Further, among the cycles with only ALU (or multiplication) operations higher priority is given to the cycle having a smaller number of ALU (or multiplication) operations. Similarly, among the cycles with mixed operations higher priority is given to cycles having smaller number of multiplications. In step 6, the initial cycle frequency is taken as the minimum operating frequency with the help of Table 6.6. In step 7, in order to fulfill time constraints, the highest priority cycle frequency is increased using Table 6.6. If needed, the process is repeated for the next higher priority cycle. In step 8, if it is found that a cycle with multiplication has the highest voltage, then the cycle having minimum number of ALU operations is eliminated and the schedule is adjusted. In step 9, voltage is assigned and energy estimates for the entire DFG are calculated. In step 10, the cycle frequency index for each cycle is calculated.

6.4.2.2 Pseudocode Description

The list of functions needed in the implementation of the algorithm is given in Table 6.4. Similarly, the data structures or the identifiers used in the algorithm description are summarized in Table 6.5. The pseudocode is presented in Algorithm 18.

In line 1, the ASAP schedule for the UDFG is obtained. The procedure CreateVertexPriorityList creates the VertexPriorityList such that the vertex with the lower operating frequency gets the higher priority to be scheduled at an earlier control step than the lower priority vertices. Table 6.7 shows such a list obtained for the DFG given in Fig. 6.5. TC-DFCSchedSteps$_{v_i}$ (line 2) is a data structure that contains the clock-cycle step for any vertex v_i. It is initialized to zero for the source vertex. ScheduledVertexList (line 2) is a data structure to maintain the list of vertices already scheduled which is initialized to the source vertex. The *while* loop (line 3)

Table 6.4 List of functions used in the TC-DFC algorithm

Functions	Description	Complexity
ASAPScheduler	: Determines the ASAP time of the vertices	$\Theta(|V| + |E|)$
CreateVertexPriorityList	: Creates a priority list of vertices such that the vertex with lower operating frequency gets the higher priority	$\Theta(|V|)$
TOP	: Finds the first vertex from priority list array	$\Theta(1)$
CheckFrequencyConstraint	: Checks the frequency constraint in a cycle	$\Theta(1)$
Maximum	: Finds the maximum value from an array	$\Theta(c)$
CreateCyclePriorityList	: Constructs the cycle priority list in an array	$\Theta(c)$
FindMinimumFrequency	: Finds the minimum available frequency	$\Theta(L_f)$
CalculateDelay	: Calculates the critical path delay	$\Theta(c)$
FindNextHigherFrequency	: Finds the next higher available frequency	$O(L_f)$
FindCycleWithMinimumALU	: Finds the control step with a minimum number of ALU operations	$\Theta(cR_T)$
Adjust Predecessor	: Adjusts time stamp of predecessor	$O(|V|)$
Adjust Successor	: Adjusts time stamp of successor	$O(|V|)$
Update *CyclePriorityList*	: Updates the array	$O(c)$
Voltage Assignment	: Assigns voltage to each vertex	$\Theta(|V|)$
Find Cycle Frequency Index	: Finds cycles frequency indices of all cycles	$\Theta(c)$

Table 6.5 List of variables and data structures used in the TC-DFC algorithm description

Data structures	Descriptions
ASAPSchedule	: An array used to store ASAP time stamp of each vertex
TC-DFCSchedStep	: An array used to store TC-DFC time stamp of each vertex
ScheduledVertexList	: An array used to store vertices already scheduled
VertexPriorityList	: An array used to store vertices in a priority order
CyclePriorityList	: An array used to store control steps in a priority order
TC-DFCNoOfSteps	: Total number of control steps of TC-DFC schedule
CycleFrequencyList	: An array used to store frequency of each cycle
cycle, ControlStepIndicator	: Temporary variables

Algorithm 18 Pseudocode for TC-DFC Scheduling Algorithm

1: ASAPScheduler(UDFG); CreateVertexPriorityList(ASAPSchedule); cycle = 1;
2: TC-DFCSchedSteps$_{v_0}$ = 0; ScheduledVertexList = v_0; /* source vertex scheduled */
3: **while** (VertexPriorityList $\neq$ NULL) **do**
4:        v_i = TOP(VertexPriorityList);
5:        **if** ($v_i \notin$ ScheduledVertexList and AllPredecessor$_{v_i} \in$ ScheduledVertexList) **then**
6:                **if** (CheckFrequencyConstraint(cycle)) **then**
7:                        **return** cycle = Maximum (TC-DFCSchedSteps) + 1;
8:                **else**
9:                        **return** Schedule in current cycle;
10:               **end if**
11:               TC-DFCSchedSteps$_{v_i}$ = cycle; VertexPriorityList = VertexPriorityList $- v_i$;
12:               ScheduledVertexList = ScheduledVertexList $\cup v_i$;
13:       **end if**
14: **end while**
15: TC-DFCNoOfSteps = Maximum(TC-DFCSchedSteps);
16: CreateCyclePriorityList(CurrentSchedule, TC-DFCNoOfSteps);
17: (CycleFrequencyList = FindMinimumFrequency(Table 6.6);
18: T_s = CalculateDelay(CycleFrequencyList); ControlStepIndicator = 1;
19: **while** (ControlStepIndicator) **do**
20:       **while** ($T_s > T_{con}$) **do**
21:               c_i = TOP(CyclePriorityList);
22:               CycleFrequencyList$_{c_i}$ = FindNextHigherFrequency(Table 6.6);
23:               T_s = CalculateDelay(CycleFrequencyList);
24:       **end while**
25:       **if** (no multiplier is operating at highest frequency) **then**
26:               **return** ControlStepIndicator = 0;
27:       **else**
28:               **return** c_i = FindCycleWithMinimumALU(for all cycle c_i);
29:               **for all** ($v_i \in c_i$) **do**
30:                       Reduce time stamp of v_i.
31:                       Adjust Predecessor$_{v_i}$ and Successor$_{v_i}$.
32:               **end for**
33:               CycleFrequencyList = FindMinimumFrequency(Table 6.6);
34:               T_s = CalculateDelay(CycleFrequencyList); Update CyclePriorityList;
35:       **end if**
36: **end while**
37: Do voltage assignment; Find cycle frequency index;

Table 6.6 TC-DFC frequency selection: from left $\rightarrow$ right

	MULT$_{Low}$	MULT$_{Med}$	ALU$_{Med}$	ALU$_{High}$
Frequency (MHz)	4.5	9	18	36
cfi$_c$	8	4	2	1

Table 6.7 Vertex priority list

v0	v1	v2	v6	v8	v3	v7	v10	v9	v11	v4	v5	v12
0	1	2	3	4	5	6	7	8	9	10	11	12

takes the highest priority vertex each time (line 4) and schedules it in an appropriate cycle checking for the frequency constraint violation provided all of its predecessors are already scheduled. The function CheckFrequencyConstraint (line 6) helps in checking the frequency constraint. It assures that two vertices operating at different frequencies are not scheduled during the same cycle. TC-DFCNoOfSteps (line 10) is the number of control steps for the schedule already generated.

Procedure CreateCyclePriorityList (line 11) creates the CyclePriorityList in which the higher priority cycles will be assigned higher frequencies. Table 6.8 shows such a list obtained for the schedule generated using lines 1–9. The data structure CycleFrequencyList (line 12) is used to store the operating frequency of each cycle. Initially, each cycle is assigned the minimum frequency from Table 6.6, and the critical delay of the schedule is found (line 12). While the time constraint is not satisfied, with the help of CyclePriorityList appropriate clock cycles are assigned to the next higher frequency and checked if time constraint is satisfied (lines 14–24). The algorithm terminates if no cycle has multipliers scheduled operating at the highest frequency (line 18). Otherwise, the cycle having the minimum number of ALUs is eliminated (line 20), CyclePriorityList is updated and lines 14–24 are repeated. Table 6.9 shows an updated CyclePriorityList. Finally, proper voltage values are assigned to the vertices. The algorithm also calculates the energy value of the schedule. It then finds the cycle frequency index using CycleFrequencyList. The final scheduled data path is shown in Figs. 6.6(a) and 6.6(b) for different time constraints.

6.4.2.3 Time Complexity

Let there be $|V|$ number of vertices and $|E|$ number of edges in the DFG. Suppose the number of control steps found out from the ASAP scheduling is c. Let L_f denote the number of frequency levels and R_T denote the number of resource types. Based on the time complexity of the different functions given in Table 6.4, the following analysis for the worst-case running time of the TC-DFC algorithm is provided. Time taken by the instruction from lines 1–2 is $\Theta\left(|V|+|E|\right)+\Theta\left(|V|\right)$. The running time of the code segment in lines 3–9 is $\Theta\left(c\,|V|\right)$. Similarly, $\Theta\left(c\right)+\Theta\left(L_f\right)$ is the running time of the code segment in lines 10–13. Assuming that the *while* loops are executed

Table 6.8 Cycle priority list : $T_{con} \approx 2 \times T_{cp}\, or\, 1.75 \times T_{cp}$

Cycles	c5	c4	c3	c2	c1	c6	c0
Priorities	0	1	2	3	4	5	6

Table 6.9 Cycle priority list : $T_{con} \approx 1.5 \times T_{cp}$

Cycles	c4	c3	c2	c1	c5	c0
Priorities	0	1	2	3	4	5

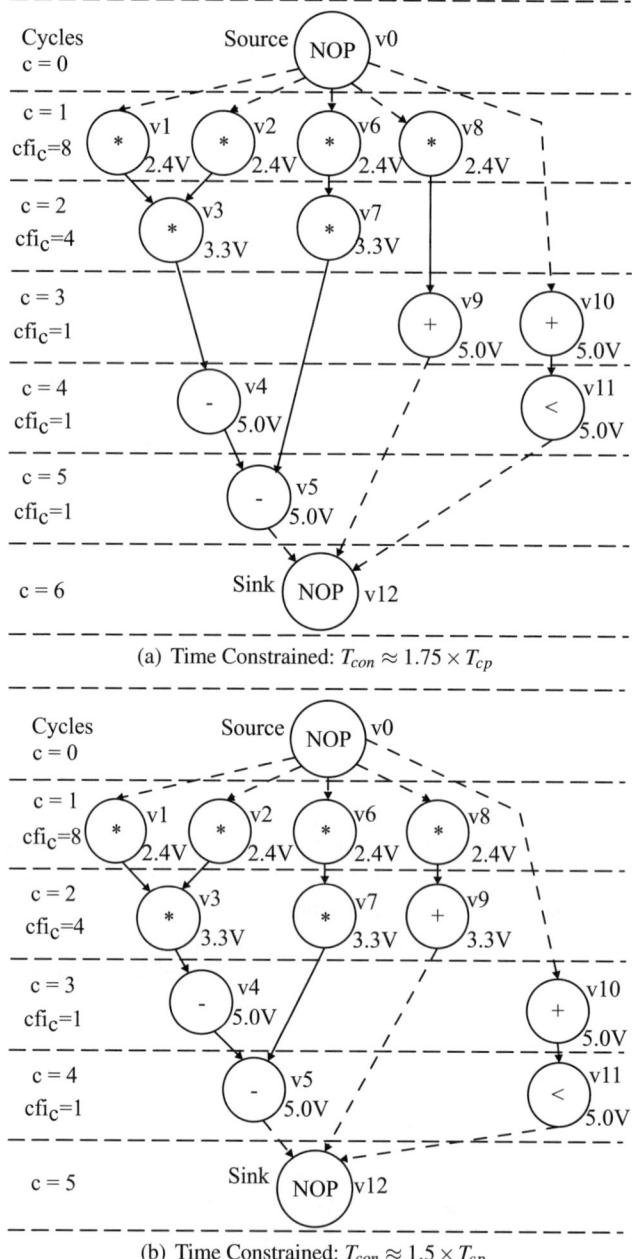

(a) Time Constrained: $T_{con} \approx 1.75 \times T_{cp}$

(b) Time Constrained: $T_{con} \approx 1.5 \times T_{cp}$

Fig. 6.6 Schedules obtained for HAL benchmark for different time constraints using TC-DFC

for a constant number of time (independent of the input size $|V|$ or $|E|$), the time complexity of the code segment lines 14–25 is $\Theta\left(cR_T\right) + \Theta\left(|V|\right) + \Theta\left(L_f\right) + \Theta\left(c\right)$. Without loss of generality, we can assume that the R_T, L_f and c are upper bounded by the number of vertices $|V|$. Based on this assumption, the overall running time of the algorithm is expressed as $\Theta\left(|V| + |E|\right) + \Theta\left(|V| \, |V|\right)$. For strong data dependency $|E| \approx |V|^2$, and for weak data dependency, $|E| << |V|^2$. In either case, *the simplified time complexity of the TC-DFC scheduling algorithm is $|V|^2$; in other words, the time complexity is polynomial to the number of vertices (operations) in the DFG.*

6.4.3 Resource-Constrained Scheduling: RC-DFC

The objective of RC-DFC is to minimize the EDP given in Equation (6.6), while assigning a schedule for the DFG. RC-DFC attempts to operate the multipliers at as low a frequency as possible, where the resulting decrease in performance is compensated by operating the ALUs at as high a frequency as possible. Depending on which FUs are active in a given cycle, the algorithm determines the frequency using a lookup table (LUT), called "frequency selection LUT," such as the one shown in Table 6.10, scanning it from left to right. In a schedule, if only multipliers are needed in a particular cycle, the frequency selection is in the order $MULT_{Low}, MULT_{Med}, MULT_{High}$. If both multipliers and ALUs are all operating in a given clock cycle, the frequency selection is in the order $MULT_{Low}, ALU_{Low}, MULT_{High}$. If only ALUs are operating in a control step, then the frequency selection is in the order $ALU_{High}, ALU_{Med}, ALU_{Low}$. Another LUT called "resource assignment LUT" constructed considering the resource constraints is used to match the selected frequency with a corresponding voltage level. The resources are assigned scanning the LUT from left to right. The scheduling algorithm uses heuristics to minimize the number of times that level conversions are needed. An example resource assignment LUT is shown in Table 6.11 with resource constraints: one MULT at 2.4 V, two MULT at 3.3 V, one MULT at 5.0 V, one ALU

Table 6.10 Frequency selection (from left to right in each step)

FUs in a cycle		Frequency priority order
MULT	–	$MULT_{Low}, MULT_{Med}, MULT_{High}$
MULT	ALU	$MULT_{Low}, ALU_{Low}, MULT_{High}$
–	ALU	$ALU_{High}, ALU_{Med}, ALU_{Low}$

Table 6.11 Resource look-up table (from left to right)

Clock	MULT			ALU		
	2.4 V	3.3 V	5.0 V	5.0 V	3.3 V	2.4 V
cycle						
c	1	2	1	1	1	0

at 3.3 V and one ALU at 5.0 V. The dimensions of this LUT depend on the total number of clock cycles of the schedule and the number of resource types. It should be noted that the arrangement of the MULTs is in the order from low to high voltage, whereas for the ALUs it is the opposite: from high to low. The LUT is updated during each assignment to ensure that the resource constraints are not violated.

6.4.3.1 Algorithm Flow

Algorithm 19 shows the flow of the proposed algorithm assuming that the DFG is modeled as a sequencing graph. The inputs to the algorithm are a UDFG, the resource constraints which include the number of resources, their corresponding operating voltages and the scaled-down operating frequencies. In step 1, the scheduler determines the ASAP and the ALAP schedules for the UDFG. In step 2, the total number of resources is calculated as the sum of each resource at different voltage levels. In step 3, the ASAP and ALAP schedules of step 1 are modified by using the number of resources found in step 2. In step 4, the total number of control steps for both ASAP and ALAP schedules are calculated, and the number of control steps for the final steps is assumed to be the maximum of the two. In step 5, the "resource assignment LUT" and "frequency selection LUT" are constructed. In step 6, the vertices having non-zero mobility and the vertices with zero mobility are determined and the current schedule is initialized as the ASAP schedule obtained in step 3. In step 7, voltage and frequency assignments are made for the current schedule using the LUTs. In step 8, the scheduler finds a proper step for each vertex having non-zero mobility such that the number of level converters needed for the execution of the whole DFG is minimum. As long as the voltage and frequency assignments follow the order in the LUTs, energy consumption is kept at a minimum. In step 9, current schedule, the LUTs are adjusted to satisfy the precedence. In step 10, cycle frequency indices are found for all cycles that would be stored in the controller

---

Algorithm 19 RC-DFC Scheduling Algorithm Flow

  1: Derive ASAP and ALAP schedules for the unscheduled DFG.
  2: Determine the number of resources at different operating voltages.
  3: Using the above number of resources, modify the schedules obtained in step 1.
  4: Calculate the total number of control steps that is the larger those of ASAP and ALAP schedules from Step 3.
  5: Construct the "resource assignment LUT" and "frequency selection LUT".
  6: Find the vertices having non-zero mobility and vertices with zero mobility and assume ASAP schedule in step 3 as the current schedule.
  7: Do voltage and frequency assignment using the current schedule and the LUTs.
  8: Taking a vertex with non-zero mobility time stamp it using LUTs such that EDP of the execution of whole DFG is minimum.
  9: Adjust current schedule, predecessor and successor time stamps, and LUTs, and repeat steps 7 and 8 to time stamp remaining non-zero mobility vertices.
 10: Determine the clock frequency index for each cycle.

---

and would be passed to the DCU for dynamic frequency generation. The algorithm terminates once all non-zero mobility vertices are scheduled.

6.4.3.2 Pseudocode of the Resource-Constrained Algorithm

The list of functions needed in the implementation of the algorithm is given in Table 6.12. Similarly, the data structures or the identifiers used in the algorithm description are summarized in Table 6.13. The pseudocode is presented in Algorithm 20.

Table 6.12 List of functions used in the RC-DFC algorithm

Functions	Description	Complexity
ASAPScheduler	: Determines ASAP time of the vertices	$\Theta(|V|+|E|)$
ALAPScheduler	: Determines ALAP time of the vertices	$\Theta(|V|+|E|)$
ModifySchedule	: Modifies the unconstrained schedules to incorporate resource constraints	$\Theta(|V|+|E|)$
ConstructResAssignmentTable	: Constructs resource assignment LUT	$\Theta(cL_vR_T)$
Maximum	: To find maximum of two control steps	$\Theta(1)$
FindResTypeForEachVertex	: Identifies the FU needed for each vertex	$\Theta(|V|)$
ConstructFreqSelectionLUT	: Constructs frequency selection LUT	$\Theta(L_f)$
FindMobileVertexList	: Finds the mobility of each vertex	$\Theta(|V|)$
AllocateVoltAndFreq	: Allocates the voltage and frequency levels using LUTs and current schedule	$\Theta(c|V|L_vR_T)$
CalculateEDP	: Calculates the EDP of the whole DFG	$\Theta(|V|)$
AdjustSchedule	: Adjusts the predecessor and successor time stamps such that the precedence is satisfied	$O(|V|)$
Update Res Assignment LUT	: Updates resource assignment LUT	$\Theta(1)$
FindEnergyAndDelay	: Determines energy and delay	$\Theta(|V|)$
FindCycleFreqIndex	: Finds cycle's frequency indices	$\Theta(c)$

Table 6.13 List of variables and data structures used in the RC-DFC algorithm description

Data structures	Descriptions
ASAPSchedule	: An array used to store ASAP time stamp of each vertex
ALAPSchedule	: An array used to store ALAP time stamp of each vertex
CurrentSchedule	: An array used to store current schedule time stamp
TempSchedule	: An array used to store temporary schedule time stamp
MULT	: Number of multipliers at all voltage levels
ALU	: Number of ALUs at all voltage levels
ASAPControlSteps	: Total number of control steps of ASAP schedule
ALAPControlSteps	: Total number of control steps of ALAP schedule
NoOfControlSteps	: Number of control steps of the schedule
ResAssignmentLUT	: Resource assignment look-up table
FreqSelectionLUT	: Frequency selection look-up table
max, start, end, cycle	: Temporary variables
CurrentEDP, TempEDP, ExtraEDP	: Temporary variables
CurrentVertex, CurrentCycle	: Temporary variables
VoltageArray	: An array used to store operating voltage for each vertex
FrequencyArray	: An array used to store operating frequency for each cycle
ZeroMobilityVertexList	: An array storing the vertices with zero mobility
NonZeroMobilityVertexList	: An array storing the vertices with non-zero mobility

---

Algorithm 20 Pseudocode for the RC-DFC Scheduler

1: ASAPScheduler(UDFG); ALAPScheduler(UDFG);
2: MULT = $\sum$ Multipliers of different voltage levels; ALU = $\sum$ ALUs of different voltage levels;
3: ModifySchedule(ASAPSchedule, MULT, ALU); ModifySchedule(ALAPSchedule, MULT, ALU);
4: NoOfControlSteps = Maximum(ASAPControlSteps, ALAPControlSteps);
5: ConstructResAssignmentLUT(NoOfControlSteps, FUs);
6: FindResTypeForEachVertex(UDFG); ConstructFreqSelectionLUT(Operating Frequency);
7: FindMobileVertexList(ASAPSchedule, ALAPSchedule); CurrentSchedule = ASAPSchedule;
8: **while** (NonZeroMobilityVertexList is NOT empty) **do**
9:       max = $-\infty$; AllocateVoltAndFreq(CurrentSchedule, LUTs);
10:      CurrentEDP = CalculateEDP (VoltageArray,FrequencyArray);
11:      **for all** ($v_i \in$ NonZeroMobilityVertexList) **do**
12:           start = CurrentSchedule[v_i]; end = ALAPSchedule[v_i];
13:           **for all** (cycle = start $\rightarrow$ end in steps of 1) **do**
14:                TempSchedule = AdjustSchedule(CurrentSchedule, v_i, cycle);
15:                AllocateVoltAndFreq(TempSchedule, LUTs);
16:                TempEDP = CalculateEDP(VoltageArray,FrequencyArray);
17:                ExtraEDP = CurrentEDP $-$ TempEDP;
18:                **if** (ExtraEDP $>$ max) **then**
19:                     **return** max = ExtraEDP; CurrentVertex = v_i; CurrentCycle = cycle;
20:                **end if**
21:           **end for**
22:      **end for**
23:      CurrentSchedule = AdjustSchedule(CurrentSchedule, CurrentVertex, Currentcycle);
24:      Update the "resource assignment LUT";
25:      ZeroMobilityVertexList = ZeroMobilityVertexList $\cup$ CurrentVertex;
26:      NonZeroMobilityVertexList = NonZeroMobilityVertexList $-$ CurrentVertex;
27: **end while**
28: AllocateVoltAndFreq(CurrentSchedule, LUTs);
29: EnergyAndDelayDetails(VoltageArray, FrequencyArray);
30: FindCycleFreqIndex(FrequencyArray);

The inputs to the algorithm are the UDFG and resource constraints, which include the number and type of FUs, the operating voltage levels and the operating frequencies. The procedures in line 1, ASAPScheduler and ALAPScheduler, find the unconstrained ASAP and ALAP schedules for the UDFG, respectively. In line 2, the total number of multiplier and ALU FUs with different voltage levels is determined. For example, if the resource constraint is two ALUs at 2.4 V, one ALU at 3.3 V, one multiplier at 2.4 V and three multipliers at 5.0 V, then the number of ALUs is three and the number of multipliers is four. Using the number of multipliers and ALUs found above as initial resource constraint (with relaxed voltage constraint), the ModifySchedule procedure (line 3) modifies the ASAP and ALAP schedules so that the resource constraints are not violated. In this process, the mobility of the vertices is restricted to a great extent, and the search space for the following steps is reduced. Next, the total number of cycles for the schedule is assumed as the maximum of the number of cycles for the ASAP and ALAP schedules (line 4). The resource assignment LUT is constructed (similar to Table 6.11) in line 5 whose size depends on (NoOfControlSteps × NoOfResourceTypes). The procedure FindResTypeForEachVertex (line 6) identifies the FUs required at each vertex of the DFG. In line 6, the frequency selection LUT similar to Table 6.10 is constructed. The FindMobileVertexList procedure (line 7) takes as input the modified ASAP and the modified ALAP schedules (line 4) to determine two lists: the list ZeroMobilityVertexList, containing the vertices with zero mobility (same ASAP and ALAP time stamps), and another, NonZeroMobilityVertexList, containing the non-zero mobility vertices (different ASAP and ALAP time stamps).

In line 7, the CurrentSchedule is initialized as the modified ASAP schedule (obtained in line 3). The procedure AllocateVoltAndFreq (lines 9 and 24) allocates the voltage levels and frequency levels to the FUs using the LUTs and the current schedule. This procedure returns two lists: one containing the assigned voltage of each vertex (VoltageArray) and the other (FrequencyArray) containing the selected frequency. FrequencyArray is in turn used to derive the cfi_c for the control steps. The procedure CalculateEDP (line 10) calculates the EDP of the whole DFG using a schedule with voltage assignment stored in VoltageArray and frequency contained in FrequencyArray. The procedure AdjustSchedule (lines 14 and 20) schedules each vertex to a specific cycle while adjusting its predecessor and successor time stamps. The *for loop* (lines 11–19) considers all the vertices from the NonZeroMobilityVertexList and finds a suitable vertex and its time stamp such that the EDP of the whole DFG with the current schedule is minimum. In line 21, the resource assignment LUT is updated. The *while* loop (lines 8–23) terminates when all the vertices with non-zero mobility have been assigned the proper time stamp. The procedure FindEnergyAndDelay (line 25) determines the energy consumption and execution time for the schedule. Line 25, FindCycleFreqIndex, finds the cycle frequency indices of all cycles, which will be useful in dynamic frequency generation. Figure 6.7 is obtained after executing the RC-DFC algorithm for the resource constraint (one MULT at 2.4 V, one MULT 3.3 V, one ALU at 3.3 V and one ALU at 5.0 V).

6.4.3.3 Time Complexity

Let there be $|V|$ number of vertices and $|E|$ number of edges in the DFG, out of which $|V_m|$ number of vertices have mobility and the maximum mobility of any mobile vertex is t_m. Let L_V denote the number of voltage levels and L_f denote the number of frequency levels. Suppose the number of control steps calculated from the ASAP scheduling is c. Assuming that L_V and L_f are upper bounded by $|V|$, the running time of the code segment in lines 1–7 is $\Theta\left(|V|+|E|\right)+\Theta\left(cL_VR_T\right)$. The time complexity of the instructions in lines 11–19 is $\Theta\left(c\,|V|\,L_VR_T\,|V_m|\,t_m\right)$. The code segment in lines 9–19 has running time $\Theta\left(c\,|V|\,L_VR_T\,|V_m|\,t_m\right)+\Theta\left(|V|\right)$ $+\Theta\left(c\,|V|\,L_VR_T\right)=\Theta\left(c\,|V|\,L_VR_T\,|V_m|\,t_m\right)$. The running time of the code segment in lines 8–19 is $\Theta\left(c\,|V|\,L_VR_T|V_m|^2t_m\right)$. The time complexity of lines 20–25 is $\Theta\left(|V|\right)$ $+\Theta\left(c\,|V|\,L_VR_T\right)+\Theta\left(c\right)=\Theta\left(c\,|V|\,L_VR_T\right)$. Therefore, the running time of the overall algorithm is $\Theta\left(|V|+|E|\right)+\Theta\left(cL_VR_T\right)+\Theta\left(c\,|V|\,L_VR_T|V_m|^2t_m\right)+\Theta\left(c\,|V|\,L_VR_T\right)=$

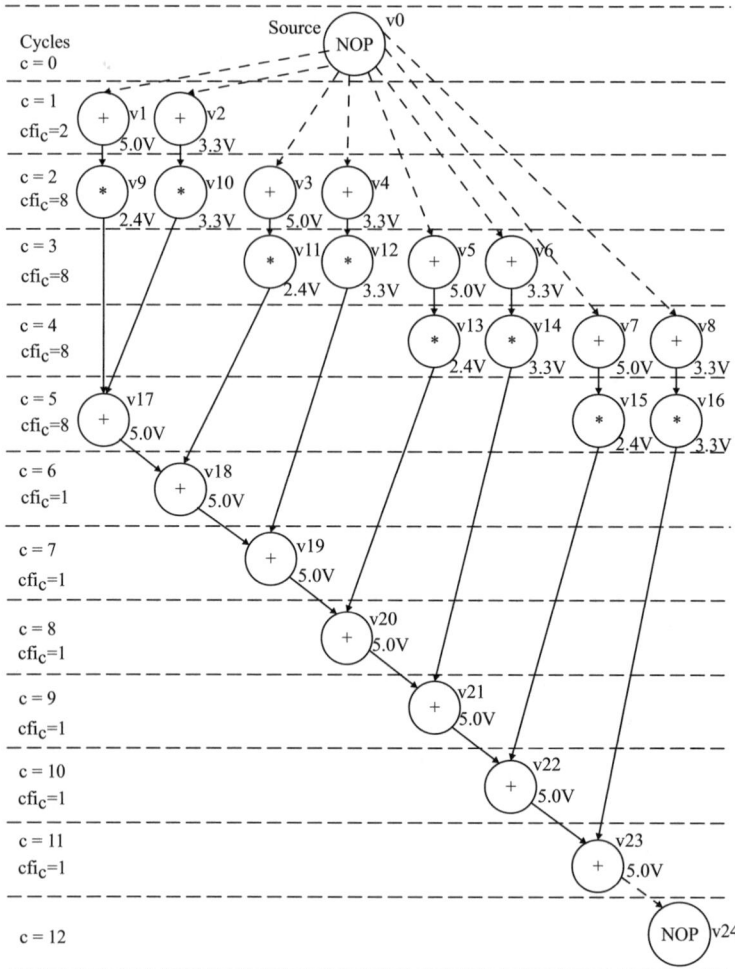

Fig. 6.7 Final schedule of FIR filter DFG (using RC-DFC)

$\Theta\left(|V|+|E|\right)+\Theta\left(c|V|L_V R_T|V_m|^2 t_m\right)$. Assuming that $|E|$ is upper bounded by $|V|^2$ and $|V_m|$ is upper bounded by $|V|$, the above expression can be simplified to $O\left(c|V|^3 L_V R_T t_m\right)$.

6.4.4 Experimental Results

Both RC-DFC and TC-DFC schedulers were implemented in C and tested with selected benchmark circuits. The benchmarks used are ARF filter, band-pass filter

(BPF), elliptic-wave filter (EWF), DCT, FIR filter and HAL differential equation solver. The FUs used are ALUs and multipliers. The energy values are computed using the data path components given in [210, 219, 221].

For the RC-DFC scheduler, the experimental results are as follows. The algorithm was tested using the different sets of resource constraints. Figure 6.8(a) shows the percentage savings averaged over all resource constraints. From the chart, it is evident that the scheduling yields approximately equal savings for all kinds of benchmark circuits. The EDP reduction averaged over all resource constraints are shown in Fig. 6.8(b). The energy estimation includes the energy consumption of the overhead units. It is assumed that each resource has equal switching activity. The results are reported for two supply voltages and for switching = 0.5. It is observed that the energy consumption is increased for higher switching and decreased for lower switching activity; but under the assumption that switching is the same for each resource, the percentage energy savings is not affected. The experiments with three supply voltage levels are also conducted, and it is found that the percentage energy savings could only increase by 5%. From the above discussion, it is concluded that the scheduling algorithm yields appreciable energy savings and EDP reduction.

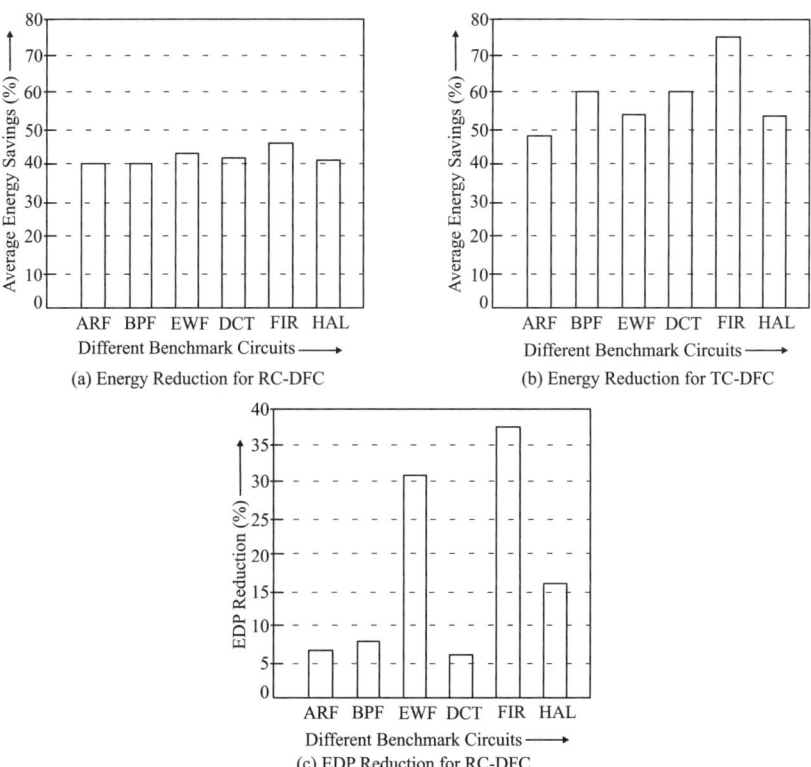

Fig. 6.8 Average energy and EDP reduction for benchmarks

The TC-DFC scheduler was tested for three different time constraints (T_{con}): 1.5, 1.75 and 2.0 times the critical path delay (T_{cp}). The voltage constraint is relaxed unlike the RC-DFC scheduler. Figure 6.8 shows the energy savings for different benchmarks averaged over all time constraints. It is observed that circuits requiring an equal number of ALU-related operations, such as addition, subtraction or comparison, and multiplier operations save more energy. The energy savings increased as the time constraints relaxed from $1.5T_{cp}$ to $2.0T_{cp}$.

The energy savings from the proposed RC-DFC scheduling algorithm are listed along with other resource-constrained multiple voltage scheduling algorithms in Table 6.14. The minimum and maximum range of energy savings is shown in the table. As is clear from the table, RC-DFC gives better energy savings for lower time penalties. The energy savings obtained using different existing multiple voltage-based time-constraint scheduling algorithms are shown in Table 6.15. In all cases, the time constraints are $1.5 \times T_{cp}$ to $2.0 \times T_{cp}$.

6.4.5 Conclusions

Energy reduction is achieved by voltage reduction and performance is maintained by using DFC along with multiple voltages. Resource-constrained and time-constrained data path scheduling algorithms are developed based on DFC. The use of DFC could generate enough slack to apply reduced voltages, which in turn saves energy. It is

Table 6.14 Savings for various resource-constrained scheduling

Benchmark circuits	Percentage energy savings and time penalties (T) in cycles							
	RC-DFC		Shiue [320]		Sarrafzadeh [311]		Johnson [144]	
	ΔE	N_D	ΔE	T	ΔE	T	ΔE	T
ARF	24–58	9–10	11–14	11–16	16–20	17–24	16–59	10–18
BPF	27–56	8–10	–	–	–	–	–	–
EWF	38–61	10–13	14–14	17–20	13–32	21–25	11–50	12–24
DCT	41–63	13–18	–	–	–	–	–	–
FIR	20–67	6–10	–	–	16–29	10–15	28–73	5–10
HAL	29–62	2–3	19–28	5–6	–	–	–	–

Table 6.15 Savings for various time-constrained scheduling cases

Benchmark Circuits	Energy Savings (%)			
	TC-DFC	Chang [70]	Shiue [320]	Manzak [199]
ARF	41–58	40–63	38–76	25–61
BPF	45–70	–	–	–
EWF	36–73	44–69	13–76	10–55
FDCT	52–75	43–69	–	–
FIR	74–74	–	–	–
HAL	43–67	41–61	22–77	19–62

observed that when using two supply voltage levels, an average energy reduction of 41% and an average reduction of 46% for three supply voltage levels is obtained for the benchmarks using the RC-DFC algorithm. Similarly, for TC-DFC, an average energy reduction of 46% (for $1.5 \times T_{\text{cp}}$) and 68% (for $2.0 \times T_{\text{cp}}$) is obtained. The processor configurations for various benchmark circuits that would result in minimum EDP can be determined through experiments. The integration of such a scheduler into a low-power data path synthesis tool will significantly benefit low-power processor design especially for data-intensive applications.

6.5 Data Path Scheduling for Energy or Average Power Reduction Using Voltage Reduction

It is known that voltage reduction is an effective method of power reduction since the power or energy consumption is quadratically dependent on the supply voltage. In this section, existing literature using multiple supply voltages during data path scheduling for minimization of energy or average power is discussed.

6.5.1 Time- or Resource-Constrained Scheduling Algorithms

Johnson and Roy [143, 145] present a method called *minimum energy schedule with voltage selection* (MESVS) based on ILP to optimize the schedule, supply voltage levels and allocation of resources. The MESVS algorithm takes a directed acyclic DFG, the allowable set of supply voltages, a limit on the number of supply voltages that can be selected, a minimum difference between the voltages that can be selected, average switching activity values for each data path operation, nominal propagation delay and average energy dissipation values for each data path resource as inputs. The objective function for MESVS is an estimate of data path energy dissipation expressed as a function of supply voltages. The outputs of the MESVS algorithm are the following: (i) a data path schedule (represented by scheduled DFG), (ii) an energy estimate, (iii) selection of optimal set of supply voltages, (iv) assignment of supply voltage to each operation and (v) allocation of resources to each supply voltage. Since different resources need to operate at different voltages, level conversion is needed. The authors claim that energy savings in the range of 46–58% are obtained compared to 5 V operation. The other observation was that the use of two supply voltages can reduce power dissipation substantially, whereas three supply voltages resulted in less than 5% reduction compared to two supply voltages.

Johnson and Roy [144] also present an algorithm called *multiple operating voltage energy reduction* (MOVER) to minimize data path energy dissipation. Energy savings ranging from 0 to 50% are obtained with an area penalty in the range 0 to 170%. MOVER generates one, two and three supply voltage designs for consideration by circuit designers. The user has control over latency constraints, resource constraints, the number of control steps, clock period and the number of power supplies. MOVER iteratively searches for the range of minimum voltage levels. MOVER also uses ILP to evaluate the feasibility of candidate supply voltage selections, to partition operations among different power supplies and to produce a minimum area schedule under latency constraints once voltages have been selected. MOVER has the following phases:

- Determining maximum and minimum bounds on the time frame in which each operation must execute
- Searching for minimum voltage
- Partitioning data path operations into two supply voltages that are either high or low
- Partitioning the lower voltage group for the three supply voltage schedule.

The MOVER algorithm [144] is similar to the MESVS algorithm [145] in the following ways:

- Both use ILP formulation.
- Behavior with respect to latency, resource and supply voltage constraints.
- Both use *differential cascade voltage switches* (DCVS).

The difference between MOVER and MESVS is that MESVS can only select a discrete set of voltages, whereas MOVER can select a continuous range of voltages. The ILP formulation handles timing and resource constraints and accounts for the cost if level shifters are used. However, MOVER and MESVS have the following drawbacks:

- They do not address conditional branches.
- They do not consider functional pipelining.
- The energy model used is data intensive, which ignores the effect of input activities on the energy dissipation of a module.
- They have exponential worst-case complexity and cannot handle large benchmarks.

Chang and Pedram [70, 69] present a dynamic programming technique for multiple supply voltage scheduling. The proposed technique handles both functionally pipelined and non-pipelined data paths and multicycling operations. The scheduling algorithm assigns a supply voltage level from a fixed set of voltage levels such that the energy consumption is minimum for given constraints. In this algorithm, level-shifters are used for both step-up and step-down of signals. It may be noted that in most of the algorithms, level shifters are used for step-up of signals only. An average saving of 40.19% is obtained using three supply voltage levels as compared to a single supply voltage level. The algorithm has pseudo-polynomial complexity and produces optimal results for trees but is suboptimal for general directed acyclic graphs. The scheduling algorithm can handle very large DFGs with results within 1% error.

Kumar and Bayoumi [175, 173, 174] proposed scheduling schemes using MVMC. The algorithms essentially have two phases, initial scheduling and rescheduling.

During initial scheduling, parallelism is exploited. Rescheduling uses an iterative approach, which is based on stochastic evolution. Level converters are used when an FU operating at lower voltage drives a FU operating at higher voltage. The time complexity of the scheduling algorithm is O $\left(n^2\right)$. The authors report power savings of up to 80% for three supply voltage levels of 5.0, 3.3 and 2.4 V. The power overhead due to the level converters is in the range of 0 to 4% and the area overhead is in the range of 0 to 6%.

6.5.2 Time- and Resource-Constrained Scheduling Algorithms

In [186], an ILP formulation and a heuristic for variable voltage scheduling are presented by Lin, Hwang and Wu. The authors have considered three different solutions to the problem, such as time-constrained, resource-constrained and time- and resource-constrained. The scheduling schemes consider variable supply voltage and multicycling. The heuristic method produces results comparable with those of the ILP method in a fraction of runtime. The time complexity of the heuristic algorithm is O $\left(n^3 \log n\right)$. The proposed heuristic is a modification over list-based algorithms with a priority function that considers three factors: the power gain of an operation, the mobility of an operation and the computation density. The authors show that using different cost and delay combinations, power consumption in a single design can differ by as much as a factor of 6 when using mixed $(3.3 \text{ and } 5.0 \text{ V})$ supply voltages.

Sarrafzadeh and Raje [311] proposed two scheduling algorithms: one is a dynamic programming algorithm and the other is a heuristic algorithm based on a geometric algorithm. The algorithms assume both time and resource constraints as inputs. The resource constraints are the number and types of each FU and their operating supply voltage. The algorithms assume only two supply voltages, such as 3.3 and 5.0 V. The aim of the algorithms is to maximize the usage of the FUs at the lower supply voltages while satisfying the time constraints. Let n be the number of nodes, k the time constraint, R a given resource constraint and β the latency of an FU that operates at a supply voltage of V_β. The runtime of the dynamic programming scheduling algorithm is O $\left(n^2 k \beta |R|^2\right)$. If C is the number of control steps, then the time complexity of the geometric algorithm is O $(nC \log nC)$ and can handle more than two supply voltages. The authors reported power reductions in the range of 13.28–31.54% for various high-level synthesis benchmarks under various resource and time constraints.

Elgamel and Bayoumi [97] use genetic algorithms to solve multiple supply voltage scheduling problems with multicycling operations. The proposed scheme assumes an unscheduled data or control flow graph, a data path component library and area and time constraints as inputs and minimizes the average power. The algorithms simultaneously solve scheduling, allocation and binding. Power reduction as high as 84% is reported. The results do not consider the power overhead due to the level converters.

Shiue and Chakrabarti [320, 318] discuss a resource-constrained and a latency-constrained list-based scheduling algorithm using multiple supply voltages. The scheduling scheme considers the effect of switching activity. The algorithms use heuristics to reduce power consumption in the level converters. The list-based algorithms assign control steps to nodes based on their priorities. The priority of a node is a function of various parameters, such as depth, mobility, switched capacitance, interconnection complexity and need for a level shifter. Level shifters are used between a low-voltage resource and a high-voltage resource for stepping up the signal. The proposed algorithms are of polynomial time complexity. The proposed schemes achieve significant power reduction when the operation voltages are 5.0 and 3.3 V or 5, 3.3 and 2.4 V.

The Lagrangian multiplier method has been used by Manzak and Chakrabarti [199] to develop resource- and time-constrained scheduling algorithms. (Others have used Lagrangian methods in circuit sizing and dual-vt selection [148].) The scheduling algorithms use the Lagrangian multiplier method iteratively, and are based on efficient distribution of slack among the nodes in the DFG. If n denotes the number of nodes and L denotes the latency, the time complexity of the two versions of the proposed algorithms are O (n^2) and O $(n^2 \log L)$. The O $(n^2 \log L)$ algorithm results in better savings in energy compared to the O (n^2) algorithm. Average power or energy reduction of 39% has been obtained when the latency constraint is 1.5 times the critical delay and is improved to 58.5% when the latency constraints are relaxed to 2 times the critical path delay. The time constraint, resource constraint consisting of the number of resource of each type operating at specific voltage and delay and energy values are given as inputs to the algorithm. The resources are allowed to operate at one of several supply voltages from 5.0, 3.3, 2.4 and 1.5 V). Level shifters are used whenever step-up of signal is necessary.

Manzak and Chakrabarti [198] proposed list-based latency and resource-constrained scheduling algorithms. The scheduling uses a priority function based on the number of available resources, the difference between the actual number of cycles left and the estimated number of cycles required to schedule the remaining nodes. The algorithms consider the switching activity of nodes. The resources are allowed to operate at one of several supply voltages from 5.0, 3.3, 2.4 and 1.5 V). The average power or energy reduction is 59.1% when the latency constraint is 1.5 times the critical delay, and the average power or energy reduction is 66.8% when the latency constraint is 2.0 times the critical delay. The time complexity of the algorithm is O $(r^2 L^2)$, where r is the number of resources and L is the latency.

A comparative view of the above-discussed algorithms which use voltage lowering for average power or energy reduction is given in Table 6.16.

6.6 Switching Activity Reduction During High-Level Synthesis

In this section, works on data path scheduling, binding and allocation which use capacitance reduction to reduce average power or energy are discussed.

Table 6.16 Data path scheduling schemes using multiple supply voltages

Proposed scheme	Optimization method used	Constraints assumed	Operating voltage levels	Time complexity
Johnson and Roy [143, 145]	ILP	Time	$(5.0\,\mathrm{V} \rightarrow 2.0\,\mathrm{V})$	Exponential
Johnson and Roy [144]	ILP	Time	$(5.0\,\mathrm{V}, 3.3\,\mathrm{V}, 2.4\,\mathrm{V})$	Exponential
Chang and Pedram [70, 69]	Dynamic programming	Time	$(5.0\,\mathrm{V}, 3.3\,\mathrm{V}, 2.4\,\mathrm{V})$	Pseudo-polynomial
Kumar and Bayoumi [175, 173, 174]	Stochastic evolution	Resource	$(5.0\,\mathrm{V}, 3.3\,\mathrm{V}, 2.4\,\mathrm{V})$	$\mathrm{O}\left(n^2\right)$
Lin, Hwang and Wu [186]	ILP and heuristic	Time and resource	$(5.0\,\mathrm{V}, 3.3\,\mathrm{V})$	Exponential $\mathrm{O}\left(n^3 \log n\right)$
Sarrafzadeh and Raje [311]	Dynamic prog geometric	Time and resource	$(5.0\,\mathrm{V}, 3.3\,\mathrm{V})$	$\mathrm{O}\left(n^2 k \beta \lvert R \rvert^2\right)$ $\mathrm{O}\left(nC \log nC\right)$
Elgamel and Bayoumi [97]	Genetic algorithms	Time and area	$(5.0\,\mathrm{V}, 3.3\,\mathrm{V}, 2.4\,\mathrm{V})$	NA
Shiue and Chakrabarti [320, 318]	List-based	Time and resource	$(5.0\,\mathrm{V}, 3.3\,\mathrm{V})$ or $(5.0\,\mathrm{V}, 3.3\,\mathrm{V}, 2.4\,\mathrm{V})$	Polynomial
Manzak and Chakrabarti [199]	Lagrangian multiplier	Time and resource	$(5.0\,\mathrm{V}, 3.3\,\mathrm{V}, 2.4\,\mathrm{V}, 1.5\,\mathrm{V})$	$\mathrm{O}\left(n^2\right)$ and $\mathrm{O}\left(n^2 \log L\right)$
Manzak and Chakrabarti [198]	List-based	Time and resource	$(5.0\,\mathrm{V}, 3.3\,\mathrm{V}, 2.4\,\mathrm{V}, 1.5\,\mathrm{V})$	$\mathrm{O}\left(r^2 L^2\right)$

6.6.1 Scheduling and/or Allocation for Switching Activity Reduction

Lee, Lee, Park and Hwang [183] propose a scheduling algorithm that reduces the switching activity of the FUs under area or time constraints and thus reduces the power consumption. The switching activity is minimized by scheduling operations such that the Hamming distance between the variables appearing in the input and output ports is minimized. The FU allocation is performed by partitioning the operations in the given behavioral description, and the switching activity is kept at a minimum. After allocation is performed, the scheduling algorithm attempts to schedule the operations using the minimum number of functional modules. The algorithm is of polynomial time complexity. The results indicate that switching reduction of 16.5% in average can be obtained.

Gupta and Katkoori [114] propose a scheduling algorithm based on the original force-directed scheduling algorithm proposed in [272]. For a given DFG and input data environment, the DFG is profiled with the representative data streams. The probability of selecting a combination among the operations that would share a resource is evaluated. Assuming that the force equation is $F = kx$, the switching capacitance inside a module is modeled as the spring constant k and the probability of selecting such a combination is modeled as the displacement x. For t number of possible time steps and n number of operations, the time complexity of the proposed algorithm is O $\left(n^4 t\right)$. It may be noted that the original force-directed scheduling algorithm has running time of O $\left(n^2\right)$. The authors have reported a power reduction of 16.4% over the conventional force-directed algorithm.

Shut-down techniques are used by Monteiro, Devadas, Ashar and Mauskar [236] to eliminate switching activity and hence power dissipation. The conditions under which the output of a module is not used for a particular cycle is identified, and the input latches for that module are disabled when the conditions are met. The proposed scheduling algorithm maximizes the shut-down period of FUs. The scheduling algorithm is time and resource constrained. Various techniques, such as multiplexer reordering and pipelining, are proposed to improve power management under these stringent constraints. Power reduction as high as 41.67% has been reported.

Kumar, Katkoori, Rader and Vemuri [176, 149] present a profile-driven approach to high-level synthesis called *profile-driven synthesis system* (PDSS). The inputs to the PDSS are a subset of VHDL and constraints in terms of clock period and area. PDSS generates a constraint-satisfying design with the least amount of estimated switching activity. In this system, the input specification is profiled to collect data for various operations and carriers using a user-specified input set of vectors. The switching activity for each module set is estimated by using this profiled data and the raw switching activity data of all modules in the library. The module set with minimum estimate of power consumption is chosen for further synthesis. The goal of profiling is to gather the following data :

• For each node (operation), the number of times the node is executed for given profiling stimuli is determined and input vectors used as profile stimuli. This number is called the event activity of the operation node.
• For each edge, the number of times the edge is traversed during execution is determined. This number is called the transaction activity of the edge.
• For each edge, the number of times the value on the edge has changed is determined. This number is called the event activity of the edge.

The authors claim that the results obtained are within an accuracy of 10% of the actual switching activity measured at the switch-level implementation of the design.

Raghunathan and Jha [288] present a comprehensive low-power data path synthesis system that performs the various high-level synthesis tasks to reduce power consumption in the synthesized data path. The authors call the system SCALP. The system considers both supply voltage and switching capacitance to reduce power consumption. The authors claim that SCALP estimates switching capacitance accurately, handles diverse module libraries and utilizes complex scheduling constructs such as multicycling, chaining and structural pipelining. The input to SCALP is a control DFG (CDFG), input sampling period and a library of components to be used for data path implementation. SCALP minimizes power consumption both by voltage scaling and switching capacitance reduction. First, the set of candidate supply voltages to a small set of supply voltages is pruned. For each supply voltage in the pruned set, a data path that has minimal capacitance is synthesized. The best solution among these data paths in terms of power consumption is then chosen.

Raghunathan and Jha [287] are also the first researchers to propose the allocation method for low power. The method is based on iterative improvement of some initial solution. The authors assume random input in a structurally pipelined design. The method can also handle non-random input sequences. The method is implemented in the framework of the *Genesis* behavioral synthesis system [44]. In this system, register and module allocations are performed simultaneously, while minimizing the amount of interconnect needed. A lifetime analysis is performed for the scheduled CDFG. Two variables are said to be compatible and can share hardware resources if they are not alive at the same time. Similarly, two operations are compatible if they are not performed at the same time. Allocation is based on a weighted graph called compatibility graph (CG). Initially, each variable and operation corresponds to a node in the CG, with undirected edges connecting compatible pairs. Weights are assigned to edges in the CG to indicate the preference on the two variables or operations for sharing the same resource. A single step of allocation selects the edge in the CG with the highest composite weight, merges the two nodes it joins and maps the corresponding variable (or operation) to the same module (register). If two or more edges have the same composite weight, the tie is broken based on the corresponding transition activity weights (or, in some cases, arbitrarily). Power reduction is achieved with the help of two factors, capacitance and transition activity. Capacitance is reduced by minimizing the number of functional modules, registers and multiplexers. The allocation scheme selects a sequence of operations (variables) for a module or register such that the transition activity is reduced.

Chiou, Muhammand and Roy [80] propose a scheduling and allocation method that reduces power consumption of data-intensive applications by minimizing the switching activity. The main idea of this synthesis technique is to reduce the signal strength difference among the inputs of shared resources. The signal strength is derived from word-level statistics. The authors have proposed a formula that relates switching power with resource sharing as follows:

$$\text{Switching increment} = \frac{\text{Difference in switching activity with and without sharing}}{\text{Switching activity without sharing}}.$$
$$(6.25)$$

It is observed that sharing resources between two operations with high signal similarity will lower switching activity and hence reduce switching power. This observation serves as the major principle behind the proposed scheduling and allocation techniques. The proposed scheduling algorithm is heuristic based and uses a greedy approach in making module selections. Average power reduction up to 49% is obtained using the proposed techniques compared to conventional ones.

A comprehensive high-level synthesis system is proposed by Khouri, Lakshminarayana and Jha [156] to synthesize both control-flow-intensive and data-intensive circuits. The system handles conventional synthesis tasks, such as scheduling, module selection and resource sharing. Moreover, power-conscious structuring of multiplexer networks, which are predominant in control-flow-intensive circuits, is the key additional feature in the system. Experimental results demonstrate power reduction of 62% for control-flow-intensive benchmarks as compared to V_{DD}-scaled area-optimized designs. The power reduction for the data-dominated benchmarks is 58% as compared to V_{dd}-scaled (delay-optimized) designs. The power reductions come with an area penalty of approximately 40%.

Henning and Chakrabarti [127, 126] propose an intuitive switching activity model to capture data characteristics in terms of statistical parameters. Then, heuristics are proposed for scheduling and allocation exploration. The novelty of the model is a relation between switching activity of data path interconnect to the fixed-point, two's complement data. The model is based on four practical parameters, which are basically the bits of the two values involved in the transition, such as sign bits, the number of intersecting sign bits, number of truncation bits in the two values and all other bits of a value that are not sign or truncation bits. Since the model is dependent on only four parameters, scheduling and allocation are efficient. The heuristic is applied to synthesize a speech codec design. It is reported that average power reduction is about 15% during encoding.

6.6.2 Scheduling and/or Binding for Switching Activity Reduction

An ILP-based resource-binding scheme is proposed by Shiue and Chakrabarti [319] that minimizes the amount of switching at the inputs of FUs. The idea of resource binding is to find n disjoint paths from a multistage graph with m stages, where m is the number of cycles in the schedule and n is the number of nodes per stage. The first step of binding is to find a multistage graph called the binding graph. The total number of nodes of such a graph is $n \times m$, plus two nodes for source and sink. If two nodes are located in two different stages and can share a resource, then the two are connected with an edge. Each edge is labeled with a cost corresponding to the switching activity. The LP objective is to find n disjoint paths such that the total cost of these paths is minimum. Power savings in the range of 8.2–34.4% are obtained using the proposed binding scheme for various resource constraints as compared to random binding scheme.

Musoll and Cortadella [246] present algorithms for scheduling and resource binding to reduce power consumption during behavioral synthesis. The algorithms reduce power consumptions by reducing the transitions of their input operands. The power consumption of an FU is divided into *useful* and *useless* power. Useful power is consumed when an operation is executed and useless power is the consumption due to an input transition while the FU is idle. The algorithms proposed reduce both useful and useless power consumption. The scheduling algorithm is list based in which the operation priority is set in such a way that operations sharing the same operand are scheduled in control steps as close as possible. For n number of operations and m number of FUs, the running time of the proposed low-power list scheduling (LPLS) is $O(n^2 m)$. The algorithm for resource binding is based on clique partition that reduces power consumption by taking the average Hamming distance (AHD) among the variables. For two operands p and q, if $H(p,q)$ is the

Hamming distance and x_i is the value of operand x in cycle i, the AHD is defined as follows:

$$\text{AHD}(x) = \lim_{n \to \infty} \left(\frac{\sum_{i=1}^{n} H(x_i, x_{i-1})}{n} \right). \tag{6.26}$$

The AHD is used as a measure of energy in nJ/operation. Power reductions in the range of 5–8% have been reported.

Shin and Lin [315] propose an efficient resource allocation algorithm that minimizes switching activity to reduce the dynamic power consumption of the DSP data path. Let X be a certain binary input sequence. Suppose that L is the length of X and S is the number of "1"s in the input sequence X. The average switching activity of X is calculated as follows:

$$\alpha_{\text{switching}} = 2 \left(\frac{S}{L} \right) \left(\frac{L-S}{L} \right). \tag{6.27}$$

For example, for a input sequence 0110110000, $L = 10$ and $S = 4$. The input to the allocation algorithm is a scheduled DFG. The algorithm executes all control steps and compares FU with low-power-consuming registers and interconnects of DSP circuits. The algorithm is of polynomial time complexity. Power reduction up to 8.5% is reported using the algorithm.

Cherabuddi and Bayoumi [78] propose partitioning and binding algorithms that minimize the switching activity of FUs and global buses for single-chip applications. Cherabuddi, Bayoumi and Krishnamurthy [79] extend the same work for multi-chip applications. The authors have used a stochastic evolution-based technique for partitioning. Power reduction up to 60% has been reported. The switching activity is computed by iteratively changing the input data pattern, and a switching activity matrix is constructed. The partition algorithms partition the DFG such that each one of them can be implemented in different chips of multi-chip modules. The stochastic evolution approach is used in the partition algorithm for faster convergence. Scheduling and binding steps are performed for each move on the partitioning. An incompatible graph is constructed from the original graph for resource allocation. To find optimal solutions for low-power binding, a multistage graph is formulated and a dynamic programming approach is used. The total switching activity of a schedule is calculated as the summation of the switching activities of the chips on the module and the switching activities on the interchip buses.

Murugavel and Ranganathan [243] describe a game theory-based algorithm for average power minimization during behavioral synthesis using low-power binding. The techniques of FU sharing, path balancing and register assignment are incorporated within the binding algorithm for power reduction. For the binding algorithm, each FU in the data path is modeled as a player bidding for executing an operation with the estimated power consumption as the bid. The operations are assigned to the FUs such that the number of inputs to the FUs that change are minimized, thus reducing switching activity. The proposed algorithm yields power reduction improvement of 13.9% without any increase in area or delay overhead.

An overview summary of the discussed methods is given in Table 6.17, where the indicated percentage power reduction is the average data.

Table 6.17 High-level synthesis schemes using switching activity reduction

Proposed work	Synthesis tasks performed	Methods used	Time complexity	Percentage power reduction
Lee, Lee, Park and Hwang [183]	Scheduling	Heuristic	Polynomial	16.5
Gupta and Katkoori [114]	Scheduling	Force-directed heuristic	$O\left(n^4 t\right)$	16.4
Monteiro, Devadas, Ashar and Mauskar [236]	Scheduling	HYPER [284]	NA	22.43
Kumar, Katkoori, Rader and Vemuri [176, 149]	Scheduling, register optimization, etc.	Simulation of DFG	NA	NA
Raghunathan and Jha [288]	Transformation, scheduling and allocation	Iterative improvement	Polynomial	4.6
Raghunathan and Jha [288]	Allocation	Simulation	NA	14.6
Chiou, Muhammand and Roy [80]	Scheduling and allocation	Heuristic based	Polynomial	30.13
Khouri, Lakshmi-narayana and Jha [156]	Scheduling and resource sharing	Heuristic	Polynomial	22
Henning and Chakrabarti [127, 126]	Scheduling and allocation	Intuitive heuristic	Polynomial	15
Shiue and Chakrabarti [319]	Resource binding	ILP	Exponential	24.08
Musoll and Cortadella [246]	Scheduling and resource binding	List-based algorithm	$O\left(n^2 m\right)$	6.67
Shin and Lin [315]	Resource allocation	Heuristic	Polynomial	7.84
Cherabuddi and Bayoumi [78]	Partitioning and binding	Stochastic evolution	Polynomial	23.89
Murugavel and Ranganathan [243]	Scheduling and binding	Game theory	Exponential	13.9

6.7 Summary and Conclusions

Energy or average power reduction is important as it has direct impact on battery life and the energy bill. Techniques for energy reduction are proposed that work during different high-level synthesis tasks, such as scheduling and binding. These techniques rely on voltage reduction, frequency adaptation or switching activity reduction to optimize the power consumption. Heuristics, ILP and game theory-based algorithms proposed in the literature are discussed. These algorithms encapsulate time and physical resource-constrained approaches for design space exploration.

Chapter 7
Peak Power Reduction

7.1 Introduction

The use of multiple supply voltages for energy and average power reduction is well researched, and several relevant works have appeared in the literature. However, in low-power design for battery-driven nanoscale circuits, the peak power, the peak power differential, the average power and the total energy are equally critical design metrics or constraints. In this chapter, data path scheduling algorithms for peak power and simultaneous peak and average power minimization are presented [210, 225, 227, 222]. Minimization schemes based on ILP are developed for the design of data paths that can function in three modes of operation: (1) single supply voltage and single frequency (SVSF), (2) multiple supply voltages and dynamic frequency clocking (MVDFC) and (3) multiple supply voltages and multicycling (MVMC). The effectiveness of the presented techniques is measured by estimating the peak power consumption, the average power consumption and the PDP of the data path circuits. Various experiments are presented on selected high-level synthesis benchmark circuits under different resource constraints.

7.2 Peak and Average Power Dissipation Modeling of a Data Path Circuit

It is assumed that the data path is represented in the form of a sequencing DFG. The target architecture for the synthesis is the same as that discussed in the previous chapter. The data path uses various resources or functional units operating at different supply voltages. The voltage-level converters are considered as resource overheads that are often needed when the voltage level needs to be stepped up in any control step. The dynamic clocking unit (DCU) that generates the different frequency levels is also taken into account as a resource that will operate during all the control steps. The relevant notation and terminology are given in Table 7.1. It is

S.P. Mohanty et al., *Low-Power High-Level Synthesis for Nanoscale CMOS Circuits*,
DOI: 10.1007/978-0-387-76474-0_7, © Springer Science+Business Media, LLC 2008

Table 7.1 Notations used in the description

N	: Total number of control steps in the DFG
c	: Any control step or clock cycle in DFG
R_c	: Number of resources active in step c
f_c	: Cycle frequency for control step c
$\alpha_{i,c}$	: Switching at resource i operating in step c
$C_{i,c}$	: Load capacitance of resource i operating in control step c
$V_{i,c}$	: Operating voltage of resource i operating in control step c
P_c	: Power consumption for the DFG for any control step c
P_{peak}	: Maximum power consumption for the DFG
P_{avg}	: Average power consumption for the DFG
T_{cp}	: Critical path delay of the DFG
PDP	: Power delay product of the DFG
SVSF	: Subscript used for single supply voltage and single-frequency operation
MVDFC	: Subscript used for multiple supply voltage and dynamic frequency operation
MVMC	: Subscript used for multiple supply voltage and multicycling operation
f_{clk}	: Operating clock frequency for single-frequency or multicycling operations
V	: Operating voltage for single voltage and single-frequency operations

assumed that for single frequency and single supply voltage mode of operation, $V_{i,c}$ and f_c are the same for any clock cycle (c) and resource (i). Similarly, for a multicycling operation, f_c is the same for any clock cycle (c).

The power consumption for any control step c in a DFG is presented as

$$P_c = \sum_{i=1}^{R_c} \alpha_{i,c} C_{i,c} V_{i,c}^2 f_c. \tag{7.1}$$

The peak power consumption of the DFG is the maximum power consumption over all the control steps, which for MVDFC mode of operation is

$$P_{\text{peak}_{\text{MVDFC}}} = \text{Max}\,(P_c)_{\forall c=1,2,\ldots,N}. \tag{7.2}$$

Rewriting Equation (7.2) using Equation (7.1), the following expression is obtained:

$$P_{\text{peak}_{\text{MVDFC}}} = \text{Max}\left(\sum_{i=1}^{R_c} \alpha_{i,c} C_{i,c} V_{i,c}^2 f_c\right)_{\forall c=1,2,\ldots,N}. \tag{7.3}$$

The average power consumption of the DFG is characterized as the mean of the cycle powers (P_c) for all control steps:

$$P_{\text{avg}_{\text{MVDFC}}} = \frac{1}{N}\sum_{i=1}^{N} P_c. \tag{7.4}$$

Again, using Equation (7.1), Equation (7.4) can be rewritten as

$$P_{\text{avg}_{\text{MVDFC}}} = \frac{1}{N}\sum_{i=1}^{N}\sum_{i=1}^{R_c} \alpha_{i,c} C_{i,c} V_{i,c}^2 f_c. \tag{7.5}$$

For peak power reduction, the objective function in Equation (7.3) needs to be optimized for various constraints. On the other hand, for the simultaneous reduction of both peak and average power, the objective function to be minimized is the sum of Equations (7.3) and (7.5) for different constraints.

The critical path delay of the DFG can be calculated as

$$T_{\text{cpMVDFC}} = \sum_{i=1}^{N} \frac{1}{f_c}. \tag{7.6}$$

It may be noted that f_c is the same for single-frequency and multicycling operations for all values of c but may be different for DFC operations.

The PDP of the DFG is defined as the product of the average power consumption and critical path delay as shown below:

$$\text{PDP} = P_{\text{avg}} \times T_{\text{cp}}. \tag{7.7}$$

Using Equations (7.4) and (7.6), the following expression for the PDP is obtained:

$$\text{PDP}_{\text{MVDFC}} = \frac{1}{N} \sum_{i=1}^{N} P_c \times \sum_{i=1}^{N} \frac{1}{f_c}. \tag{7.8}$$

Similarly, the following expression for the PDP is obtained using Equations (7.5) and (7.6):

$$\text{PDP}_{\text{MVDFC}} = \frac{1}{N} \sum_{i=1}^{N} \sum_{i=1}^{R_c} \alpha_{i,c} C_{i,c} V_{i,c}^2 f_c \times \sum_{i=1}^{N} \frac{1}{f_c}. \tag{7.9}$$

To study the impact of the scheduling algorithms on the performance of the data path, the PDP of the scheduled DFGs using the above expression will be estimated.

For MVMC operation, f_c is the same for all control steps and will be denoted as f_{clk}. Following the same steps as above, the power and delay of the DFG for the MVMC operation are modeled by the following equations:

$$P_{\text{peakMVMC}} = \text{Max} \left[\left(\sum_{i=1}^{R_c} \alpha_{i,c} C_{i,c} V_{i,c}^2 \right) \times f_{\text{clk}} \right]_{\forall c=1,2,\dots,N}, \tag{7.10}$$

$$P_{\text{avgMVMC}} = \frac{1}{N} \left(\sum_{i=1}^{N} \sum_{i=1}^{R_c} \alpha_{i,c} C_{i,c} V_{i,c}^2 \right) \times f_{\text{clk}} \text{ and} \tag{7.11}$$

$$\text{PDP}_{\text{MVMC}} = \frac{1}{N} \sum_{i=1}^{N} \sum_{i=1}^{R_c} \alpha_{i,c} C_{i,c} V_{i,c}^2 f_c \times \sum_{i=1}^{N} \frac{1}{f_c},$$

$$= \frac{1}{N} \left(\sum_{i=1}^{N} \sum_{i=1}^{R_c} \alpha_{i,c} C_{i,c} V_{i,c}^2 \right) \times f_{\text{clk}} \times \sum_{i=1}^{N} \frac{1}{f_{\text{clk}}},$$

$$= \frac{1}{N} \sum_{i=1}^{N} \sum_{i=1}^{R_c} \alpha_{i,c} C_{i,c} V_{i,c}^2. \tag{7.12}$$

For an SVSF mode of operation, $V_{i,c}$ and f_c are the same for any clock cycle (c) and any operation (i). Similarly, for SVSF, the following equations express power and delay:

$$P_{\text{peak}_{\text{SVSF}}} = \text{Max} \left[\left(\sum_{i=1}^{R_c} \alpha_{i,c} C_{i,c} \right) \times V^2 f_{\text{clk}} \right]_{\forall c = 1, 2, \ldots, N}, \quad (7.13)$$

$$P_{\text{avg}_{\text{SVSF}}} = \frac{1}{N} \left(\sum_{i=1}^{N} \sum_{i=1}^{R_c} \alpha_{i,c} C_{i,c} \right) \times V^2 f_{\text{clk}} \text{ and} \quad (7.14)$$

$$\text{PDP}_{\text{SVSF}} = \left(\frac{1}{N} \sum_{i=1}^{N} \sum_{i=1}^{R_c} \alpha_{i,c} C_{i,c} \right) \times V^2. \quad (7.15)$$

As can be observed from Equations (7.12) and (7.15), the PDP for MVMC and SVSF mode operation is independent of the operating frequency. Switching capacitance and operating voltage are two important parameters that also affect the PDP.

7.3 ILP-Based Scheduling for Peak Power Reduction

In this section, ILP models are formulated for peak power minimization for both MVDFC and MVMC scenarios. The ILP models ensure that the dependency constraints and resource constraints are satisfied. Level converters are considered as resources operating in the control step in which they are needed to step up the signal. The DCU that generates the dynamic frequency is considered as a resource operating in all control steps. The power dissipation of the level converters and the DCU is included. In the formulation of an ILP-based model for Equation (7.3), and hence a scheduling scheme for the DFG, the notations given in Table 7.2 are used.

Table 7.2 Notations used in ILP formulations

N_v	: Total number of vertices in the DFG excluding the source and sink
L_V	: Number of voltage levels for multiple supply voltage operation
L_f	: Number of frequency levels for dynamic frequency operation
v_i	: Any vertex or operation i, $1 \le i \le N_v$
F_{k,L_V}	: Functional unit of type k operating at voltage level L_V
M_{k,L_V}	: Maximum number of units of type k operating at voltage level L_V
$C_S[i]$	: As soon as possible (ASAP) time stamp for the operation v_i
$C_L[i]$	: As late as possible (ALAP) time stamp for the operation v_i
$P(i, L_V, L_f)$	: Power consumption of vertex v_i at voltage level L_V and frequency level L_f
x_{i,c,L_V,L_f}	: Decision variable that takes the value of 1 if vertex v_i is scheduled in control step c using the functional unit F_{k,L_V}
$y_{i,L_V,l,m}$	: Decision variable that takes the value of 1 if v_i is using the functional unit F_{k,L_V} and scheduled in control steps $l \to m$
L_{i,L_V}	: Latency in terms of the number of clock cycles for vertex v_i using resource operating at voltage level L_V

7.3.1 ILP Formulations

7.3.1.1 Multiple Supply Voltages and Dynamic Frequency Clocking (MVDFC)

In this case, the objective is to minimize the peak power consumption of the whole DFG over all control steps described in Equation (7.3). Thus the objective function changes into the equation given below:

$$\text{Minimize} : P_{\text{peak}}, \text{ or} \tag{7.16}$$

$$\text{Minimize} : \text{Max}\left(\sum_{i=1}^{R_c} \alpha_{i,c}C_{i,c}V_{i,c}^2 f_c\right)_{\forall c=1 \rightarrow N}. \tag{7.17}$$

It should be noted that P_{peak} is an unknown that has to be minimized. It may be the power consumption of any control step in the DFG depending on the scheduled operations and hence is later used as a constraint. The constraints of the formulation, such as uniqueness constraints, precedence constraints, resource constraints, frequency constraints and peak power constraints, are necessary as presented below.

Uniqueness Constraints: These constraints ensure that each vertex v_i is scheduled to one unique control step within the mobility range $\left(C_S[i], C_L[i]\right)$ with a particular supply voltage and operating frequency. They are represented as $\forall i, 1 \leq i \leq N_v$,

$$\sum_c \sum_{L_V} \sum_{L_f} x_{i,c,L_V,L_f} = 1. \tag{7.18}$$

Precedence Constraints: These constraints ascertain that for a vertex v_i, all its predecessors are scheduled in an earlier control step and its successors are scheduled in a later control step. These are modeled as $\forall i, j, v_i \in \text{Pred}_{v_j}$,

$$\sum_{L_V} \sum_{L_f} \sum_{d=C_S[i]}^{C_L[i]} d \times x_{i,d,L_V,L_f} - \sum_{L_V} \sum_{L_f} \sum_{e=C_S[j]}^{C_L[j]} e \times x_{j,e,L_V,L_f} \leq -1. \tag{7.19}$$

Resource Constraints: These constraints establish that no control step contains more than F_{k,L_V} operations of type k operating at voltage level L_V. These can be enforced as $\forall c, 1 \leq c \leq N$ and $\forall L_V$,

$$\sum_{i \in F_{k,L_V}} \sum_{L_f} x_{i,c,L_V,L_f} \leq M_{k,L_V}. \tag{7.20}$$

Frequency Constraints: This set ensures that if an FU is operating at a higher voltage level, then it can be scheduled in a lower frequency control step; whereas if an FU is operating a lower voltage level, then it cannot be scheduled in a higher frequency control step. These constraints are written as $\forall i, 1 \leq i \leq N_v, \forall c, 1 \leq c \leq N$, if $L_f < L_V$, then $x_{i,c,L_V,L_f} = 0$.

Peak Power Constraints: These constraints ensure that the maximum power consumption of the DFG does not exceed P_{peak} for any control step. These constraints are applied as follows: $\forall c, 1 \leq c \leq N$ and $\forall L_V$,

$$\sum_{i \in F_{k,L_V}} \sum_{L_f} x_{i,c,L_V,L_f} \times P(i,L_V,L_f) \le P_{\text{peak}}. \tag{7.21}$$

7.3.1.2 Multiple Supply Voltages and Multicycling (MVMC)

The ILP formulation for peak power minimization using MVMC is now presented. In this case, the objective is to minimize the peak power consumption of the whole DFG over all control steps. Thus, the ILP formulation becomes the following:

$$\text{Minimize} : P_{\text{peak}}, \text{ or} \tag{7.22}$$

$$: \text{Max} \left(\sum_{i=1}^{R_c} \alpha_{i,c} C_{i,c} V_{i,c}^2 f_{\text{clk}} \right)_{\forall c=1 \to N}. \tag{7.23}$$

It should be noted that P_{peak} is an unknown that has to be minimized. It may be power consumption of any control step in the DFG depending on the scheduled operations and hence is later used as a constraint. The constraints of the formulation, such as uniqueness constraints, precedence constraints, resource constraints and peak power constraints, are presented below.

Uniqueness Constraints: These constraints confirm that every vertex v_i is scheduled in appropriate control steps within the mobility range $\left(C_S[i], C_L[i] \right)$ with a particular supply voltage. It may be operated at more than one clock cycle depending on the supply voltage. These constraints are represented as $\forall i, 1 \le i \le N_v$,

$$\sum_{L_V} \sum_{l=C_S[i]}^{C_S[i]+C_L[i]+1-L_{i,L_V}} y_{i,L_V,l,(l+L_{i,L_V}-1)} = 1. \tag{7.24}$$

When the operators are operating at the highest voltage, they are scheduled in one unique control step; whereas when they are to be operated at lower voltages, they need more than one clock cycle for completion. Thus, for lower voltage, the mobility is restricted.

Precedence Constraints: These constraints guarantee that for a vertex v_i, all its predecessors are scheduled in an earlier control step and its successors are scheduled in a later control step. These constraints should also take care of the multicycling operations. These are modeled as $\forall i, j, v_i \in \text{Pred}_{v_j}$,

$$\sum_{L_V} \sum_{l=C_S[i]}^{C_L[i]} (l+L_{i,L_V}-1) \times y_{i,L_V,l,(l+L_{i,L_V}-1)}$$

$$-\sum_{L_V} \sum_{l=C_S[j]}^{C_L[j]} l \times y_{j,L_V,l,(l+L_{j,L_V}-1)} \le -1. \tag{7.25}$$

Resource Constraints: These constraints make sure that no control step contains more than F_{k,L_V} operations of type k operating at voltage level L_V. These can be enforced as $\forall L_V$ and $\forall l, 1 \le l \le N$,

$$\sum_{i \in F_{k,L_V}} \sum_l y_{i,L_V,l,(l+L_{i,L_V}-1)} \leq M_{k,L_V}. \tag{7.26}$$

Peak Power Constraints: These constraints ensure that the maximum power consumption of the DFG does not exceed P_{peak} for any control step. These constraints are enforced as follows: $\forall l,\ 1 \leq l \leq N$,

$$\sum_{i \in F_{k,L_V}} \sum_{L_V} y_{i,L_V,l,(l+L_{i,L_V}-1)} \times P(i,L_V,L_{f_{\text{clk}}}) \leq P_{\text{peak}}. \tag{7.27}$$

7.3.2 ILP-Based Scheduler

In this section, the solutions for the ILP formulations obtained in the previous section are discussed. The same target architecture and the same characterized data path components used in [210, 225, 222] are assumed. The ILP-based scheduler, which minimizes peak power consumption of the DFG is presented in Algorithm 21. The first step is to determine the ASAP time stamp of each operation. The second step is the determination of the ALAP time stamp of each vertex for the DFG. The ASAP time stamp is the start time, and the ALAP time stamp is the finish time of each operation. These two times provide the mobility of an operation, and the operation must be scheduled in this mobility range. This mobility graph needs to be modified for the MVMC scheme. After the ILP formulation is solved (using LP-Solve), the scheduled DFG is obtained. The scheduler determines the cycle frequencies for the scheduled DFG for the MVDFC scheme.

7.3.2.1 Scheduling for MVDFC

The solution for the ILP formulation in the MVDFC case, with the help of the DFG shown in Fig. 7.1, is illustrated. The ASAP schedule is shown in Fig. 7.1(a), and

---

Algorithm 21 ILP-Based Scheduling for Peak Power Minimization

---

1: Find ASAP and ALAP schedules of the UDFG.
2: Determine the mobility graph of each vertex in the DFG.
3: Modify the mobility graph for MVDFC and MVMC.
4: Construct the ILP formulations for peak power.
5: Solve the ILP formulations using LP-Solve.
6: Find the scheduled DFG.
7: Determine the cycle frequencies for MVDFC scheme.
8: Estimate the power consumptions of the DFG.

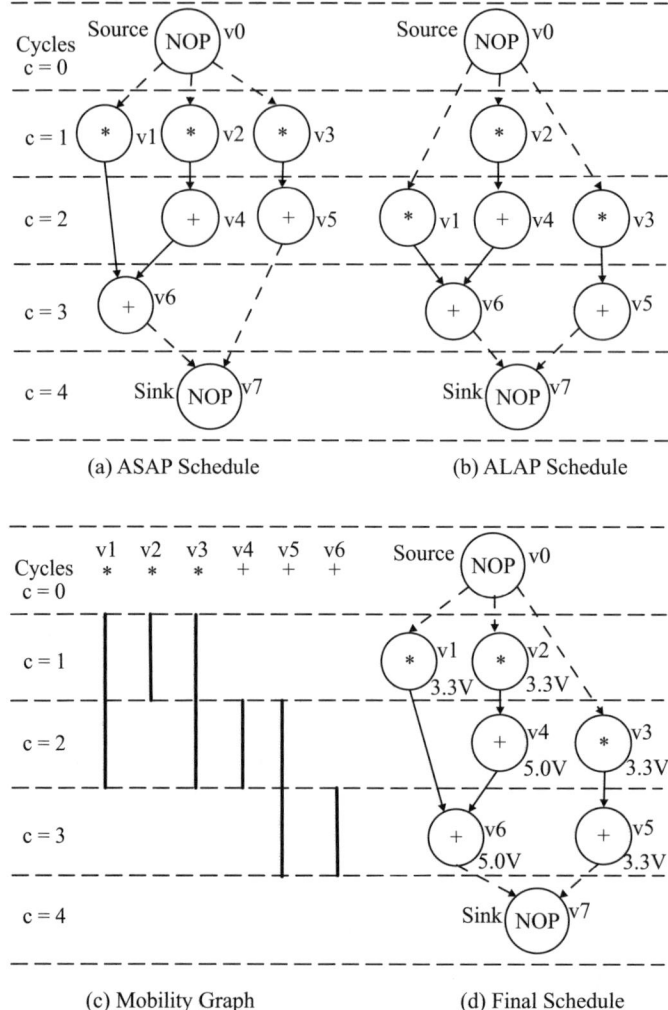

(a) ASAP Schedule   (b) ALAP Schedule

(c) Mobility Graph   (d) Final Schedule

Fig. 7.1 Example of DFG for MVDFC operation for resource constraint RC1: two multipliers at 3.3 V, one multiplier at 5.0 V, one ALU at 3.3 V and one ALU operating at 5.0 V

the ALAP schedule is shown in Fig. 7.1(b). From the ASAP and ALAP schedules, the mobility graph is obtained as in Fig. 7.1(c). Using this mobility graph, the ILP formulations are shown in Algorithm 22 for the resource constraint (RC1): two multipliers at 3.3 V, one multiplier at 5.0 V, one ALU at 3.3 V and one ALU operating at 5.0 V. The formulation is solved using LP-Solve; based on the results, the scheduled DFG shown is Fig. 7.1(d) is obtained. In Algorithm 22, the following additional notations are used. PP: peak power, Mmult1: number of multipliers at voltage level 1, Mmult2: number of multipliers at voltage level 2, Malu1: number of ALUs at voltage level 1 and Malu2: number of ALUs at voltage level 2 are used. The corresponding formulation expressed in AMPL [101] is given in Algorithm 23.

Algorithm 22 ILP Formulation for the Example of DFG for the MVDFC Scheme

1: /* Objective Function */
2: min: PP;
3: /* Uniqueness Constraints */
4: $x1111 + x1112 + x1121 + x1122 + x1211 + x1212 + x1221 + x1222 = 1$;
5: $x2111 + x2112 + x2121 + x2122 = 1$;
6: $x3111 + x3112 + x3121 + x3122 + x3211 + x3212 + x3221 + x3222 = 1$;
7: $x4211 + x4212 + x4221 + x4222 = 1$;
8: $x5211 + x5212 + x5221 + x5222 + x5311 + x5312 + x5321 + x5322 = 1$;
9: $x6311 + x6312 + x6321 + x6322 = 1$;
10: /* Precedence Constraints */
11: $3 x6311 + 3 x6312 + 3 x6321 + 3 x6322 - 2 x1211 - 2 x1212 - 2 x1221 - 2 x1222 - x1111 - x1112 - x1121 - x1122 \geq 1$;
12: $2 x4211 + 2 x4212 + 2 x4221 + 2 x4222 - x2111 - x2112 - x2121 - x2122 \geq 1$;
13: $3 x6311 + 3 x6312 + 3 x6321 + 3 x6322 - 2 x4211 - 2 x4212 - 2 x4221 - 2 x4222 \geq 1$;
14: $3 x5311 + 3 x5312 + 3 x5321 + 3 x5322 + 2 x5211 + 2 x5212 + 2 x5221 + 2 x5222 - 2 x3211 - 2 x3212 - 2 x3221 - 2 x3222 - x3111 - x3112 - x3121 - x3122 \geq 1$;
15: /* Resource Constraints */
16: $x1111 + x2111 + x3111 + x1112 + x2112 + x3112 \leq 1$; /* Mmult1 */
17: $x1121 + x2121 + x3121 + x1122 + x2122 + x3122 \leq 2$; /* Mmult2 */
18: $x1211 + x3211 + x1212 + x3212 \leq 1$; /* Mmult1 */
19: $x1221 + x3221 + x1222 + x3222 \leq 2$; /* Mmult2 */
20: $x4211 + x5211 + x4212 + x5212 \leq 1$; /* Malu1 */
21: $x4221 + x5221 + x4222 + x5222 \leq 1$; /* Malu2 */
22: $x5311 + x6311 + x5312 + x6312 \leq 1$; /* Malu1 */
23: $x5321 + x6321 + x5322 + x6322 \leq 1$; /* Malu2 */
24: /* Frequency Constraints */
25: $x1121 = 0$; $x1221 = 0$; $x2121 = 0$; $x3121 = 0$; $x3221 = 0$; $x4221 = 0$; $x5221 = 0$; $x5321 = 0$; $x6321 = 0$;
26: /* Peak Power Constraints */
27: $39.6 x1111 + 19.8 x1112 + 17.3 x1121 + 8.6 x1122 + 39.6 x2111 + 19.8 x2112 + 17.3 x2121 + 8.6 x2122 + 39.6 x3111 + 19.8 x3112 + 17.3 x3121 + 8.6 x3122 \leq PP$;
28: $39.6 x1211 + 19.8 x1212 + 17.3 x1221 + 8.6 x1222 + 39.6 x3211 + 19.8 x3212 + 17.3 x3221 + 8.6 x3222 + 1.0 x4211 + 0.5 x4212 + 0.5 x4221 + 0.2 x4222 + 1.0 x5211 + 0.5 x5212 + 0.5 x5221 + 0.2 x5222 \leq PP$;
29: $1.0 x5311 + 0.5 x5312 + 0.5 x5321 + 0.2 x5322 + 1.0 x6311 + 0.5 x6312 + 0.5 x6321 + 0.2 x6322 \leq PP$;
30: /* Integer Constraints */
31: INT $x1111, x1112, x1121, x1122, x1211, x1212, x1221, x1222, x2111, x2112, x2121, x2122,$ $x3111, x3112, x3121, x3122, x3211, x3212, x3221, x3222, x4211, x4212, x4221, x4222,$ $x5211, x5212, x5221, x5222, x5311, x5312, x5321, x5322, x6311, x6312, x6321, x6322$;

7.3.2.2 Scheduling for MVMC

The solution for the ILP formulation of the MVMC case is illustrated with the help of the DFG shown in Fig. 7.2. The ASAP schedule is shown in Fig. 7.2(a), and the ALAP schedule is shown in Fig. 7.2(b). From the ASAP and ALAP schedules, the mobility graph shown in Fig. 7.2(c) is obtained. This mobility graph is different from that shown in Fig. 7.2(c). In the MVMC case, the mobility graph considers

---

Algorithm 23 AMPL Modeling of ILP Formulation for the Sample of the DFG for the MVDFC Scheme

---

1: param TASK;   # number of Tasks
2: param LEVEL;   # number of levels in DFG
3: param VOLT;   # number of voltage levels
4: param FREQ;   # number of frequency levels
5: param ASAP {1..TASK} > 0, ≤ LEVEL;   #ASAP Schedule for Each Task
6: param ALAP {1..TASK} > 0, ≤ LEVEL;   #ALAP Schedule for Each Task
7: param OP {1..TASK};   #Type of FU
8: param POWER {1..2, 1..VOLT, 1..FREQ};   #Power Consumption of Each Unit
9: param M {1..2, 1..VOLT};   #Resource Constraints
10: var PP;
11: var X {i in 1..TASK, j in ASAP[i]..ALAP[i], v in 1..VOLT, f in 1..FREQ} binary;
12: #Objective Function
13: minimize peak_power : PP;
14: # Uniqueness Constraints
15: subject to uniq_cons {i in 1..TASK}: sum {j in ASAP[i]..ALAP[i], v in 1..VOLT, f in 1..FREQ} X[i, j, v, f] = 1;
16: # Precedence Constraints
17: subject to pred_cons1: sum {j in ASAP[6]..ALAP[6], v in 1..VOLT, f in 1..FREQ} j * X[6, j, v, f] − sum {j in ASAP[1]..ALAP[1], v in 1..VOLT, f in 1..FREQ} j * X[1, j, v, f] ≥ 1;
18: subject to pred_cons2: sum {j in ASAP[4]..ALAP[4], v in 1..VOLT, f in 1..FREQ} j * X[4, j, v, f] − sum {j in ASAP[2]..ALAP[2], v in 1..VOLT, f in 1..FREQ} j * X[2, j, v, f] ≥ 1;
19: subject to pred_cons3: sum {j in ASAP[6]..ALAP[6], v in 1..VOLT, f in 1..FREQ} j * X[6, j, v, f] − sum {j in ASAP[4]..ALAP[4], v in 1..VOLT, f in 1..FREQ} j * X[4, j, v, f] ≥ 1;
20: subject to pred_cons4: sum {j in ASAP[5]..ALAP[5], v in 1..VOLT, f in 1..FREQ} j * X[5, j, v, f] − sum {j in ASAP[3]..ALAP[3], v in 1..VOLT, f in 1..FREQ} j * X[3, j, v, f] ≥ 1;
21: # Resource Constraints
22: subject to res_cons_mult {j in 1..LEVEL, v in 1..VOLT}: sum {f in 1..FREQ, i in 1..TASK: ASAP[i] ≤ j ≤ ALAP[i] && OP[i] = 2} X[i, j, v, f] ≤ M[2, v];
23: subject to res_cons_alu j in 1..LEVEL, v in 1..VOLT: sum {f in 1..FREQ, i in 1..TASK: ASAP[i] ≤ j ≤ ALAP[i] && OP[i] = 1} X[i, j, v, f] ≤ M[1, v];
24: # Peak Power Constraints
25: subject to pp_cons {j in 1..LEVEL}: sum {v in 1..VOLT, f in 1..FREQ, i in 1..TASK: ASAP[i] ≤ j ≤ ALAP[i]} POWER[OP[i], v, f] * X[i, j, v, f] ≤ PP;
26: #Frequency Constraints
27: subject to freq_cons {i in 1..TASK, j in ASAP[i]..ALAP[i]}: X[i, j, 2, 1] = 0;

---

multicycle operations. It is assumed that there are two operating voltage levels, and when the multipliers are operated at a lower voltage, they take two clock cycles. For the characterized cells used in the experiments, the operating clock frequency, f_{clk} is 9 MHz. Using this mobility graph, the ILP formulation shown in Algorithm 24 is obtained for the resource constraint (RC1) of two multipliers at 3.3 V, one multiplier at 5.0 V, one ALU at 3.3 V and one ALU operating at 5.0 V. The corresponding formulation expressed in AMPL is given in Algorithm 25. The formulation using LP-Solve and based on the results obtained from the scheduled DFG is shown in Fig. 7.2(d). In Algorithm 24, the additional notations, such as PP, Mmult1, Mmult2, Malu1 and Malu2 have the same meaning as that of the MVDFC case shown in Algorithm 22.

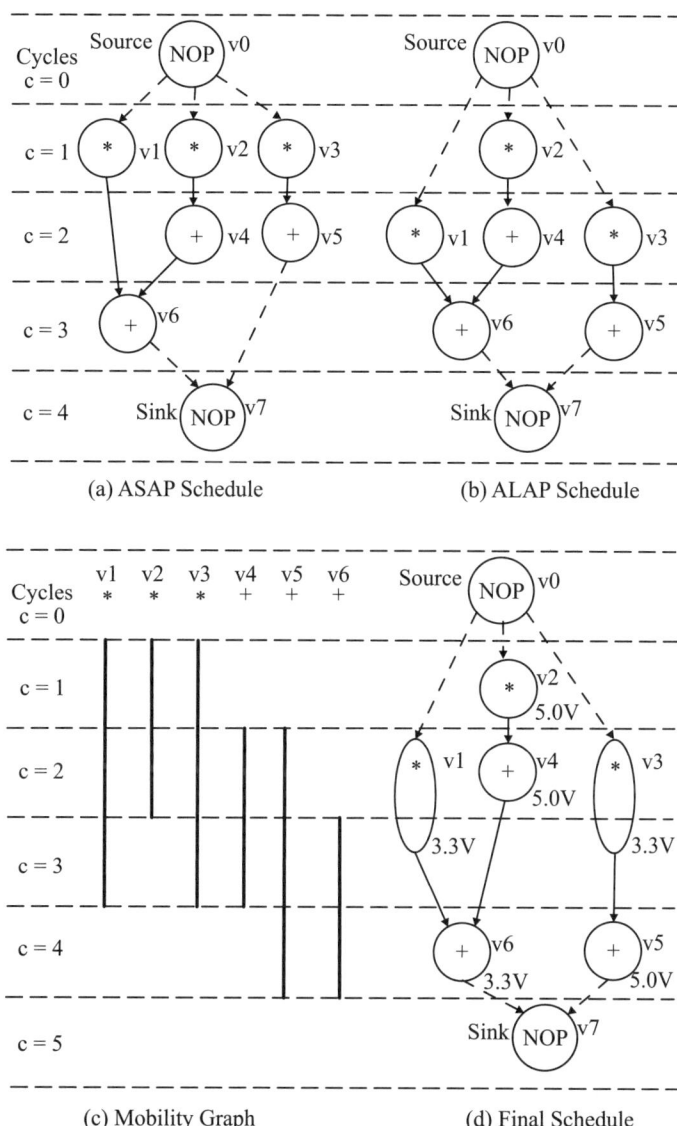

(a) ASAP Schedule   (b) ALAP Schedule

(c) Mobility Graph   (d) Final Schedule

Fig. 7.2 Example of DFG for MVMC operation for resource constraint RC1: two multipliers at 3.3 V, one multiplier at 5.0 V, one ALU at 3.3 V and one ALU operating at 5.0 V

7.3.3 Experimental Results

The ILP-based MVDFC and MVMC peak power minimization schedulers were tested with selected benchmark circuits: Example circuit (EXP), FIR filter, IIR filter, HAL differential equation solver and ARF. The FUs used are ALUs and multipliers.

Algorithm 24 ILP Formulation for Peak Power Minimization for MVMC Scheme

1: /* Objective Function */
2: min: PP;
3: /* Uniqueness Constraints */
4: x1212 + x1223 + x1111 + x1122 + x1133 = 1;
5: x2212 + x2111 + x2122 = 1;
6: x3111 + x3122 + x3133 + x3212 + x3223 = 1;
7: x4122 + x4133 + x4222 + x4233 = 1;
8: x5122 + x5133 + x5144 + x5222 + x5233 + x5244 = 1;
9: x6133 + x6144 + x6233 + x6244 = 1;
10: /* Peak Power Constraints */
11: 39.6 x1111 + 8.6 x1212 + 39.6 x2111 + 8.6 x2212 + 39.6 x3111 + 8.6 x3212 ≤ PP;
12: 39.6 x1122 + 8.6 x1212 + 8.6 x1223 + 39.6 x2122 + 8.6 x2212 + 39.6 x3122 + 8.6 x3212 +
    8.6 x3223 + 1.0 x4122 + 0.5 x4222 + 1.0 x5122 + 0.5 x5222 ≤ PP;
13: 39.6 x1133 + 8.6 x1223 + 39.6 x3133 + 8.6 x3223 + 1.0 x4133 + 0.5 x4233 + 1.0 x5133 + 0.5
    x5233 + 1.0 x6133 + 0.5 x6233 ≤ PP;
14: 1.0 x5144 + 0.5 x5244 + 1.0 x6144 + 0.5 x6244 ≤ PP;
15: /* Resource Constraints */
16: x1111 + x2111 + x3111 ≤ 1; /* Mmult1 */
17: x1212 + x2212 + x3212 ≤ 2; /* Mmult2 */
18: x1122 + x2122 + x3122 ≤ 1; /* Mmult1 */
19: x1212 + x1223 + x2212 + x3212 + x3223 ≤ 2; /* Mmult2 */
20: x1133 + x3133 ≤ 1; /* Mmult1 */
21: x1223 + x3223 ≤ 2; /* Mmult2 */
22: x4122 + x5122 ≤ 1; /* Malu1 */
23: x4222 + x5222 ≤ 1; /* Malu2 */
24: x4133 + x5133 + x6133 ≤ 1; /* Malu1 */
25: x4233 + x5233 + x6233 ≤ 1; /* Malu2 */
26: x5144 + x6144 ≤ 1; /* Malu1 */
27: x5244 + x6244 ≤ 1; /* Malu2 */
28: /* Precedence Constraints */
29: 4 x6144 + 4 x6244 + 3 x6133 + 3 x6233 − 3 x1133 − 3 x1223 − 2 x1122 − 2 x1212 − x1111
    ≥ 1;
30: 4 x6144 + 4 x6244 + 3 x6133 + 3 x6233 − 3 x4133 − 3 x4233 − 2 x4122 − 2 x4222 ≥ 1;
31: 3 x4133 + 3 x4233 + 2 x4122 + 2 x4222 − 2 x2122 − 2 x2212 − x2111 ≥ 1;
32: 4 x5144 + 4 x5244 + 3 x5133 + 3 x5233 + 2 x5122 + 2 x5222 − 3 x3133 − 3 x3223 − 2
    x3122 − 2 x3212 − x3111 ≥ 1;
33: /* Integer Constraints */
34: INT x1111, x1122, x1133, x1212, x1223, x2111, x2122, x2212, x3111, x3122, x3133, x3212,
    x3223, x4122, x4133, x4222, x4233, x5122, x5133, x5144, x5222, x5233, x5244, x6133,
    x6144, x6233, x6244;

To obtain a visual picture of the experimental results, the peak power reductions
and the PDP reductions averaged over all resource constraints are plotted. Fig. 7.3
shows the average reductions for different benchmarks, averaged over all resource
constraints. The power estimation includes the power consumption of the overhead,
such as level converters (used in both MVDFC and MVMC schemes) and DCUs
(needed in the MVDFC case). It is assumed that each resource has equal switching
activity ($\alpha_{i,c}$). The results are reported for two supply voltages and for switching
$= 0.5$. It is obvious from the figure that the reductions are appreciable. It is also

---

Algorithm 25 AMPL Models for ILP Formulation for Peak Power Minimization for the MVMC Scheme

---

1: param TASK;   # Number of Tasks
2: param LEVEL;   # Number of Levels in DFG
3: param VOLT;   # Number of Voltage Levels
4: param ASAP {1..TASK} > 0;   #ASAP Schedule for Each Task
5: param ALAP {1..TASK} > 0;   #ALAP Schedule for Each Task
6: param OP {1..TASK};   #Type of FU
7: param M {1..2, 1..VOLT};   #Resource Constraints
8: param POWER {1..2, 1..VOLT};   #Power consumption of the FUs
9: var PP;
10: var X {i in 1..TASK, v in 1..VOLT, j in ASAP[i]..ALAP[i], k in ASAP[i]..ALAP[i]} binary;
11: #Objective Function
12: minimize peak_power: PP;
13: # Uniqueness Constraints
14: subject to uniq_cons {i in 1..TASK}: sum{j in ASAP[i]..ALAP[i]} X[i, 1, j, j] + (if OP[i] = 2 then sum{j in ASAP[i]..ALAP[i]-1} X[i, 2, j, j+1] else sum{j in ASAP[i]..ALAP[i]} X[i, 2, j, j]) = 1;
15: # Precedence Constraints
16: subject to pred_cons1: sum {v in 1..VOLT, j in ASAP[6]..ALAP[6]} j * X[6, v, j, j] - sum {j in ASAP[1]..ALAP[1]} j * X[1, 1, j, j] - sum {j in ASAP[1]..ALAP[1]-1} (j+1) * X[1, 2, j, j+1] ≥ 1;
17: subject to pred_cons2: sum {v in 1..VOLT, j in ASAP[6]..ALAP[6]} j * X[6, v, j, j] - sum {v in 1..VOLT, j in ASAP[4]..ALAP[4]} j * X[4, v, j, j] ≥ 1;
18: subject to pred_cons3: sum {v in 1..VOLT, j in ASAP[4]..ALAP[4]} j * X[4, v, j, j] - sum {j in ASAP[2]..ALAP[2]} j * X[2, 1, j, j] - sum {j in ASAP[2]..ALAP[2]-1} (j+1) * X[2, 2, j, j+1] ≥ 1;
19: subject to pred_cons4: sum {v in 1..VOLT, j in ASAP[5]..ALAP[5]} j * X[5, v, j, j] - sum {j in ASAP[3]..ALAP[3]} j * X[3, 1, j, j] - sum {j in ASAP[3]..ALAP[3]-1} (j+1) * X[3, 2, j, j+1] ≥ 1;
20: # Resource Constraints subject to res_cons_mult {j in 1..LEVEL, v in 1..VOLT}: if v = 1 then sum {i in 1..TASK: ASAP[i] ≤ j ≤ ALAP[i] && OP[i] = 2} X[i, 1, j, j] else sum {i in 1..TASK: ASAP[i] < j < ALAP[i] && OP[i] = 2} (X[i, 2, j-1, j] + X[i, 2, j, j+1]) + sum {i in 1..TASK: ALAP[i] = j && OP[i] = 2} X[i, 2, j-1, j] + sum {i in 1..TASK: ASAP[i] = j && OP[i] = 2} X[i, 2, j, j+1] ≤ M[2, v];
21: subject to res_cons_alu {j in 1..LEVEL, v in 1..VOLT}: sum {i in 1..TASK: ASAP[i] ≤ j ≤ ALAP[i] && OP[i] = 1} X[i, v, j, j] ≤ M[1, v];
22: # Peak Power Constraints
23: subject to pp_cons {j in 1..LEVEL-1}: sum {i in 1..TASK: ASAP[i] ≤ j ≤ ALAP[i]} X[i, 1, j, j] * POWER[OP[i], 1] + sum {i in 1..TASK: ASAP[i] < j < ALAP[i] && OP[i] = 2} (X[i, 2, j-1, j] * POWER[OP[i], 2] + X[i, 2, j, j+1] * POWER[OP[i], 2]) + sum {i in 1..TASK: j = ALAP[i] && OP[i] = 2} X[i, 2, j-1, j] * POWER[OP[i], 2] + sum {i in 1..TASK: ASAP[i] = j && OP[i] = 2} X[i, 2, j, j+1] * POWER[OP[i], 2] + sum {i in 1..TASK: ASAP[i] ≤ j ≤ ALAP[i] && OP[i] = 1} X[i, 2, j, j] * POWER[OP[i], 2] ≤ PP;

---

observed that the power consumption increases for higher switching activity and decreases for lower switching activity. The peak power reductions for the proposed scheduling schemes are listed along with other scheduling algorithms dealing with peak power reduction in Table 7.3. The table is not meant to provide an exact comparison, but rather to provide a general idea of relative performances.

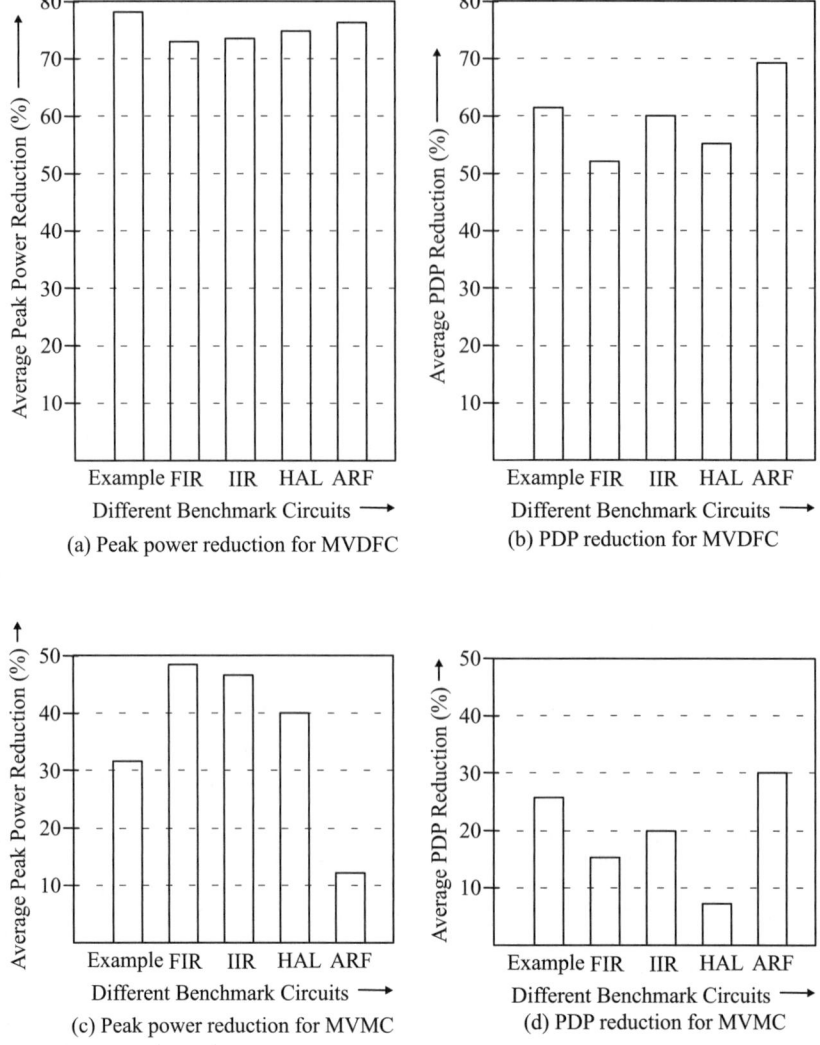

Fig. 7.3 Average peak power and PDP reductions for benchmark circuits

Table 7.3 Peak power reduction for various scheduling schemes

Benchmark Circuits	% Percentage Estimated Average Reduction in Peak Power						
	This Work				Shiue [316]	Martin [202]	Raghunathan [291]
	MVDFC		MVMC				
	ΔP_{DFC}	$\Delta\mathrm{PDP}_{\mathrm{DFC}}$	ΔP_{MC}	$\Delta\mathrm{PDP}_{\mathrm{MC}}$	ΔP	ΔP	ΔP
(1) EXP	77.9	61.8	32.2	25.3	–	–	–
(2) FIR	72.3	52.2	48.5	15.3	63.0	40.3	23.1
(3) IIR	73.4	60.0	47.2	20.3	–	–	–
(4) HAL	75.2	55.4	40.8	7.9	28.0	–	–
(5) ARF	77.8	68.9	11.1	29.8	50.0	–	–

7.3.4 Conclusions

Reduction of peak power consumption is important in a CMOS circuit. Two data path scheduling schemes, one using MVDFC and another using MVMC, have been introduced using ILP-based optimizations for peak power reduction. In both cases, the presented scheduling schemes could achieve significant amount of peak power reduction over the single supply voltage and single-frequency scenario. It is observed that for the MVDFC case, using two supply voltage levels, an average peak power reduction of 75% is achieved, and the average PDP reduction is 60%. Similarly, for the MVMC case, the average peak power reduction is 36%, and the average PDP reduction is 20%. The results clearly indicate that dynamic frequency clocking is a better scheme than the multicycling approach for peak power minimization.

7.4 ILP-Based Scheduling for Simultaneous Peak and Average Power Reduction

7.4.1 ILP Formulations

In this section, the ILP formulations to minimize the peak and average power consumption of a data path circuit are discussed. First, the formulations for MVDFC-based circuits are discussed followed by MVMC-based circuits.

7.4.1.1 Formulations for MVDFC

In this section, the ILP formulation for simultaneous peak (Equation 7.3) and average power (Equation 7.5) minimization using MVDFC is described [227, 222].

Objective Function: The objective is to simultaneously minimize the peak power and the average power consumption of the whole DFG over all control steps simultaneously. Thus, the objective function can be represented as follows:

$$\text{Minimize} : P_{\text{peak}} + P_{\text{avg}}. \tag{7.28}$$

Using decision variables, the objective function can be rewritten as follows:

$$\text{Minimize} : P_{\text{peak}} + \frac{1}{N} \sum_{c} \sum_{L_V} \sum_{i \in F_{k,L_V}} \sum_{L_f} x_{i,c,L_V,L_f} \times P(i,L_V,L_f). \tag{7.29}$$

It should be noted that P_{peak} is unknown and has to be minimized. It may be the power consumption of any control step in the DFG depending on the scheduled operations and hence is later used as a constraint. The constraints of the formulation, such as uniqueness constraints, precedence constraints, resource constraints, frequency constraints and peak power constraints, are the same as the formulations used for peak power reduction minimization using MVDFC.

7.4.1.2 Formulations for MVMC

In this section, the ILP formulations for simultaneous minimization of both peak and average power consumption of the DFG using MVMC are discussed [227, 222].

Objective function : The objective is to minimize the peak and average power consumption of the whole DFG over all control steps. Thus, the objective function can be represented as follows:

$$\text{Minimize} : P_{\text{peak}} + P_{\text{avg}}. \tag{7.30}$$

In terms of decision variables, Equation (7.30) is written as

$$\text{Minimize} : P_{\text{peak}} + \frac{1}{N} \sum_{l} \sum_{i \in F_{k,L_V}} \sum_{L_V} y_{i,L_V,l,(l+L_{i,L_V}-1)} \times P(i, L_V, L_{f_{\text{clk}}}). \tag{7.31}$$

P_{peak} is used as a constraint later. The constraints of the formulation, such as uniqueness constraints, precedence constraints, resource constraints and peak power constraints, are the same as the formulations used for peak power reduction minimization using MVMC.

7.4.2 ILP-Based Scheduler

The solutions for the ILP formulations are now discussed for the same target architecture and the characterized data path components of the previous section. Simultaneous peak and average power consumption of the DFG is minimized by the ILP-based scheduler outlined in Algorithm 26. The first step is to determine the ASAP time stamp of each operation. The second step is the determination of the ALAP time stamp of each vertex for the DFG. The ASAP time stamp is the start time, and

---

Algorithm 26 ILP-Based Scheduling for Simultaneous Peak and Average Power Minimization

---

1: Find ASAP and ALAP schedules of the UDFG.
2: Determine the mobility graph of each vertex in the DFG.
3: Modify the mobility graph for MVDFC and MVMC.
4: Construct the ILP formulations for simultaneous peak and average power.
5: Solve the ILP formulations using LP-Solve.
6: Find the scheduled DFG.
7: Determine the cycle frequencies for MVDFC scheme.
8: Estimate the power consumptions of the DFG.

---

the ALAP time stamp is the finish time of each operation. These two times provide the mobility of an operation, and the operation must be scheduled in this mobility range. This mobility graph needs to be modified for the multicycling scheme. At this point, the operating frequency of an FU is assumed as the inverse of its operational delay determined by using the delay model. The ILP formulations are solved to derive the scheduled DFG. Finally, the power consumption of the scheduled DFG is estimated.

7.4.2.1 Scheduler for MVDFC

The steps in the solution for the ILP formulations for the MVDFC is illustrated using the DFG shown in Fig. 7.4. From the ASAP and ALAP schedules, the mobility graph shown in Fig. 7.4(a) is determined. One ILP formulation for the resource constraint (RC3), with two multipliers at 2.4 V and two ALUs operating at 3.3 V using switching activity of 0.5 is presented in Algorithm 27. The following additional notations are used: (i) PP: peak power, (ii) Mmult1: number of multipliers at voltage level 1, (iii) Mmult2: number of multipliers at voltage level 2, (iv) Malu1: number of ALUs at voltage level 1 and (v) Malu2: number of ALUs at voltage level 2. The ILP formulations are solved using LP-Solve and the scheduled DFG is shown in Fig. 7.4(b).

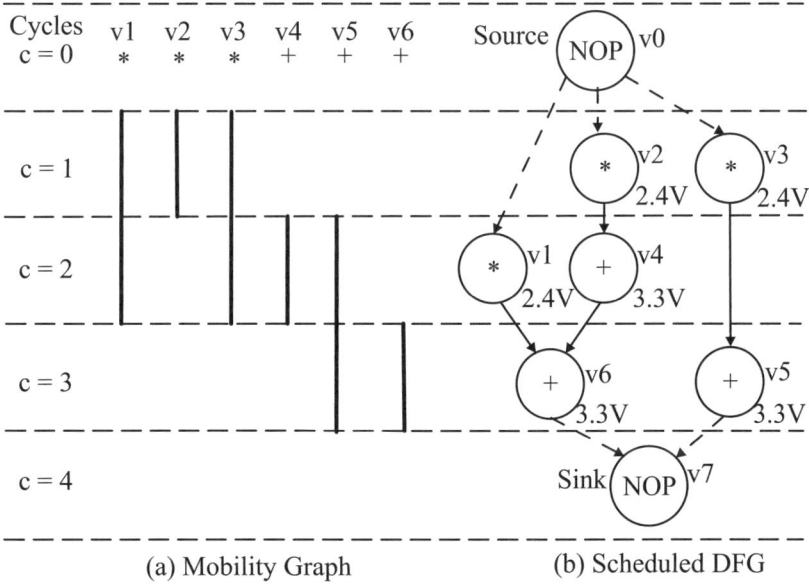

(a) Mobility Graph   (b) Scheduled DFG

Fig. 7.4 Example of DFG for resource constraint RC3 using MVDFC

Algorithm 27 ILP Formulation for Simultaneous Peak and Average Power Minimization for the MVDFC Scheme

1: /* Objective function */
2: min : 2.89 x1111 + 1.44 x1112 + 1.52 x1121 + 0.76 x1122 + 2.89 x2111 + 1.44 x2112 + 1.52
   x2121 + 0.76 x2122 + 2.89 x3111 + 1.44 x3112 + 1.52 x3121 + 0.76 x3122 + 2.89 x1211 +
   1.44 x1212 + 1.52 x1221 + 0.76 x1222 + 2.89 x3211 + 1.44 x3212 + 1.52 x3221 + 0.76 x3222
   + 0.08 x4211 + 0.04 x4212 + 0.04 x4221 + 0.02 x4222 + 0.08 x5211 + 0.04 x5212 + 0.04
   x5221 + 0.02 x5222 + 0.08 x5311 + 0.04 x5312 + 0.04 x5321 + 0.02 x5322 + 0.08 x6311 +
   0.04 x6312 + 0.04 x6321 + 0.02 x6322 + PP;
3: /* Uniqueness Constraints */
4: x1111 + x1112 + x1121 + x1122 + x1211 + x1212 + x1221 + x1222 = 1;
5: x2111 + x2112 + x2121 + x2122 = 1;
6: x3111 + x3112 + x3121 + x3122 + x3211 + x3212 + x3221 + x3222 = 1;
7: x4211 + x4212 + x4221 + x4222 = 1;
8: x5211 + x5212 + x5221 + x5222 + x5311 + x5312 + x5321 + x5322 = 1;
9: x6311 + x6312 + x6321 + x6322 = 1;
10: /* Precedence Constraints */
11: 3 x6311 + 3 x6312 + 3 x6321 + 3 x6322 − 2 x1211 − 2 x1212 − 2 x1221 − 2 x1222 − x1111
    − x1112 − x1121 − x1122 ≥ 1;
12: 2 x4211 + 2 x4212 + 2 x4221 + 2 x4222 − x2111 − x2112 − x2121 − x2122 ≥ 1;
13: 3 x6311 + 3 x6312 + 3 x6321 + 3 x6322 − 2 x4211 − 2 x4212 − 2 x4221 − 2 x4222 ≥ 1;
14: 3 x5311 + 3 x5312 + 3 x5321 + 3 x5322 + 2 x5211 + 2 x5212 + 2 x5221 + 2 x5222 − 2 x3211
    − 2 x3212 − 2 x3221 − 2 x3222 − x3111 − x3112 − x3121 − x3122 ≥ 1;
15: /* Resource Constraints */
16: x1111 + x2111 + x3111 + x1112 + x2112 + x3112 ≤ 0; /* Mmult1 */
17: x1121 + x2121 + x3121 + x1122 + x2122 + x3122 ≤ 2; /* Mmult2 */
18: x1211 + x3211 + x1212 + x3212 ≤ 0; /* Mmult1 */
19: x1221 + x3221 + x1222 + x3222 ≤ 2; /* Mmult2 */
20: x4211 + x5211 + x4212 + x5212 ≤ 2; /* Malu1 */
21: x4221 + x5221 + x4222 + x5222 ≤ 0; /* Malu2 */
22: x5311 + x6311 + x5312 + x6312 ≤ 2; /* Malu1 */
23: x5321 + x6321 + x5322 + x6322 ≤ 0; /* Malu2 */
24: /* Frequency Constraints */
25: x1121 = 0; x1221 = 0; x2121 = 0; x3121 = 0; x3221 = 0; x4221 = 0; x5221 = 0; x5321 = 0;
    x6321 = 0;
26: /* Peak Power Constraints */
27: 8.64 x1111 + 4.32 x1112 + 4.56 x1121 + 2.28 x1122 + 8.64 x2111 + 4.32 x2112 + 4.56 x2121
    + 2.28 x2122 + 8.64 x3111 + 4.32 x3112 + 4.56 x3121 + 2.28 x3122 ≤ PP;
28: 8.64 x1211 + 4.32 x1212 + 4.56 x1221 + 2.28 x1222 + 8.64 x3211 + 4.32 x3212 + 4.56 x3221
    + 2.28 x3222 + 0.23 x4211 + 0.11 x4212 + 0.12 x4221 + 0.06 x4222 + 0.23 x5211 + 0.11
    x5212 + 0.12 x5221 + 0.06 x5222 ≤ PP;
29: 0.23 x5311 + 0.11 x5312 + 0.12 x5321 + 0.06 x5322 + 0.23 x6311 + 0.11 x6312 + 0.12 x6321
    + 0.06 x6322 ≤ PP;
30: /* Integer Constraints */
31: INT x1111, x1112, x1121, x1122, x1211, x1212, x1221, x1222, x2111, x2112, x2121, x2122,
    x3111, x3112, x3121, x3122, x3211, x3212, x3221, x3222, x4211, x4212, x4221, x4222,
    x5211, x5212, x5221, x5222, x5311, x5312, x5321, x5322, x6311, x6312, x6321, x6322;

7.4.2.2 Scheduler for MVMC

The solution for the ILP formulation for MVDFC is illustrated by using the DFG shown in Fig. 7.5. From the ASAP and ALAP schedules, the mobility graph shown

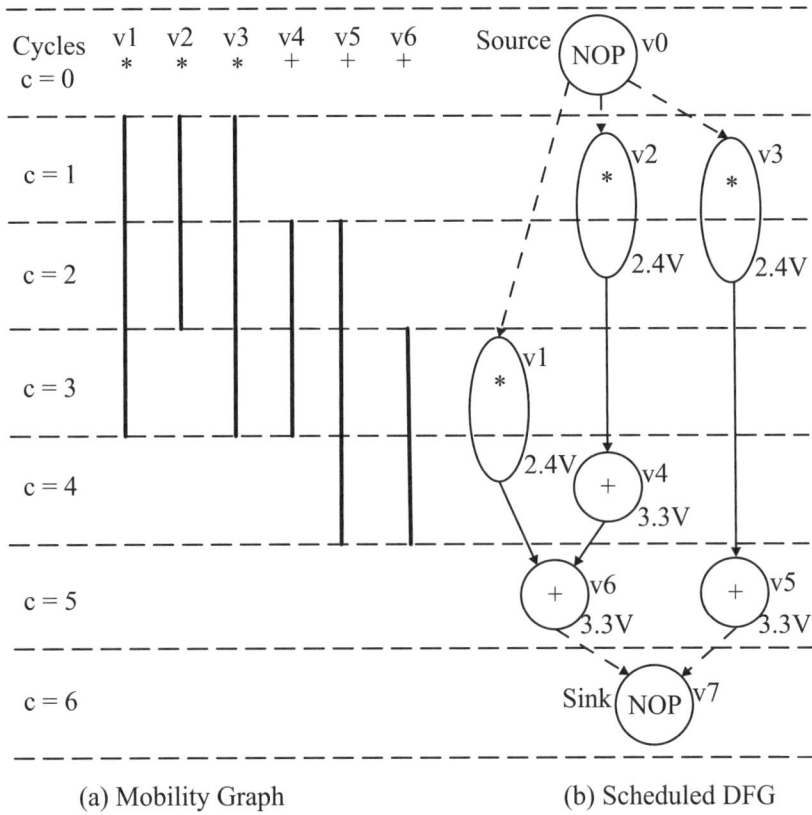

(a) Mobility Graph   (b) Scheduled DFG

Fig. 7.5 Example of DFG for resource constraint RC3 using MVMC

in Fig. 7.5(a) is obtained. This mobility graph is different from that shown in Fig. 7.4(a), as it considers multicycle operations. Two operating voltage levels are assumed in Fig. 7.5(a). The multipliers take two clock cycles when operated at a low-voltage level. For the characterized cells used in the experiment, the operating clock frequency is f_{clk} is 9 MHz. The ILP formulations are obtained by using this mobility graph. One such ILP formulation in Algorithm 28 is for the resource constraint (RC3), with two multipliers at 2.4 V two ALUs at 3.3 V and switching activity = 0.5. The additional notations, such as PP, Mmult1, Mmult2, Malu1 and Malu2 have the same meaning as that of the MVDFC case. The ILP formulations are solved by using LP-Solve, and the scheduled DFG is shown in Fig. 7.5(b).

7.4.3 Experimental Results

The ILP-based schedulers for both MVDFC and MVMC schemes were tested with selected high-level synthesis benchmark circuits: (1) Example circuit (EXP), (2) FIR filter, (3) IIR filter, (4) HAL differential equation solver and (5) ARF. The schedulers were tested using different sets of resource constraints for each benchmark circuit. The experimental results for power included the overhead, such as level converters

Algorithm 28 ILP Formulation for Simultaneous Peak and Average Power Minimization for the MVMC Scheme

1: /* Objective function */
2: min: 1.7 x1111 + 0.9 x1212 + 1.7 x2111 + 0.9 x2212 + 1.7 x3111 + 0.9 x3212 + 1.7 x1122 +
    0.9 x1212 + 0.9 x1223 + 1.7 x2122 + 0.9 x2212 + 0.9 x2223 + 1.7 x3122 + 0.9 x3212 + 0.9
    x3223 + 0.05 x4122 + 0.02 x4222 + 0.05 x5122 + 0.02 x5222 + 1.7 x1133 + 0.9 x1223 + 0.9
    x1234 + 1.7 x2133 + 0.9 x2223 + 1.7 x3133 + 0.9 x3223 + 0.9 x3234 + 0.05 x4133 + 0.02
    x4233 + 0.05 x5133 + 0.02 x5233 + 0.05 x6133 + 0.02 x6233 + 1.7 x1144 + 0.9 x1234 + 1.7
    x3144 + 0.9 x3234 + 0.05 x4144 + 0.02 x4244 + 0.05 x5144 + 0.02 x5244 + 0.05 x6144 +
    0.02 x6244 + 0.05 x5155 + 0.02 x5255 + 0.05 x6155 + 0.02 x6255 + PP;
3: /* Uniqueness Constraints */
4: x1111 + x1122 + x1133 + x1144 + x1212 + x1223 + x1234 = 1; x2111 + x2122 + x2133 +
    x2212 + x2223 = 1; x3111 + x3122 + x3133 + x3144 + x3212 + x3223 + x3234 = 1; x4122
    + x4133 + x4144 + x4222 + x4233 + x4244 = 1; x5122 + x5133 + x5144 + x5155 + x5222 +
    x5233 + x5244 + x5255 = 1; x6133 + x6144 + x6155 + x6233 + x6244 + x6255 = 1;
5: /* Peak Power Constraints */
6: 8.6 x1111 + 4.6 x1212 + 8.6 x2111 + 4.6 x2212 + 8.6 x3111 + 4.6 x3212 ≤ PP;
7: 8.6 x1122 + 4.6 x1212 + 4.6 x1223 + 8.6 x2122 + 4.6 x2212 + 4.6 x2223 + 8.6 x3122 + 4.6
    x3212 + 4.6 x3223 + 0.2 x4122 + 0.1 x4222 + 0.2 x5122 + 0.1 x5222 ≤ PP;
8: 8.6 x1133 + 4.6 x1223 + 4.6 x1234 + 8.6 x2133 + 4.6 x2223 + 8.6 x3133 + 4.6 x3223 + 4.6
    x3234 + 0.2 x4133 + 0.1 x4233 + 0.2 x5133 + 0.1 x5233 + 0.2 x6133 + 0.1 x6233 ≤ PP;
9: 8.6 x1144 + 4.6 x1234 + 8.6 x3144 + 4.6 x3234 + 0.2 x4144 + 0.1 x4244 + 0.2 x5144 + 0.1
    x5244 + 0.2 x6144 + 0.1 x6244 ≤ PP;
10: 0.2 x5155 + 0.1 x5255 + 0.2 x6155 + 0.1 x6255 ≤ PP;
11: /* Resource Constraints */
12: x1111 + x2111 + x3111 ≤ 0; /* Mmult1 */ x1212 + x2212 + x3212 ≤ 2; /* Mmult2 */
13: x1122 + x2122 + x3122 ≤ 0; /* Mmult1 */
14: x1212 + x1223 + x2212 + x2223 + x3212 + x3223 ≤ 2; /* Mmult2 */
15: x1133 + x2133 + x3133 ≤ 0; /* Mmult1 */
16: x1223 + x1234 + x2223 + x3223 + x3234 ≤ 2; /* Mmult2 */
17: x1144 + x3144 ≤ 0; /* Mmult1 */ x1234 + x3234 ≤ 2; /* Mmult2 */
18: x4122 + x5122 ≤ 2; /* Malu1 */ x4222 + x5222 ≤ 0; /* Malu2 */
19: x4133 + x5133 + x6133 ≤ 2; /* Malu1 */ x4233 + x5233 + x6233 ≤ 0; /* Malu2 */
20: x4144 + x5144 + x6144 ≤ 2; /* Malu1 */ x4244 + x5244 + x6244 ≤ 0; /* Malu2 */
21: x5155 + x6155 ≤ 2; /* Malu1 */ x5255 + x6255 ≤ 0; /* Malu2 */
22: /* Precedence Constraints */
23: 5 x6155 + 5 x6255 + 4 x6144 + 4 x6244 + 3 x6133 + 3 x6233 - 4 x1144 - 4 x1234 - 3 x1133 -
    3 x1223 - 2 x1122 - 2 x1212 - x1111 ≥ 1;
24: 5 x6155 + 5 x6255 + 4 x6144 + 4 x6244 + 3 x6133 + 3 x6233 - 4 x4144 - 4 x4244 - 3 x4133 -
    3 x4233 - 2 x4122 - 2 x4222 ≥ 1;
25: 4 x4144 + 4 x4244 + 3 x4133 + 3 x4233 + 2 x4122 + 2 x4222 - 3 x2133 - 3 x2223 - 2 x2122 -
    2 x2212 - x2111 ≥ 1;
26: 5 x5155 + 5 x5255 + 4 x5144 + 4 x5244 + 3 x5133 + 3 x5233 + 2 x5122 + 2 x5222 - 4 x3144
    - 4 x3234 - 3 x3133 - 3 x3223 - 2 x3122 - 2 x3212 - x3111 ≥ 1;
27: /* Integer Constraints */
28: INT x1111, x1122, ..., x6244, x6255;

(used in both the schemes) and DCUs (needed for DFC case). It is assumed that
each resource has equal switching activity ($\alpha_{i,c}$). The results are reported for two
supply voltages and for switching $= 0.5$.

To derive a visual picture of the experimental results, the peak power reduction, the average power reduction and the PDP reduction are averaged over the different sets of resource constraints. Figure 7.6 shows the average reductions for different benchmarks averaged over all resource constraints. It is obvious from the figure

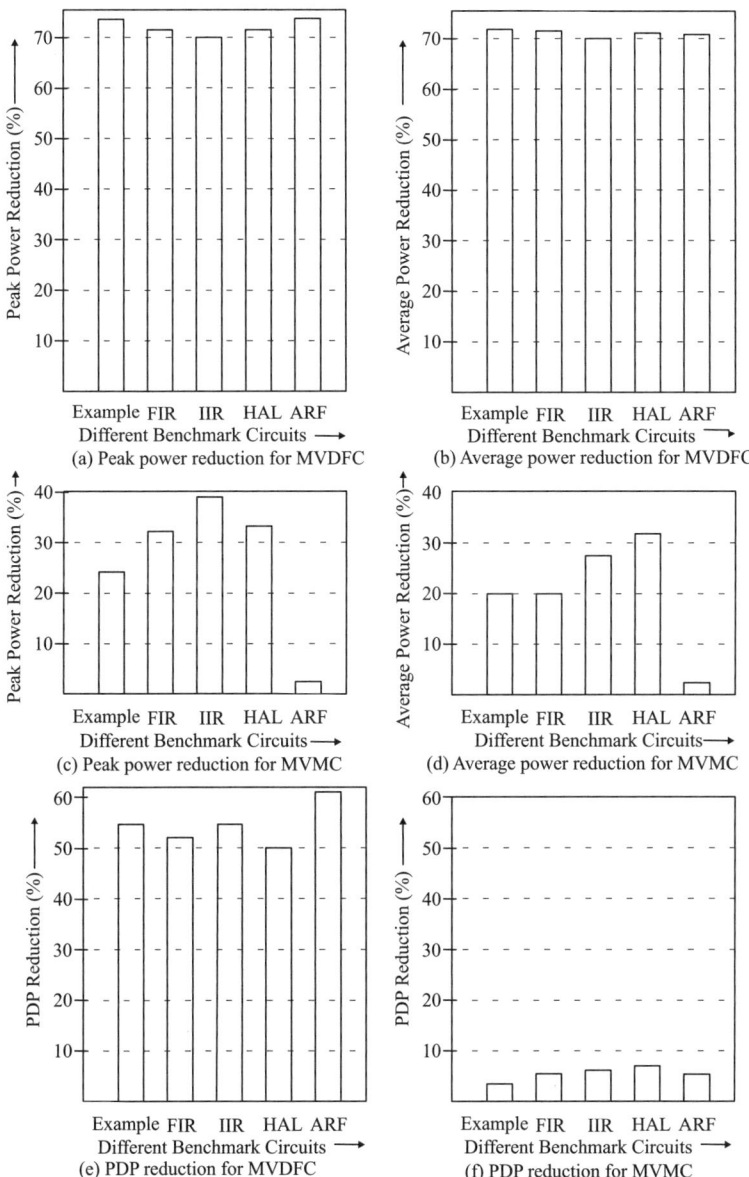

Fig. 7.6 Average reduction for different benchmarks circuits

Table 7.4 Peak and average power reduction for various scheduling schemes

Benchmark Circuits	Percentage Average Data for Various Schemes									
	DFC-Based		Shiue [316]		Martin [222]		Raghunathan [291]		Mohanty [218]	
	ΔP_{p}	ΔP_{a}	ΔP_{p}	ΔP_{a}	ΔP_{p}	ΔP_{a}	ΔP_{p}	ΔP_{a}	ΔP_{p}	ΔP_{a}
(1) EXP	73	72	–	–	–	–	–	–	–	–
(2) FIR	71	72	63	NA	40	NO	23	38	71	53
(3) IIR	69	69	–	–	–	–	–	–	–	–
(4) HAL	71	71	28	NA	–	–	–	–	73	70
(5) ARF	73	71	50	NA	–	–	–	–	68	67

that the reductions using combined MVDFC are appreciable. It is observed that the power consumption increases for higher switching activity and decreases for lower switching activity. The power reductions for the proposed scheduling scheme are listed along with other scheduling algorithms dealing with peak power reduction in Table 7.4. The table is not meant to provide an exact comparison, but rather to provide a general idea of relative performance.

7.4.4 Conclusions

Reduction of both peak power and average power consumption of a CMOS circuit is important. The previous sections addressed simultaneous reduction of peak power and average power at behavioral level using low-power data path scheduling techniques. Data path scheduling schemes, one using MVDFC and another using MVMC have been introduced. ILP-based optimization techniques were used for the above two modes of data path operations. A significant amount of peak and average power reduction over the single supply voltage and single-frequency scenario could be achieved in both the cases by the presented scheduling algorithm. The reductions attained in peak power, average power and PDP by using combined MVDFC were noteworthy. *The results clearly indicate that DFC is a better scheme than the multicycling approach for power minimization.*

7.5 Scheduling or Binding for Peak Power Reduction

Few background research works have addressed peak power minimization at the behavioral level. In this section, those works are briefly discussed, and an overview of their relative performance is presented in Table 7.5.

7.5.1 Scheduling Algorithms

Shiue [316, 317], Shiue and Chakrabarti [319], and Shiue et al. [321] propose different data path scheduling schemes to minimize peak power at the behavioral level.

Table 7.5 Relative performance of various schemes proposed for peak power minimization

Proposed Work	Synthesis Tasks Performed	Methods Used	Time Complexity	% Power Reduction
Shiue and et al. [316, 317, 321, 319]	Scheduling	ILP Force Directed	Exponential $O\left(cn^3\right)$	50.0–75.0
Nielsen and Madsen [254]	Scheduling Allocation	ASAP Heuristic	$O(|V| + |E|)$	NA
Martin and Knight [202, 201]	Scheduling Assignment	Genetic Algorithms	NA	40.3–60.0

In [319, 321, 317], ILP formulations are proposed, and [316] also includes a modified force-directed scheduling algorithm. The running time of the proposed modified force-directed scheduling algorithm is $O\left(cn^3\right)$, if c is the number of control steps and n is the number of nodes. The scheduling schemes in [316] minimize peak power while satisfying time constraint. The scheduling algorithms in [319, 321, 317] minimize both peak power and peak area while satisfying latency constraints. The simultaneous minimization is performed with the help of a multi-cost objective function using user-defined weighting factors. The formulation considers multicycling and pipelining and single supply voltage design. Peak power reductions in the range of 50%–75% have been reported after scheduling and pipelining. The reduction in peak area is also in the range of 50%–75%.

A heuristic high-level synthesis algorithm that solves scheduling, allocation and assignment simultaneously under both a time and power constraint is presented by Nielsen and Madsen [254]. The proposed approach enables exploration of the design space to include different types of FUs to trade area versus power. The underlying idea of the proposed algorithm is to elongate the classical ASAP schedule to fit the power constraint using a heuristic approach.

7.5.2 Binding Algorithms

Martin and Knight [202, 201] have proposed a scheme that combines SPICE simulations with a behavioral synthesis tool to estimate and optimize digital ASICs' peak power consumption. SPICE is used to measure the power consumption accurately. The behavioral synthesis tool is used for simultaneous assignment and scheduling such that the use of power in each clock cycle is minimal. Genetic algorithms are used in the behavioral synthesis tool for optimization. The authors claim that genetic algorithms have advantages over other conventional optimization tools since they never get stuck in local minima and do not need fine-tuning. The proposed synthesis tool can minimize the following parameters:

- average power with area, delay and peak power constraints
- peak power with area, delay and average power constraints

- delay with area and peak or average power constraints
- area with delay, average and/or peak power constraints
- any combination of area and power as weighted formula.

The optimizer searches for the best combination of architecture and schedule while satisfying all given constraints. They reported peak power reduction in the range of 40%–60%, which comes at a cost of 0.3%–2.7% penalty in average power. The work also considers mixed supply voltage scenario $(3.3\,V, 5.0\,V)$. It is reported that the time penalty is large if the circuit is operated at low voltage, but significant power reduction is achieved.

7.6 Summary and Conclusions

Peak power in a circuit, which is the maximum instantaneous power dissipation, is becoming increasingly important. The goal of reducing peak power is to improve the battery efficiency and circuit reliability. The current literature is sparse in this area of research. Only a few approaches are available that use voltage or frequency scaling for peak power reduction, and they tend to use ILP-based optimization.

The next chapter deals with fluctuation in power during circuit operation.

Chapter 8
Transient Power Reduction

8.1 Introduction

In this chapter, a framework for simultaneous reduction of the energy and transient power during behavioral synthesis is presented. A new parameter called "cycle power function" (CPF) is defined, which captures the transient power characteristics as an equally weighted sum of normalized mean cycle power and normalized mean cycle differential power [220, 218, 210]. Minimizing this parameter using MVDFC results in the reduction of both energy and transient power. Based on these ideas, a new data path scheduling algorithm called CPF-scheduler is developed, which attempts to optimize power and energy via minimization of the CPF parameter by the scheduling process. Further, another modified version of the non-linear CPF, denoted as CPF*, is presented so that ILP can be used for its minimization during data path scheduling [228, 210, 231, 232]. The first scheme, CPF-MVDFC, combines both MV and DFC for CPF* minimization, whereas the second scheme [228, 210, 232], CPF-MVMC, uses MV and MC [228, 210, 231].

8.2 Modeling for Power Transience or Fluctuation of a Data Path Circuit

The CPF is defined in a way such that it can capture simultaneously the average power, the peak power and the peak power differential of the data path. The peak power and peak power differential determine the transient power characteristics of the circuit. Minimization of the CPF using MV results in the minimization of energy as well. The data path is represented as a sequencing DFG. The notations and terminology needed for the presented models are given in Table 8.1.

The CPF consists of two main components: the normalized mean cycle power and the normalized mean cycle difference power. The normalized mean cycle power (P_{norm}) is the mean cycle power (P) normalized with respect to the peak power

S.P. Mohanty et al., *Low-Power High-Level Synthesis for Nanoscale CMOS Circuits*,
DOI: 10.1007/978-0-387-76474-0_8, © Springer Science+Business Media, LLC 2008

Table 8.1 List of notations and terminology used in CPF modeling

N	: Total number of control steps in the DFG
L_V	: Number of voltage levels for multiple supply voltage operation
L_f	: Number of frequency levels for dynamic frequency operation
N_v	: Total number of vertices in the DFG
c	: A control step or a clock cycle in the DFG
v_i	: Any vertex i, where $1 \leq i \leq N_v$
P_c	: The total power consumption in control step c (cycle power consumption)
P_{peak}	: Peak power consumption for the DFG equal to $\max(P_c)_{\forall c}$
P	: Mean power consumption of the DFG (average P_c over all control steps)
P_{norm}	: Normalized mean power consumption of the DFG
PD_c	: Cycle difference power (for cycle c; a measure of cycle power fluctuation)
PD_{peak}	: Peak differential power consumption for the DFG equal to $\max(PD_c)_{\forall c}$
PD	: Mean of the cycle difference powers for all control steps in DFG
PD_{norm}	: Normalized mean of the mean difference powers for all steps in DFG
CPF	: Cycle power function
FU_{k,L_V}	: Any functional unit of type k operating at voltage level L_V
FU_i	: Any functional unit FU_{k,L_V} needed by o_i for its execution ($v_i \in FU_{k,L_V}$)
$FU_{i,c}$	: Any functional unit FU_i active in control step c
R_c	: Total number of functional units active in step c (same as the number of operations scheduled in c)
$\alpha_{i,c}$	: Switching activity of resource $FU_{i,c}$
$V_{i,c}$	: Operating voltage of resource $FU_{i,c}$
$C_{i,c}$	: Load capacitance of resource $FU_{i,c}$
f_c	: Frequency of control step c

consumption (P_{peak}) of the DFG. The normalized mean cycle difference power (PD_{norm}) is the mean cycle difference power (PD) normalized with respect to the peak power differential of the DFG. The second component varies between the two models. The mean difference power is the mean of the cycle difference power PD_c over the control steps. In model 1, the cycle difference power PD_c is defined as the absolute deviation of the cycle power from the mean cycle power. Then, the mean cycle difference power PD is the mean deviation of the cycle power from the mean cycle power. In model 2, the cycle difference power PD_c of a current cycle is modeled as the cycle-to-cycle power gradient. In other words, the cycle difference power PD_c of a current control step c is the difference (or gradient) of the current cycle power and the previous cycle power. This difference can be expressed mathematically as $PD_c = P_c - P_{c-1}$ or $DP_{c+1} = P_{c+1} - P_c$. In this case, the mean cycle difference power DP is the mean difference (or the gradient). The two models are further elaborated and used in defining the CPF.

8.2.1 Model 1: CPF Using Mean Deviation

For a set of n observations, $x_1, x_2, ..., x_n$ from a given distribution, the sample mean (which is an unbiased estimator for the population mean, μ) is $m = \frac{1}{n}\sum_{i=1}^{n} x_i$. The absolute deviation of these observations is defined as $\Delta x_i = |x_i - m|$. The mean deviation of the observations is given by $MD = \frac{1}{n}\sum_{i=1}^{n}|x_i - m|$. In this case, we model the cycle difference power PD_c as the absolute deviation of cycle power P_c from the mean cycle power P. Similarly, the mean difference power PD is modeled as the mean deviation of the cycle power P_c. The mean cycle power P is an unbiased estimate of the average power consumption of the DFG.

The power consumption for any control step c, given by Equation 8.1, is the total power consumption of all FUs active in control step c. It also includes the power consumption of the level converters, which are considered as resources operating in a cycle c if the current resource is driven by a resource operating at lower voltage.

$$P_c = \sum_{i=1}^{R_c} \alpha_{i,c} C_{i,c} V_{i,c}^2 f_c. \tag{8.1}$$

The peak power consumption of the DFG is the maximum power consumption over all the N control steps, which can be expressed as

$$P_{\text{peak}} = \max\left(P_c\right)_{\forall c=1,2,\ldots,N} \tag{8.2}$$

$$= \max\left(\sum_{i=1}^{R_c} \alpha_{i,c} C_{i,c} V_{i,c}^2 f_c\right)_{\forall c=1,2,\ldots,N} \tag{8.3}$$

The mean cycle power consumption of the DFG (P) is defined as

$$P = \frac{1}{N}\sum_{c=1}^{N} P_c, \tag{8.4}$$

$$= \frac{1}{N}\sum_{c=1}^{N}\left(\sum_{i=1}^{R_c} \alpha_{i,c} C_{i,c} V_{i,c}^2 f_c\right). \tag{8.5}$$

The mean cycle power P is an unbiased estimate of the average power consumption of the DFG. The true average power consumption of the DFG is the total energy consumption of the DFG per clock cycle or per second. The normalized mean cycle power (P_{norm}) is obtained by dividing P by the maximum cycle power (P_{peak}):

$$P_{\text{norm}} = \frac{P}{P_{\text{peak}}}, \tag{8.6}$$

$$= \frac{\frac{1}{N}\sum_{c=1}^{N}\sum_{i=1}^{R_c} \alpha_{i,c} C_{i,c} V_{i,c}^2 f_c}{\max\left(\sum_{i=1}^{R_c} \alpha_{i,c} C_{i,c} V_{i,c}^2 f_c\right)_{\forall c=1,2,\ldots,N}}. \tag{8.7}$$

Thus, the normalized mean cycle power (P_{norm}) is a quantity without units in the range of $[0, 1]$.

The cycle difference power (PD_c) for any control step can be defined as follows: It is the absolute deviation of the cycle power from the mean cycle power consumption of the DFG and a measure of the cycle power fluctuation of the DFG.

$$PD_c = |P - P_c|, \tag{8.8}$$

$$= \left| \frac{1}{N} \sum_{c=1}^{N} \left(\sum_{i=1}^{R_c} \alpha_{i,c} C_{i,c} V_{i,c}^2 f_c \right) - \sum_{i=1}^{R_c} \alpha_{i,c} C_{i,c} V_{i,c}^2 f_c \right|. \tag{8.9}$$

The peak differential power which characterizes the maximum power fluctuation of the DFG is given by PD_{peak}. It characterizes the maximum power fluctuation or the transient of the DFG over the entire set of control steps.

$$PD_{peak} = \max \left(|P - P_c| \right)_{\forall c=1,2,\dots,N}, \tag{8.10}$$

$$= \max \left(\left| \frac{1}{N} \sum_{c=1}^{N} \left(\sum_{i=1}^{R_c} \alpha_{i,c} C_{i,c} V_{i,c}^2 f_c \right) \right. \right.$$

$$\left. \left. - \sum_{i=1}^{R_c} \alpha_{i,c} C_{i,c} V_{i,c}^2 f_c \right| \right)_{\forall c=1,2,\dots,N}. \tag{8.11}$$

The mean cycle difference power (PD) is calculated as the sample mean of PD_c. It is a measure of the power spread or distribution of the cycle power over all control steps of the DFG.

$$PD = \frac{1}{N} \sum_{c=1}^{N} DP_c, \tag{8.12}$$

$$= \frac{1}{N} \sum_{c=1}^{N} |P - P_c|, \tag{8.13}$$

$$= \frac{1}{N} \sum_{c=1}^{N} \left(\left| \frac{1}{N} \sum_{c=1}^{N} \left(\sum_{i=1}^{R_c} \alpha_{i,c} C_{i,c} V_{i,c}^2 f_c \right) - \sum_{i=1}^{R_c} \alpha_{i,c} C_{i,c} V_{i,c}^2 f_c \right| \right). \tag{8.14}$$

The normalized mean cycle difference power (DP_{norm}) can be written as

$$PD_{norm} = \frac{PD}{PD_{peak}}, \tag{8.15}$$

$$= \frac{\frac{1}{N} \sum_{c=1}^{N} \left(\left| \frac{1}{N} \sum_{c=1}^{N} \left(\sum_{i=1}^{R_c} \alpha_{i,c} C_{i,c} V_{i,c}^2 f_c \right) - \sum_{i=1}^{R_c} \alpha_{i,c} C_{i,c} V_{i,c}^2 f_c \right| \right)}{\max \left(\left| \frac{1}{N} \sum_{c=1}^{N} \left(\sum_{i=1}^{R_c} \alpha_{i,c} C_{i,c} V_{i,c}^2 f_c \right) - \sum_{i=1}^{R_c} \alpha_{i,c} C_{i,c} V_{i,c}^2 f_c \right| \right)_{\forall c}}. \tag{8.16}$$

The above normalized mean cycle difference power DP_{norm} is a quantity without units in the range of $[0,1]$.

The CPF, which is modeled as the equally weighted sum of the normalized mean cycle power (P_{norm}), and the normalized mean cycle difference power (PD_{norm}) are given below:

$$CPF(P_{norm}, DP_{norm}) = P_{norm} + DP_{norm}. \tag{8.17}$$

Thus, the CPF will have a value in the range of $[0,2]$. The CPF can be affected by various constraints, including the resource constraints. In terms of peak cycle power

(P_{peak}) and peak cycle difference power (PD_{peak}), the CPF can be expressed as

$$\text{CPF} = \frac{P}{P_{\text{peak}}} + \frac{PD}{PD_{\text{peak}}}, \tag{8.18}$$

$$= \frac{\frac{1}{N}\sum_{c=1}^{N} P_c}{P_{\text{peak}}} + \frac{\frac{1}{N}\sum_{c=1}^{N}|P - P_c|}{DP_{\text{peak}}}. \tag{8.19}$$

Using Equations 8.7 and 8.16, the cycle power function (CPF) can be written as follows:

$$\text{CPF} = \frac{\frac{1}{N}\sum_{c=1}^{N}\sum_{i=1}^{R_c} \alpha_{i,c}C_{i,c}V_{i,c}^2 f_c}{\max\left(\sum_{i=1}^{R_c} \alpha_{i,c}C_{i,c}V_{i,c}^2 f_c\right)_{\forall c}}$$

$$+ \frac{\frac{1}{N}\sum_{c=1}^{N}\left(\left|\frac{1}{N}\sum_{c=1}^{N}\left(\sum_{i=1}^{R_c} \alpha_{i,c}C_{i,c}V_{i,c}^2 f_c\right) - \sum_{i=1}^{R_c} \alpha_{i,c}C_{i,c}V_{i,c}^2 f_c\right|\right)}{\max\left(\left|\frac{1}{N}\sum_{c=1}^{N}\left(\sum_{i=1}^{R_c} \alpha_{i,c}C_{i,c}V_{i,c}^2 f_c\right) - \sum_{i=1}^{R_c} \alpha_{i,c}C_{i,c}V_{i,c}^2 f_c\right|\right)_{\forall c}}. \tag{8.20}$$

8.2.2 Model 2: CPF Using Cycle-to-Cycle Gradient

For a set $x_1, x_2, \ldots, x_n$ of n observations from a given distribution, the observation-to-observation gradient can be defined as, $|x_{i+1} - x_i|$, where $1 \le i \le n - 1$. The mean gradient is given by $\frac{1}{n-1}\sum_{i=1}^{n-1}|x_{i+1} - x_i|$. It should be noted that there are $n - 1$ gradients for n observations. In this case, the cycle difference power DP_c is modeled as the cycle-to-cycle power gradient and the mean difference power DP as the mean gradient. The models for the mean cycle power or the average power (Equations 8.1–8.5) remain the same as in the previous subsection.

The cycle difference power (PD_c) for any control step is defined as the difference in the power consumption of the current versus the previous control step, as given below:

$$PD_{c+1} = |P_{c+1} - P_c|, \tag{8.21}$$

$$= \left|\sum_{i=1}^{R_{c+1}} \alpha_{i,c+1}C_{i,c+1}V_{i,c+1}^2 f_{c+1} - \sum_{i=1}^{R_c} \alpha_{i,c}C_{i,c}V_{i,c}^2 f_c\right|. \tag{8.22}$$

The peak differential power (PD_{peak}) is characterized by

$$PD_{\text{peak}} = \max\left(|P_{c+1} - P_c|\right)_{\forall c=1,2,\ldots,N-1}, \tag{8.23}$$

$$= \max\left(\left|\sum_{i=1}^{R_{c+1}} \alpha_{i,c+1}C_{i,c+1}V_{i,c+1}^2 f_{c+1} - \sum_{i=1}^{R_c} \alpha_{i,c}C_{i,c}V_{i,c}^2 f_c\right|\right)_{\forall c=1\to N-1}. \tag{8.24}$$

The mean cycle difference power (PD) is calculated as

$$PD = \frac{1}{N-1} \sum_{c=1}^{N-1} DP_{c+1},$$

(8.25)

$$= \frac{1}{N-1} \sum_{c=1}^{N-1} |P_{c+1} - P_c| \text{ and}$$

(8.26)

$$= \frac{1}{N-1} \sum_{c=1}^{N-1} \left(\left| \sum_{i=1}^{R_{c+1}} \alpha_{i,c+1} C_{i,c+1} V_{i,c+1}^2 f_{c+1} - \sum_{i=1}^{R_c} \alpha_{i,c} C_{i,c} V_{i,c}^2 f_c \right| \right).$$

(8.27)

The normalized mean cycle difference power (DP_{norm}) can be written as given below:

$$PD_{norm} = \frac{PD}{PD_{peak}},$$

(8.28)

$$= \frac{\frac{1}{N-1} \sum_{c=1}^{N-1} \left(\left| \sum_{i=1}^{R_{c+1}} \alpha_{i,c+1} C_{i,c+1} V_{i,c+1}^2 f_{c+1} - \sum_{i=1}^{R_c} \alpha_{i,c} C_{i,c} V_{i,c}^2 f_c \right| \right)}{\max \left(\left| \sum_{i=1}^{R_{c+1}} \alpha_{i,c+1} C_{i,c+1} V_{i,c+1}^2 f_{c+1} - \sum_{i=1}^{R_c} \alpha_{i,c} C_{i,c} V_{i,c}^2 f_c \right| \right)_{\forall c=1 \to N-1}}.$$

(8.29)

Using Equations 8.7 and 8.29, the CPF can be written as follows:

$$CPF = P_{norm} + PD_{norm},$$

(8.30)

$$= \frac{P}{P_{peak}} + \frac{PD}{PD_{peak}},$$

(8.31)

$$= \frac{\frac{1}{N} \sum_{c=1}^{N} P_c}{P_{peak}} + \frac{\frac{1}{N-1} \sum_{c=1}^{N-1} |P_{c+1} - P_c|}{PD_{peak}},$$

(8.32)

$$= \frac{\frac{1}{N} \sum_{c=1}^{N} \sum_{i=1}^{R_c} \alpha_{i,c} C_{i,c} V_{i,c}^2 f_c}{\max \left(\sum_{i=1}^{R_c} \alpha_{i,c} C_{i,c} V_{i,c}^2 f_c \right)_{\forall c}}$$

$$+ \frac{\frac{1}{N-1} \sum_{c=1}^{N-1} \left(\left| \sum_{i=1}^{R_{c+1}} \alpha_{i,c+1} C_{i,c+1} V_{i,c+1}^2 f_{c+1} - \sum_{i=1}^{R_c} \alpha_{i,c} C_{i,c} V_{i,c}^2 f_c \right| \right)}{\max \left(\left| \sum_{i=1}^{R_{c+1}} \alpha_{i,c+1} C_{i,c+1} V_{i,c+1}^2 f_{c+1} - \sum_{i=1}^{R_c} \alpha_{i,c} C_{i,c} V_{i,c}^2 f_c \right| \right)_{\forall c=1 \to N-1}}.$$

(8.33)

8.2.3 Minimization of CPF as an Objective Function

The previously derived functions (Equation 8.20 or 8.33) can be used as the objective function for low-power data path scheduling. *The minimization of this objective function using MV, DFC and MC will lead to the reduction of energy and power parameters.* From Equations 8.19, 8.20 and 8.33, the following observations are made about the CPF. The CPF is a *non-linear* function. It is a function of four parameters: average power (P), peak power (P_{peak}), average difference power (PD) and peak difference power (PD_{peak}). Each of these power parameters dependents on switching activity, capacitance, operating voltage and operating frequency. The absolute function (abs or $|\ |$) in the numerator (of Equation 8.20 or 8.33) contributes to the non-linearity. The complex behavior of the function is also contributed by the denominator parameters, P_{peak} and PD_{peak}.

The power models expressed in Equations 8.33 and 8.20 for the CPF use generic parameters, such as $\alpha_{i,c}, C_{i,c}, V_{i,c}$, and f_c. The intent of using such parameters is to make the CPF model a general one, independent of any specific energy or power model. It can accommodate both LUT-based energy (power) models and energy (power) macro-models. The generic model can also help in easy integration of the CPF model in a behavioral synthesis tool that uses both a behavioral power estimator and a data path scheduler. Using the dynamic energy model proposed in [70], we can express the effective switching capacitance as

$$\alpha_i C_i = C_{\text{sw}i}(\alpha_i^1, \alpha_i^2). \tag{8.34}$$

Here, α_i and C_i are the parameters corresponding to the functional unit FU_i. $C_{\text{sw}i}$ is a measure of the effective switching capacitance of resource (functional unit) FU_i, which is a function of α_i^1 and α_i^2; α_i^1 and α_i^2 are the average switching activity values on the first and second input operands of resource FU_i. It should be noted that the above switching model (in Equation 8.34) handles input pattern dependencies. Moreover, the generic CPF model can be easily tuned to handle any of the four modes of data path circuit operation, such as (i) single supply voltage and single frequency, (ii) multiple supply voltages and single frequency, (iii) MVDFC and (iv) MVMC. For example, for single supply voltage and single frequency scheme, $V_{i,c}$ and f_c are the same for all c; for multiple supply voltage and multicycling, f_c is the same for all c. Using Equation 8.34, we rewrite Equation 8.20 as (Model: 1)

$$\text{CPF} = \frac{\frac{1}{N}\sum_{c=1}^{N}\sum_{i=1}^{R_c} C_{\text{sw}i,c} V_{i,c}^2 f_c}{\max\left(\sum_{i=1}^{R_c} C_{\text{sw}i,c} V_{i,c}^2 f_c\right)_{\forall c}}$$

$$+ \frac{\frac{1}{N}\sum_{c=1}^{N}\left(\left|\frac{1}{N}\sum_{c=1}^{N}\left(\sum_{i=1}^{R_c} C_{\text{sw}i,c} V_{i,c}^2 f_c\right) - \sum_{i=1}^{R_c} C_{\text{sw}i,c} V_{i,c}^2 f_c\right|\right)}{\max\left(\left|\frac{1}{N}\sum_{c=1}^{N}\left(\sum_{i=1}^{R_c} C_{\text{sw}i,c} V_{i,c}^2 f_c\right) - \sum_{i=1}^{R_c} C_{\text{sw}i,c} V_{i,c}^2 f_c\right|\right)_{\forall c}}. \tag{8.35}$$

Similarly, using Equation 8.34, we rewrite Equation 8.33 as (Model: 2)

$$\text{CPF} = \frac{\frac{1}{N}\sum_{c=1}^{N}\sum_{i=1}^{R_c} C_{\text{sw}i,c} V_{i,c}^2 f_c}{\max\left(\sum_{i=1}^{R_c} C_{\text{sw}i,c} V_{i,c}^2 f_c\right)_{\forall c}}$$

$$+ \frac{\frac{1}{N-1}\sum_{c=1}^{N-1}\left(\left|\sum_{i=1}^{R_{c+1}} C_{\text{sw}i,c+1} V_{i,c+1}^2 f_{c+1} - \sum_{i=1}^{R_c} C_{\text{sw}i,c} V_{i,c}^2 f_c\right|\right)}{\max\left(\left|\sum_{i=1}^{R_{c+1}} C_{\text{sw}i,c+1} V_{i,c+1}^2 f_{c+1} - \sum_{i=1}^{R_c} C_{\text{sw}i,c} V_{i,c}^2 f_c\right|\right)_{\forall c=1\rightarrow N-1}}. \tag{8.36}$$

The notation $C_{swi,c}$ represents C_{swi} for the functional unit FU_i active in control step c. The above two functions (Equations 8.35 and 8.36) are used as objective functions for the scheduling algorithm. The combinations α_i^1 and α_i^2 are estimated using behavioral simulation of a DFG [180, 312, 297]. A LUT is constructed to store the C_{sw} values for different combinations of (α^1 and α^2) for different types of FUs, such as multipliers and ALUs. An interpolation technique is used to determine the C_{sw} values for the (α^1 and α^2) combinations that are not available in the LUT.

8.3 Heuristic-Based Scheduling Algorithm for CPF Minimization

8.3.1 Introduction

In this section, a heuristic-based scheduling algorithm is presented that minimizes the objective functions (Equation 8.35 or 8.36) using MVDFC to reduce energy and power. It should be noted that frequency scaling helps in reducing power but not energy. Moreover, frequency reduction facilitates the operation of the different FUs at different voltages, which in turn helps in energy reduction. The target architecture model assumed for the scheduling is the same as in the previous chapters.

8.3.2 Algorithm Flow

The inputs to the algorithm are an unscheduled data flow graph (UDFG), the resource constraints, the number of allowable voltage levels (L_V), the number of allowable frequencies (L_f), the delay of each resource such as an FU (d_{FU}), multiplexer (d_{Mux}) and register (d_{Reg}) at different voltage levels. The delays of level converters (d_{Conv}) are represented in the form of a matrix that shows the delay for converting one voltage level V_i to another voltage level V_j (where both $V_i, V_j \in V_{L_V}$). The resource constraint includes the number of ALUs and multipliers at different voltage levels V_i (where $V_i \in V_{L_V}$). The scheduling algorithm determines and outputs the proper time stamp (c) for each operation, f_{base}, cfi_c and the voltage level such that the objective function in Equations 8.35 or 8.36 as well as the time penalty is minimum. To reduce the time penalty, the low-energy-consuming resources are used at as maximum a frequency as possible.

The flow of the proposed algorithm is outlined in Algorithm 29. In step 1, the switching activities at the inputs of each node are determined by using behavioral simulation of the DFG. For this purpose, different sets of application-specific input vectors (having different correlations) are given at the primary inputs of the DFG, and the average switching activity at each node is calculated. In step 2, the scheduler constructs a LUT with effective switching capacitance and the average switching activity pair as described in Equation 8.34. The size of the LUT affects the accuracy of the results. If the LUT is large enough to contain the switching capacitance for

---

Algorithm 29 The CPF-Scheduler Algorithm Flow

1: Calculate the switching activity at the inputs of each node through behavioral simulation of the DFG.
2: Construct a LUT of effective switching capacitance, switching activity pairs.
3: Find ASAP and ALAP schedules of the UDFG.
4: Determine the number of multipliers and ALUs at different operating voltages.
5: Modify both ASAP and ALAP schedules obtained in step 1 using the number of resources found in step 2 as initial resource constraint.
6: Calculate the total number of control steps, which is the maximum of ASAP and ALAP schedules from step 5.
7: Find the vertices having non-zero mobility and zero mobility.
8: Use the CPF-scheduler Heuristic to assign the time stamp and operating voltage for the vertices, and the cycle frequencies such that CPF and time penalty are minimum.
9: Find base frequency f_{base} and cycle frequency index cfi_c.
10: Calculate power, energy and delay details.

---

all estimated average switching activities in step 1, then the power model accuracy is highest. The scheduler uses interpolation techniques to find the switching capacitance for a pair of input average switching activities that do not exist in the LUT. The algorithm determines the ASAP and the ALAP schedules for the UDFG in step 3. The ASAP schedule is unconstrained, and the ALAP schedule uses the number of clock steps found in the ASAP schedule as the latency constraint. In step 4, the number of resources of each type and voltage level is determined. For example, if the resource constraint is one multiplier at 2.4 V, two multipliers at 3.3 V, two ALUs at 2.4 V and three ALUs at 3.3 V, then the relaxed voltage initial resource constraint is found to be three multipliers and five ALUs. In step 5, the scheduler uses the above relaxed voltage resource constraints and modifies the ASAP and ALAP schedules to take into account the resource constraints. This helps in restricting the mobility of vertices to a great extent and to reduce the solution search space for the heuristic. Because of the resource constraints, the number of control steps of modified ASAP and modified ALAP schedules may be different from that of the ASAP and ALAP schedule in step 3. In step 6, the scheduler fixes the total number of control steps of the schedule, which is the maximum of the control steps of the modified ASAP or modified ALAP in step 5. In step 7, the vertices are marked as having zero mobility or non-zero mobility. The zero mobility vertices are those having the same modified ASAP time stamp and modified ALAP time stamp, and non-zero mobility vertices are those having different modified ASAP and modified ALAP time stamps. On determining the vertices having zero mobility and vertices having non-zero mobility, the proper time stamp and operating voltage for mobile vertices and operating voltages for non-mobile vertices are calculated. Furthermore, operating clock frequencies are established such that the CPF as well as the time penalty is minimum. The CPF-scheduler uses a heuristic algorithm for the same. In step 9, the scheduler determines the base frequency (f_{base}) and cycle frequency index (cfi_c). In step 10, the scheduler calculates the peak power, average power, peak power differential, energy estimates of the scheduled DFG and the critical path delay.

8.3.3 Pseudocode of the Algorithm Heuristic

Algorithm 30 shows the heuristic algorithm used by the CPF-scheduler. The inputs to the CPF-scheduler heuristic are the modified ASAP time stamp of each vertex ($C_S[i]$), the modified ALAP time stamp of each vertex ($C_L[i]$), the resource constraints, the number of allowable voltage levels (L_V) and the number of allowable frequencies (L_f). The delays of each functional unit (d_{FU}), multiplexor (d_{Mux}) and register (d_{Reg}) at different voltage levels are also given as inputs. The delays of level converters (d_{Conv}) are represented in the form of a matrix. The heuristic has to find the time stamp c (in the range $[C_S[i], C_L[i]]$) and the operating voltage $V_{i,c}$ for each

---

Algorithm 30 The CPF-Scheduler Algorithm Heuristic

```
 1:  Initialize CurrentSchedule as modified ASAPSchedule;
 2:  while (All mobile vertices are not time-stamped) do
 3:      for all (Vertex v_i in the CurrentSchedule ) do
 4:          if ( v_i is a multiplication ) then
 5:              return Find the lowest available voltage for multipliers;
 6:          end if
 7:          if ( v_i is add/sub/comparison ) then
 8:              return Find the highest available operating voltage for ALUs;
 9:          end if
10:      end for
11:      Find CPF for CurrentSchedule and denote it as CurrentCPF ;
12:      Find R_T for CurrentSchedule and denote it as CurrentR_T;
13:      Maximum = −∞ ;
14:      for all (Mobile vertex v_i) do
15:          c_1 = CurrentSchedule[v_i];
16:          c_2 = ALAPSchedule[v_i];
17:          for all (c = c_1 to c_2 in steps of 1) do
18:              Find a TempSchedule by adjusting CurrentSchedule in which v_i is scheduled
                 in step c ;
19:              Find next higher operating voltage for multiplication vertex for the Temp-
                 Schedule (next lower for ALU operation) ;
20:              Find CPF for TempSchedule, denoted by TempCPF ;
21:              Find R_T for TempSchedule, denoted TempR_T;
22:              Difference = (CurrentCPF + CurrentR_T) − (TempCPF + TempR_T);
23:              if ( Difference > Maximum ) then
24:                  Maximum = Difference ;
25:                  CurrentVertex = v_i ;
26:                  CurrentCycle = c ;
27:                  CurrentVoltage = Operating voltage of v_i;
28:              end if
29:          end for
30:      end for
31:      Adjust CurrentSchedule to accommodate CurrentVertex in Currentcycle operating at
         voltage assigned above ;
32:  end while
```

---

vertex v_i. The aim of the heuristic is to minimize CPF as described in Equations 8.35 and 8.36 while keeping the time penalty at a minimum. The heuristic minimizes the time ratio R_T along with CPF to minimize the time penalty. The time ratio (R_T) is defined as the ratio between the critical path delay when the vertices of the DFG are operating in MVDFC mode $\left(T_{cp_{MVDFC}}\right)$ and when each of the vertices of the DFG is operated at the highest voltage in SVSF mode $\left(T_{cp_{SVSF}}\right)$. Expressed mathematically, $\left(R_T = \frac{T_{cp_{MVDFC}}}{T_{cp_{SVSF}}}\right)$. These two objectives, minimization of CPF (minimization of energy and power) and minimization of time penalty, are conflicting because if the operating voltage is reduced to minimize energy power consumption, this results in increase of critical path delay and hence increase of time penalty. The heuristic operates the energy hungry FUs at the highest possible voltage (frequency) and the least energy-consuming FUs at the lowest voltage (frequency) to achieve the simultaneous minimization of the mutually conflicting objectives. The heuristic fixes the operating voltages of the non-mobile vertices according to this order depending on the types of resource they need. The heuristic attempts to find suitable time stamps and operating voltages for the mobile vertices using exhaustive search. The mobile vertices are attempted to be placed in each of the time stamps within their mobile range $([C_S[i], C_L[i]])$; when each placement and voltage assignment is done, the CPF and R_T values are calculated. The predecessor and successor time stamps are adjusted accordingly to maintain the precedence. For this purpose, the heuristic maintains a matrix of dimension $(N * |k|V_{L_V})$ having the number of resources of different types (k) as entries row-wise over all control steps. $|k|$ is the type of resource available; for example, if only multiplier and ALUs are the available resources, then $|k| = 2$. If a voltage is assigned for a vertex, then the matrix entry of the corresponding type and operating voltage is decremented. A particular vertex is placed in a cycle for which the sum of CPF and R_T is minimum. The heuristic initially assumes the modified ASAP schedule (with relaxed voltage resource constrained) as the current schedule (line 01). If a vertex is a multiplication operation, the initial voltage assignment is the minimum available operating voltage depending on the number of multipliers, whereas for ALU operations vertex, it is the maximum available operating voltage (lines 04–08). Then the CPF and R_T values for the current schedule are calculated (lines 09 and 10). The heuristic finds CPF (and R_T) values for each allowable control step of each mobile vertex and for each available operating voltage, denoted as TempCPF (and TempR_T) (lines 17–20). The statement in line 17 adjusts the current schedule by adjusting the time stamps of successor vertices while maintaining the resource constraint (using the matrix) and guaranteeing that the precedence is satisfied. In line 12, the vertices are visited in ASAP manner. Another possible way of visiting the mobile vertices is to prioritize them in some manner: say, vertices with lower mobility are visited first. The heuristic fixes the time step and operating voltage for a vertex and hence the cycle frequency for which CPF + R_T is minimum (lines 22–26). For CPF computation, the heuristic uses $(1/d_c)$ as a temporary measure for f_c. The above steps are repeated until all mobile vertices are time-stamped.

8.3.4 Algorithm Time Complexity

Suppose there are $|V|$ number of vertices in the DFG, out of which $|V_m|$ number of vertices have mobility and the maximum mobility of any mobile vertex is t_m. It should be noted that the total number of vertices in the DFG is the total number of operations in DFG and the total number of NO–OPs. The run time of finding an operating voltage from the matrix for particular type of operation is $O(L_V)$. The statements from lines 04 to 08 have a run time of $\Theta(|V|L_V)$. The worst-case run time of the statement in line 17 (or line 31) that adjusts the current schedule is $O(|V_m|)$. The run time of the code segment between lines 17 and 28 is $O(|V_m|) + O(L_V) + \Theta(|V|) + \Theta(|V|)$, which is $\Theta(|V|)$, since it is always true that $|V_m|, L_V < |V|$. So, the run time of the code segment from lines 15 to 29 is $\Theta(t_m|V|)$. Thus, the run time of the code segment lines 12–30 is $\Theta(t_m|V_m||V|)$. The other statements of the pseudocode have a constant run time. So, the run time or time complexity of the code segment in lines 03–29 is $\Theta(|V||L_V|) + \Theta(t_m|V_m||V|) + O(|V_m|)$. This can be simplified to a weak upper bound on worst-case run of the code segment (lines 03–31) under the assumption that $|V_m| \approx |V|$, but in practice $|V_m| << |V|$. Under the above assumption, we conclude that the worst-case upper bound on the run time of the code segment in lines 03–31 is $\Theta\left(t_m|V|^2\right)$. Considering the *while* loop in line 02, the overall run time of the algorithm can be written as $\Theta\left(t_m|V|^2|V_m|\right)$. Again, under the assumption that $|V_m| \approx |V|$, the worst-case upper bound on the run time of the algorithm is $\Theta\left(t_m|V|^3\right)$. In other words, *the heuristic runs in time cubic to the number of vertices in the DFG. It can be noted that the time complexity of the algorithm is independent of the number of operating voltage levels.*

8.3.5 Experimental Results

The CPF-scheduler algorithm was implemented in C and tested with selected benchmark circuits. The benchmarks used are ARF, BPF, DCT filter, EWF, FIR filter and HAL differential equation solver. The presented algorithm can handle large DFGs and can find solutions in a reasonable time.

For an average switching capacitance calculation, a LUT method is used. The LUT construction consists of two phases: input pattern generation and cell characterization. The primary input signals of different correlations are generated using the autoregressive moving average (ARMA) model [297]. The characterization of the physical implementations of the library modules available in [219, 210] follows by applying the input patterns generated as discussed above for the values of (α_i^1, α_i^2) pairs in the table. To find the average switching capacitance for any of (α_i^1, α_i^2) pairs that do not exist in the LUT, the interpolation method is used. It should be noted that the larger the size of the LUT, the better the accuracy of the algorithm. The LUT has 100 pairs of entries for (α_i^1, α_i^2). The signals are propagated through different operators in the DFG, and the average switching activities are calculated for each node.

The first set of experiments was performed for the CPF model 1 (Equation 8.35) in which the cycle difference power is based on the absolute deviation. The scheduling algorithm uses several sets of resource constraints (RC1, RC2, RC3 and RC4) [220, 210]. The sets of resource constraints were chosen to cover resources at different operating voltages. The number of allowable voltage levels was assumed to be two (2.4 V and 3.3 V), and the maximum number of allowable frequencies are three. The CPF-scheduler determines the frequencies (in this case, 4.5 MHz, 9.0 MHz and 18.0 MHz). The results averaged over all the resource constraints are shown in Figure 8.1 for various benchmarks. The results take into account the power or energy consumption in overhead, such as level converters and DCU. This indicates that the scheduling scheme could achieve significant reductions in peak power, peak power differential, average power and total energy with reasonable time penalties. The time penalties for the benchmark circuits (ARF and HAL) were relatively high. *For many cases, the CPF-scheduler could reduce energy and power even without any time penalty or even with gain in time.* This happens when the

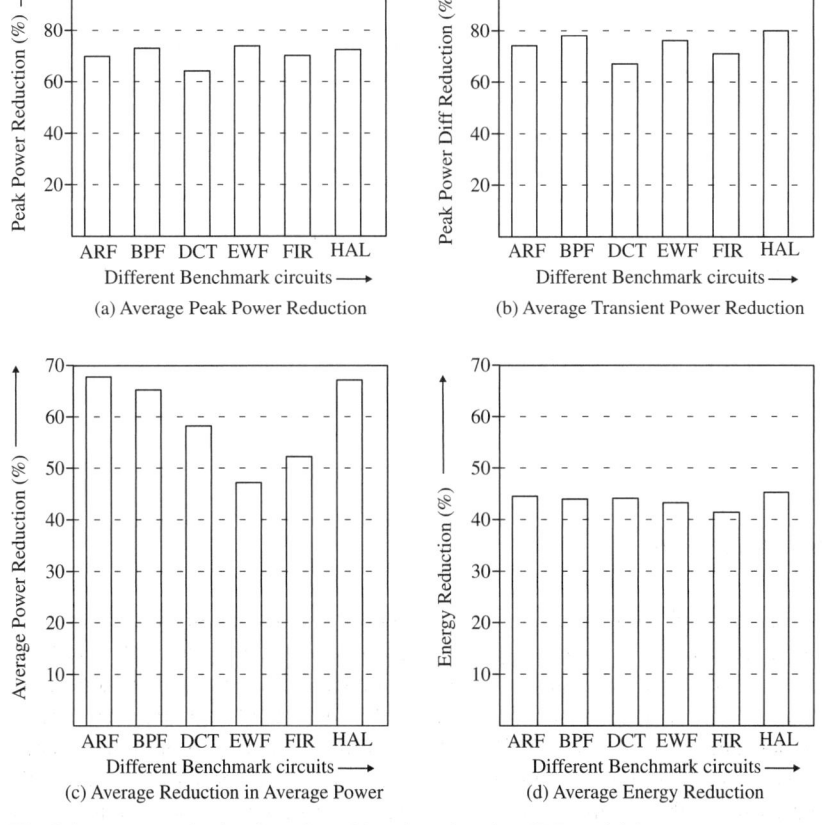

Fig. 8.1 Average reduction for selected benchmarks using CPF model 1

performance degradation due to multiplications in the critical path is adequately compensated by the number of ALU operations in the critical path. For this to happen, the ALU operations should be larger than or equal to the number of multiplications in the critical path. This is the case for most of the schedules obtained for the EWF and FIR benchmarks indicated by a time ratio (R_T) of less than or equal to one.

For the above experimental setup, the power consumption per cycle, over all the control steps (clock steps), is plotted for different benchmark circuits in Figs. 8.2(a), 8.2(b), 8.3(a) and 8.3(b) for resource constraints RC1, RC2, RC3 and RC4, respectively. The curves labeled as "SVSF" correspond to the profile when the schedule is operated at a single frequency (which is the maximum frequency of the slowest operator, the multiplier) and single voltage (SVSF operation). The profiles labeled as "MVDFC" correspond to the case when the MVDFC scheme is used. The effectiveness of the proposed scheduling scheme is obvious from the figures. Since the CPF is a complex function consisting of several parameters, it is difficult to accurately quantify the impact of a specific parameter.

The experiments are also performed with three voltage levels (1.5 V, 2.4 V and 3.3 V) and four frequency levels. The results could improve within the range of 5–10% in terms of power or energy reduction. However, the time penalty increases by 15%. It is to be noted that the number of allowable frequency levels should be as close to the number of allowable voltages in order to keep the time penalty within a reasonable limit.

The same set of experiments is also performed for the CPF model 2 Equation 8.36) in which the cycle difference power is modeled as cycle-to-cycle power gradient. The experimental results for different benchmarks averaged for different resource constraints are presented in Fig. 8.4. The results take into account the power or energy consumptions due to overhead. The results indicate that the energy and power reductions were similar, with small differences, but there were no changes in terms of time penalty. It can be concluded that the minor difference is due to the fact that the quantitative differences between the values of $\left(\frac{1}{N}\sum_{c=1}^{N}|P - P_c|\right)$ and $\left(\frac{1}{N-1}\sum_{c=1}^{N-1}|P_{c+1} - P_c|\right)$ are not significant. The cycle power plot for this model was almost similar to that of model 1.

8.3.6 Conclusions

For nanometer technology designs for low-power battery-driven systems, simultaneous minimization of total energy and transient power is beneficial. The CPF parameter defined and used in this work essentially facilitates such simultaneous optimization. The data path scheduling algorithm described is particularly useful for synthesizing data-intensive ASICs. The algorithm attempts to optimize energy and power while keeping the time penalty at a minimum. The CPF-scheduler algorithm assumes the number of different types of resources at each voltage level and the number of allowable frequencies as resource constraints. This work pro-

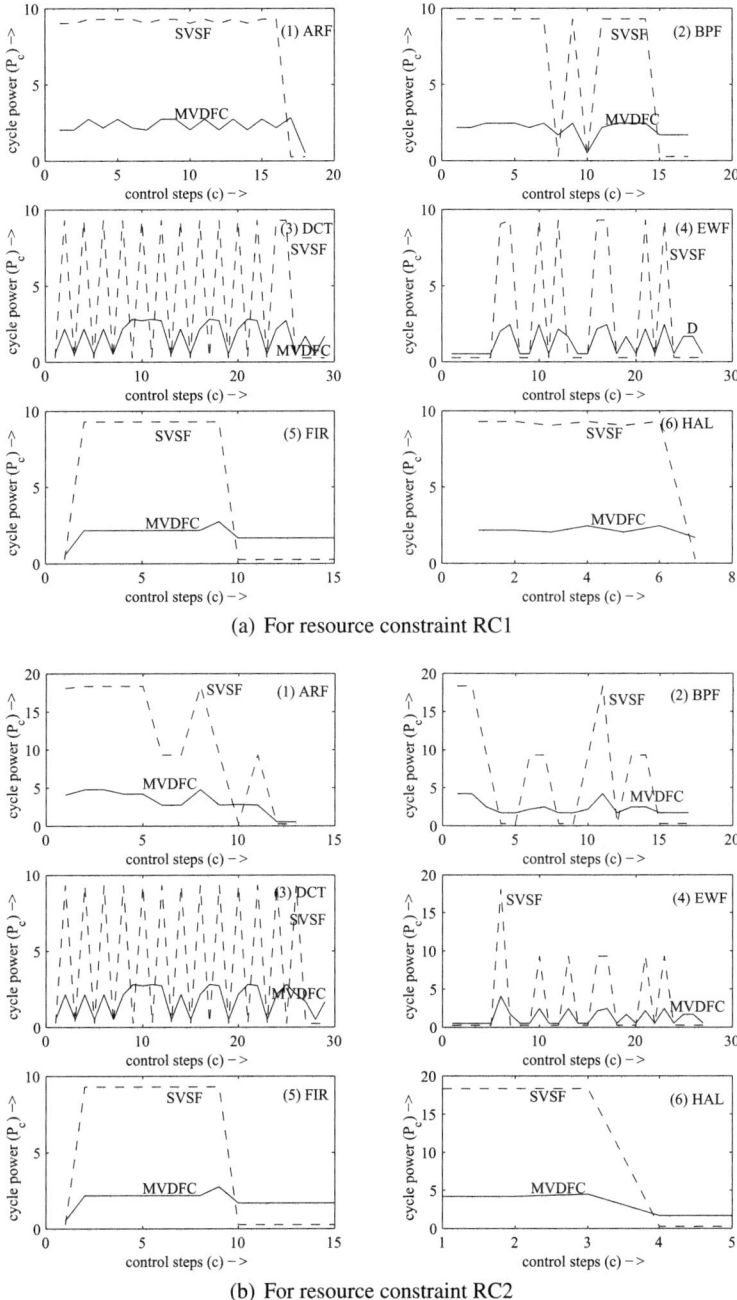

(a) For resource constraint RC1

(b) For resource constraint RC2

Fig. 8.2 Cycle power consumptions for selected resource constraints – RC1 and RC2

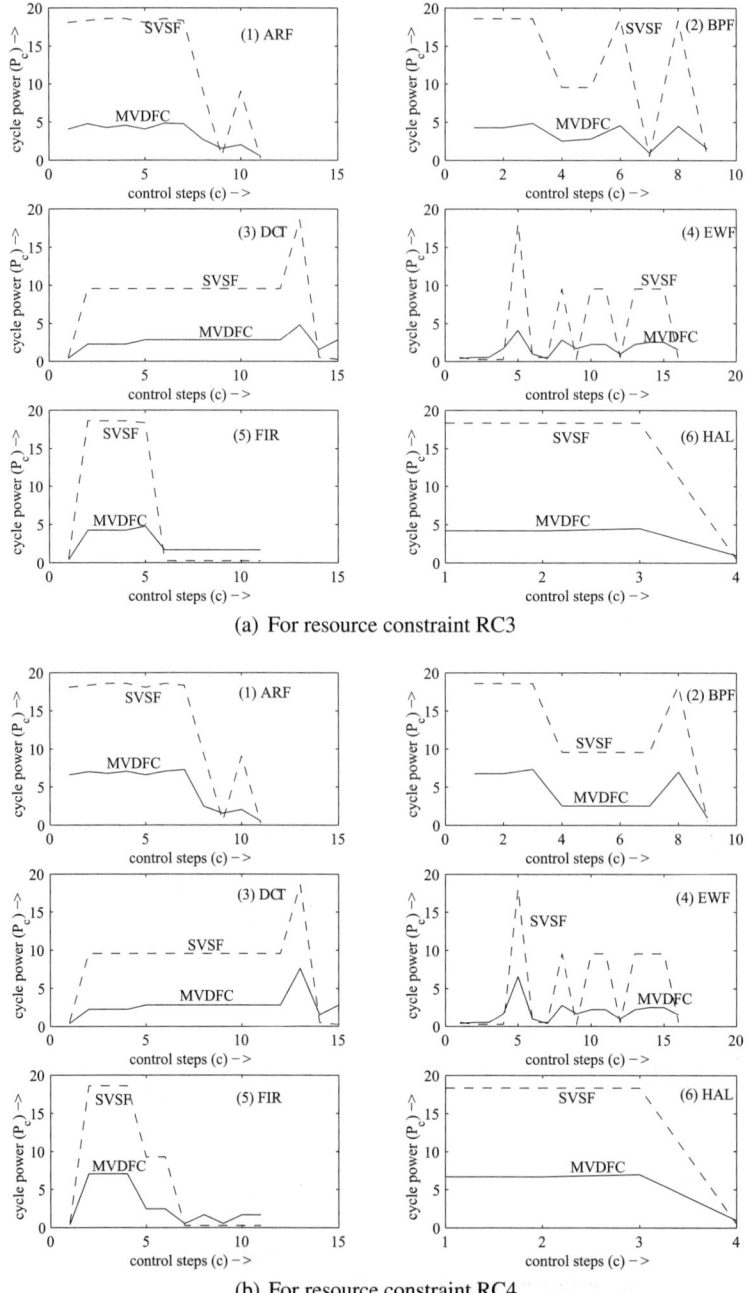

(a) For resource constraint RC3

(b) For resource constraint RC4

Fig. 8.3 Cycle power consumptions for selected resource constraints – RC3 and RC4

vides a unified framework for simultaneous multi-cost space metric optimization of different energy and power components in CMOS circuit design. Future work could address interconnect issues. The effectiveness of the CPF in the context of a pipelined data path and control-intensive applications also needs to be investigated.

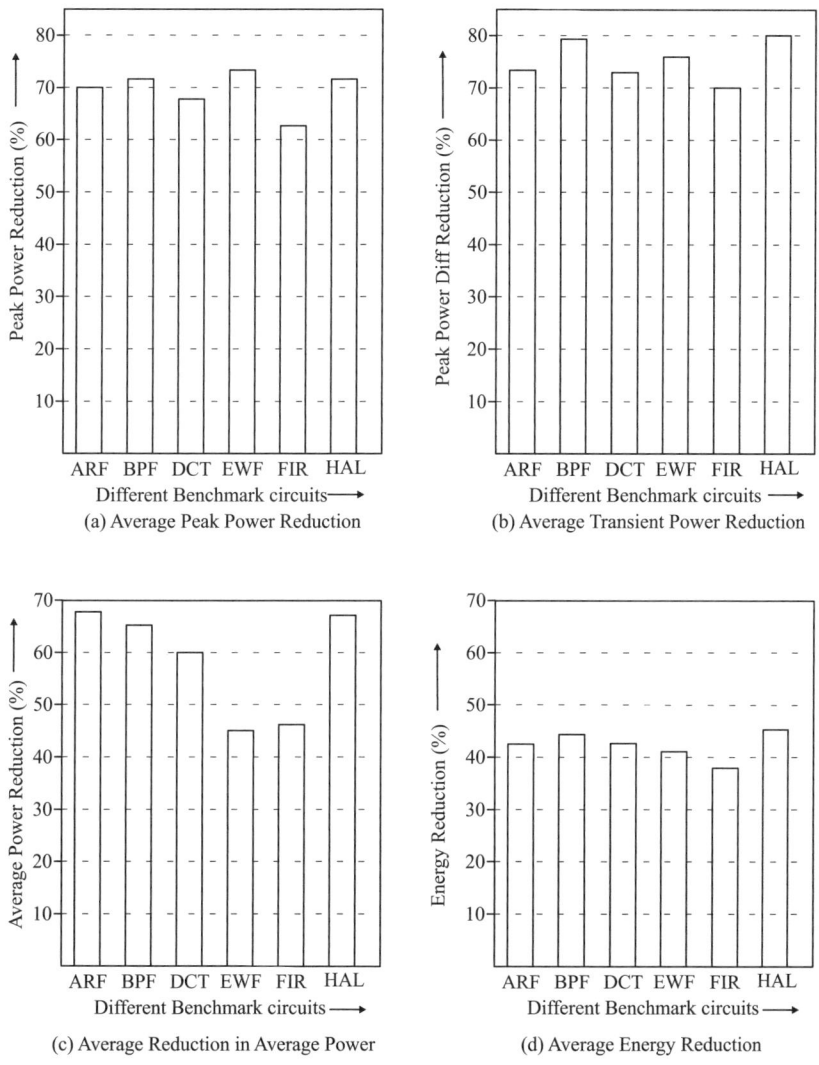

Fig. 8.4 Average reduction for selected benchmarks using CPF model 2

8.4 Modified Cycle Power Function (CPF*)

In this section, the parameter called cycle power function (CPF), which captures
the peak power, the peak power differential and the average power of the data path
circuit, is redefined. It should be noted that CPF captures the transient power characteristics of the circuit, and the minimization of CPF using multiple voltages could
lead to a reduction of energy. The non-linear behavior of CPF is studied, and it is
then modified so that ILP can be used for its minimization. The data path is represented as a sequencing DFG. The definitions and notations used are the same as that
of the previous section (Table 8.1).

Following the same steps as in the previous section, the CPF is modeled as an
equally weighted sum of the normalized mean cycle power (P_{norm}) and the normalized mean cycle difference power (PD_{norm}) as given below:

$$CPF(P_{\text{norm}}, PD_{\text{norm}}) = P_{\text{norm}} + PD_{\text{norm}}. \tag{8.37}$$

CPF has a value in the range $[0, 2]$. In terms of peak cycle power (P_{peak}) and peak
cycle difference power (PD_{peak}), CPF can be expressed as

$$CPF = \frac{P}{P_{\text{peak}}} + \frac{PD}{PD_{\text{peak}}} = \frac{\frac{1}{N}\sum_{c=1}^{N} P_c}{P_{\text{peak}}} + \frac{\frac{1}{N}\sum_{c=1}^{N} |P - P_c|}{PD_{\text{peak}}}. \tag{8.38}$$

Thus, the CPF can be written as follows:

$$CPF = \frac{\frac{1}{N}\sum_{c=1}^{N} \sum_{i=1}^{R_c} \alpha_{i,c} C_{i,c} V_{i,c}^2 f_c}{\max\left(\sum_{i=1}^{R_c} \alpha_{i,c} C_{i,c} V_{i,c}^2 f_c\right)_{\forall c}}$$

$$+ \frac{\frac{1}{N}\sum_{c=1}^{N} \left(\left|\frac{1}{N}\sum_{c=1}^{N} \left(\sum_{i=1}^{R_c} \alpha_{i,c} C_{i,c} V_{i,c}^2 f_c\right) - \sum_{i=1}^{R_c} \alpha_{i,c} C_{i,c} V_{i,c}^2 f_c\right|\right)}{\max\left(\left|\frac{1}{N}\sum_{c=1}^{N} \left(\sum_{i=1}^{R_c} \alpha_{i,c} C_{i,c} V_{i,c}^2 f_c\right) - \sum_{i=1}^{R_c} \alpha_{i,c} C_{i,c} V_{i,c}^2 f_c\right|\right)_{\forall c}}. \tag{8.39}$$

This function (Equation 8.39) can serve as the objective function for low-power
data path scheduling. *The minimization of this objective function using multiple supply voltages, DFC and multicycling can reduce both power and energy.* From Equations 8.38 and 8.39, we can make the following observations about the CPF. The
CPF is a *non-linear* function. It is a function of four parameters: average power
(P), peak power (P_{peak}), average difference power (PD) and peak difference power
(PD_{peak}). The absolute function (abs or | |) in the numerator (of Equation 8.39)
contributes to the non-linearity. The complex behavior of the function is also affected by the denominator parameters, P_{peak} and PD_{peak}.

It is needed to develop scheduling algorithms that accept an unscheduled DFG,
the resource/time constraints, the switching activity information, the load capacitance, the voltage levels and the number of allowable frequency levels as input
parameters. For optimum minimization of the function, such an algorithm has to
be based on non-linear optimization techniques, which are of large time and space

complexity. However, ILP-based models for minimization of the CPF can provide an optimal solution. The CPF alteration is necessary in order to simplify the ILP-based formulations. It is known that for the denominator parameters, P_{peak} equals to $\max(P_c)_{\forall c}$ and PD_{peak} equals to $\max(|P - P_c|)_{\forall c}$. It is evident that $|P - P_c|$ is upper bounded by P_c for all control steps c, since $|P - P_c|$ is a measure of mean difference error of P_c. Thus, it can be concluded that PD_{peak} is upper bounded by P_{peak}. CPF is modified by substituting PD_{peak} with P_{peak} and the modified CPF, denoted as CPF*, is defined as follows:

$$\text{CPF}^* = \frac{P}{P_{\text{peak}}} + \frac{\text{PD}}{P_{\text{peak}}}, \tag{8.40}$$

$$= \frac{P + \text{PD}}{P_{\text{peak}}}, \tag{8.41}$$

$$= \frac{\frac{1}{N}\sum_{c=1}^{N} P_c + \frac{1}{N}\sum_{c=1}^{N} |P - P_c|}{P_{\text{peak}}}, \tag{8.42}$$

$$= \frac{\frac{1}{N}\sum_{c=1}^{N} \sum_{i=1}^{R_c} \alpha_{i,c} C_{i,c} V_{i,c}^2 f_c}{\max\left(\sum_{i=1}^{R_c} \alpha_{i,c} C_{i,c} V_{i,c}^2 f_c\right)_{\forall c}}$$

$$+ \frac{\frac{1}{N}\sum_{c=1}^{N}\left(\left|\frac{1}{N}\sum_{c=1}^{N}\left(\sum_{i=1}^{R_c} \alpha_{i,c} C_{i,c} V_{i,c}^2 f_c\right) - \sum_{i=1}^{R_c} \alpha_{i,c} C_{i,c} V_{i,c}^2 f_c\right|\right)}{\max\left(\sum_{i=1}^{R_c} \alpha_{i,c} C_{i,c} V_{i,c}^2 f_c\right)_{\forall c}}. \tag{8.43}$$

Unlike CPF, CPF* is dependent on three factors, P, P_{peak} and PD. The absence of PD_{peak} in the denominator helps in reducing the complexity of the ILP formulations (which will be discussed in the next section) to a great extent. We minimize the "modified cycle power function" (CPF*) instead of CPF using an ILP-based formulation and optimization.

Let $C_{\text{sw}i}$ be a measure of the effective switching capacitance of the functional unit FU_i, which is a function of α_i^1 and α_i^2; α_i^1 and α_i^2 are the average switching activities on the first and second input operands of resource FU_i. Equation 8.43 can then be rewritten as:

$$\text{CPF}^* = \frac{\frac{1}{N}\sum_{c=1}^{N} \sum_{i=1}^{R_c} C_{\text{sw}i,c} V_{i,c}^2 f_c}{\max\left(\sum_{i=1}^{R_c} C_{\text{sw}i,c} V_{i,c}^2 f_c\right)_{\forall c}}$$

$$+ \frac{\frac{1}{N}\sum_{c=1}^{N}\left(\left|\frac{1}{N}\sum_{c=1}^{N}\left(\sum_{i=1}^{R_c} C_{\text{sw}i,c} V_{i,c}^2 f_c\right) - \sum_{i=1}^{R_c} C_{\text{sw}i,c} V_{i,c}^2 f_c\right|\right)}{\max\left(\sum_{i=1}^{R_c} C_{\text{sw}i,c} V_{i,c}^2 f_c\right)_{\forall c}}. \tag{8.44}$$

The notation $C_{\text{sw}i,c}$ represents $C_{\text{sw}i}$ for the functional unit FU_i active in control step c. Equation 8.44 is used as the objective function for the scheduling algorithm. The generic CPF* model can be easily tuned to handle any of the four modes of data path circuit operation: (i) single supply voltage and single frequency, (ii) multiple supply voltages and single frequency, (iii) multiple supply voltages and dynamic frequency and (iv) multiple supply voltage and multicycling. For the single supply voltage and single frequency scheme, $V_{i,c}$ and f_c are the same for all c, while for multiple supply voltages and multicycling, only f_c is same for all c.

8.5 Linear Programming Modeling of Non-linearities

The "modified cycle power function" (CPF*) discussed in the previous section is a *non-linear function*. The non-linearity arises because of the *absolute function* (abs or | |) and also because of the *fractional form* of the function itself. ILP formulations need to handle these two forms of non-linearity. The transformations required to derive linear models of the non-linear functions are first addressed. Let us represent the general linear programming model as follows [302]:

$$\text{Minimize}: \sum_j c_j \times x_j, \tag{8.45}$$

$$\text{Subject to}: \sum_j a_{ij} \times x_j \le b_i, \quad \forall i, \tag{8.46}$$

$$x_j \ge 0, \quad \forall j, \tag{8.47}$$

where c_j, a_{ij}, b_i are known constants and x_j is the decision variable.

8.5.1 Linear Programming Formulation Involving the Sum of Absolute Deviations

The general form of linear programming involving absolute deviations can be represented as given below [260, 205]:

$$\text{Minimize}: \sum_i |y_i|,$$

$$\text{Subject to}: y_i + \sum_j a_{ij} \times x_j \le b_i, \quad \forall i, \tag{8.48}$$

$$x_j \ge 0, \quad \forall j, \tag{8.49}$$

where y_i is the deviation between the prediction and observation. The $|y_i|$ is non-linear because of the absolute function but can be linearized by using the following transformation.

Let y_i be represented as the difference of two non-negative variables,

$$y_i = y_i^1 - y_i^2. \tag{8.50}$$

Using these variables, the LP formulation in Equation 8.49 can be rewritten as follows :

$$\text{Minimize}: \sum_i \left| y_i^1 - y_i^2 \right|, \tag{8.51}$$

$$\text{Subject to}: y_i^1 - y_i^2 + \sum_j a_{ij} \times x_j \le b_i, \quad \forall i, \tag{8.52}$$

$$x_j \ge 0, \quad \forall j, \tag{8.53}$$

$$y_i^1, y_i^2 \ge 0, \quad \forall i. \tag{8.54}$$

If the product of y_i^1 and y_i^2 is zero, then

$$\left| y_i^1 - y_i^2 \right| = \left| y_i^1 \right| + \left| y_i^2 \right| = y_i^1 + y_i^2. \tag{8.55}$$

Using the above, the linear programming formulation expressed in Equation 8.54 can be written as:

$$\text{Minimize}: \sum_i y_i^1 + y_i^2, \tag{8.56}$$

$$\text{Subject to}: y_i^1 - y_i^2 + \sum_j a_{ij} \times x_j \le b_i, \quad \forall i, \tag{8.57}$$

$$x_j \ge 0, \quad \forall j, \tag{8.58}$$

$$y_i^1, y_i^2 \ge 0, \quad \forall i. \tag{8.59}$$

The formulations in Equations 8.49 and 8.59 are equivalent, and the minimization of Equation 8.59 results in the minimization of Equation 8.49.

8.5.2 Linear Programming Formulation Involving Fractions

The general expression for the linear programming formulation involving fractions is considered below [205]:

$$\text{Minimize}: \frac{\sum_j c_j \times x_j}{\sum_j d_j \times x_j}, \tag{8.60}$$

$$\text{Subject to}: \sum_j a_{ij} \times x_j \le b_i, \quad \forall i, \tag{8.61}$$

$$x_j \ge 0, \quad \forall j, \tag{8.62}$$

where c_j and d_j are known constants and the denominator $\sum_j d_j \times x_j$ is strictly positive. Let us define new variables as follows :

$$z_0 = \left| d_0 + \sum_j d_j \times x_j \right|^{-1}, \tag{8.63}$$

$$x_j = \frac{z_j}{z_0}. \tag{8.64}$$

Using the above transformation, the original formulation in Equation 8.62 can be modified to the following:

$$\text{Minimize}: \quad c_0 \times z_0 + \sum_j c_j \times z_j, \tag{8.65}$$

$$\text{Subject to}: \quad \sum_j a_{ij} \times z_j - b_i \times z_0 \leq b_i, \quad \forall i, \tag{8.66}$$

$$\sum_j d_j \times z_j + d_0 \times z_0 = 1, \tag{8.67}$$

$$z_0, z_j \geq 0, \quad \forall j. \tag{8.68}$$

The problems defined in Equations 8.62 and 8.68 are equivalent. On solving the problem in Equation 8.68, $z_j = x_j \times z_0$ is substituted to obtain the results for x_j.

Although the ILP formulations become complicated as the objective function described in Equation 8.43 consists of both types of non-linearities, it is much simpler than the ILP formulation of Equation 8.39. It is observed that the cycle power fluctuation (DP_c) corresponds to $|y_i|$ in Equation 8.49. PD_c is a measure of the absolute deviation of cycle power from average power and DP is a measure of mean deviation of the cycle power.

8.6 ILP Formulations to Minimize (CPF^*)

In this section, the ILP models for minimization of the (CPF^*) are discussed. The ILP models for two different scenarios of ASIC design are given. The first one targets design with MVDFC. The other targets MVMC-based designs. The ILP models formulated ensure that the dependency constraints and the resource constraints are satisfied. In order to formulate an ILP-based model for Equation 8.44 and the scheduling schemes for the DFG, the notations presented in Table 8.2 are used.

8.6.1 For MVDFC Operation

In this subsection, the ILP formulation for minimization of CPF* using MVDFC is discussed.

Objective Function: The objective is to minimize the modified CPF described in Equation 8.43 of the whole DFG over all control steps:

Table 8.2 List of variables used in ILP formulations

M_{k,L_V}	: Maximum number of functional units of type k operating at voltage level L_V (FU_{k,L_V})
$C_S[i]$	: As soon as possible (ASAP) time stamp for the operation v_i
$C_L[i]$	: As late as possible (ALAP) time stamp for the operation v_i
$P(C_{swi}, L_V, L_f)$	: Power consumption of functional unit FU_i at voltage level L_V and operating frequency level L_f used by v_i for its execution
x_{i,c,L_V,L_f}	: Decision variable which takes the value of 1 if v_i is scheduled in control step c using the functional unit F_{k,L_V}
$y_{i,L_V,l,m}$	: Decision variable which takes the value of 1 if vertex v_i is using the functional unit F_{k,L_V} and scheduled in control steps $l \rightarrow m$
L_{i,L_V}	: Latency for vertex v_i using functional unit operating at voltage level L_V (in terms of number of clock cycles)

$$\text{Minimize}: \quad CPF^*. \tag{8.69}$$

Using Equation 8.42, Equation 8.69 can be restated as

$$\text{Minimize}: \quad \frac{\frac{1}{N}\sum_{c=1}^{N} P_c + \frac{1}{N}\sum_{c=1}^{N}|P - P_c|}{P_{\text{peak}}}. \tag{8.70}$$

This objective function has the two types of non-linearities mentioned in the previous section. The non-linearity introduced because of the fraction is removed by putting the denominator as a constraint. Then, the problem in Equation 8.70 is transformed to the one given below:

$$\text{Minimize}: \quad \frac{1}{N}\sum_{c=1}^{N} P_c + \frac{1}{N}\sum_{c=1}^{N}|P - P_c|, \tag{8.71}$$

$$\text{Subject to}: \quad \text{Peak power constraints}. \tag{8.72}$$

However, this transformed problem still has a non-linearity in it because of the absolute function. This can be converted to an equivalent problem using the transformation suggested in the previous section.

$$\text{Minimize}: \quad \frac{1}{N}\sum_{c=1}^{N} P_c + \frac{1}{N}\sum_{c=1}^{N}(P + P_c), \tag{8.73}$$

$$\text{Subject to}: \quad \text{Modified peak power constraints}. \tag{8.74}$$

The "peak power constraint" in Equation 8.72 and the "modified peak power constraint" in Equation 8.74 will be discussed in a later part of this subsection. The problem expressed in Equation 8.74 is simplified to the following:

$$\text{Minimize}: \quad \left(\frac{3}{N}\right)\sum_{c=1}^{N} P_c, \tag{8.75}$$

$$\text{Subject to}: \quad \text{Modified peak power constraints}. \tag{8.76}$$

Using the decision variables, the objective function is formulated as

$$\text{Minimize}: \sum_c \sum_{i \in F_{k,L_V}} \sum_{L_V} \sum_{L_f} x_{i,c,L_V,L_f} \times \left(\frac{3}{N}\right) \times P(C_{swi}, L_V, L_f), \qquad (8.77)$$

$$\text{Subject to}: \text{ Modified peak power constraints.} \qquad (8.78)$$

This can be rewritten as

$$\text{Minimize}: \sum_c \sum_{i \in F_{k,L_V}} \sum_{L_V} \sum_{L_f} x_{i,c,L_V,L_f} \times P^*(C_{swi}, L_V, L_f), \qquad (8.79)$$

$$\text{Subject to}: \text{ Modified peak power constraints,} \qquad (8.80)$$

where $P^*(C_{swi}, L_V, L_f)$ is given by $P(C_{swi}, L_V, L_f) \times \left(\frac{3}{N}\right)$.

Uniqueness Constraints: These constraints ensure that every operation o_i is scheduled to a single unique control step within the mobility range $(C_S[i], C_L[i])$ with a particular supply voltage and operating frequency. This is represented as $\forall i$, $1 \leq i \leq N_v$,

$$\sum_c \sum_{L_V} \sum_{L_f} x_{i,c,L_V,L_f} = 1. \qquad (8.81)$$

Precedence Constraints: These constraints guarantee that for a vertex v_i, all its predecessors are scheduled in an earlier control step, and its successors are scheduled in a later control step. These are modeled as $\forall i, j, v_i \in \text{Pred}_{v_j}$,

$$\sum_{L_V} \sum_{L_f} \sum_{d=C_S[i]}^{C_L[i]} d \times x_{i,d,L_V,L_f} - \sum_{L_V} \sum_{L_f} \sum_{e=C_S[j]}^{C_L[j]} e \times x_{j,e,L_V,L_f} \leq -1. \qquad (8.82)$$

Resource Constraints: These constraints ensure that no control step contains more than F_{k,L_V} operations of type k operating at voltage level L_V. These can be enforced as $\forall c, 1 \leq c \leq N$ and $\forall L_V$

$$\sum_{i \in F_{k,L_V}} \sum_f x_{i,c,L_V,L_f} \leq M_{k,L_V}. \qquad (8.83)$$

Frequency Constraints: This set ensures that if an FU is operating at a higher voltage level, then it can be scheduled in a lower frequency control step, whereas an FU that is operating at a lower voltage level cannot be scheduled in a higher frequency control step. These constraints are written as $\forall i, 1 \leq i \leq N_v, \forall c, 1 \leq c \leq N$, if $L_f < L_V$, then $x_{i,c,L_V,L_f} = 0$.

Peak Power Constraints: As discussed before, with reference to Equations 8.70 and 8.72, these constraints are introduced to eliminate the fractional non-linearity of the objective function. These constraints ensure that the maximum power consumption of the DFG does not exceed P_{peak} for any control step. These constraints are enforced as follows: $\forall c, 1 \leq c \leq N$

$$\sum_{i \in F_{k,L_V}} \sum_{L_V} \sum_{L_f} x_{i,c,L_V,L_f} \times P(C_{swi}, L_V, L_f) \le P_{\text{peak}}. \tag{8.84}$$

Modified Peak Power Constraints: To eliminate the non-linearity introduced due to the absolute function, we modify the above constraints, as outlined in Equations 8.72 and 8.74. The peak power constraints in Equation 8.84 is modified as $\forall c,\ 1 \le c \le N$

$$\frac{1}{N} \sum_{c} \sum_{i \in F_{k,L_V}} \sum_{L_V} \sum_{f} x_{i,c,L_V,L_f} \times P(C_{swi}, L_V, L_f)$$

$$- \sum_{i \in F_{k,L_V}} \sum_{L_V} \sum_{L_f} x_{i,c,L_V,L_f} \times P(C_{swi}, L_V, L_f) \le P_{\text{peak}}^*. \tag{8.85}$$

P_{peak}^* is a modified peak constraint which is added to the objective function and minimized along with it.

8.6.2 For MVMC Operation

In this subsection, the ILP formulations based on the CPF* using MVMC are described. In this scheme, the FUs are operated at multiple supply voltages. The FUs lower voltages may need to be active in more than one consecutive control step to complete execution.

Objective Function: The objective is to minimize the CPF* for the entire DFG. Using Equation 8.43, this can be represented as

$$\text{Minimize}: \quad \text{CPF}^*, \tag{8.86}$$

$$= \frac{\frac{1}{N} \sum_{c=1}^{N} P_c + \frac{1}{N} \sum_{c=1}^{N} |P - P_c|}{P_{\text{peak}}}. \tag{8.87}$$

As discussed in the previous subsection, this objective function has two types of non-linearities because of the absolute function and the fractional form. The fractional non-linearity is removed by introducing the denominators as a constraint. The corresponding constraints are known as "peak power constraints." The absolute function non-linearity is removed by modifying the peak power constraints, which gives rise to "modified peak power constraints." Thus, the problem in Equation 8.87 is transformed to the following:

$$\text{Minimize}: \quad \frac{1}{N} \sum_{c=1}^{N} P_c + \frac{1}{N} \sum_{c=1}^{N} (P + P_c), \tag{8.88}$$

$$\text{Subject to}: \quad \text{Modified peak power constraints.} \tag{8.89}$$

The "peak power constraint" and the "modified peak power constraint" are discussed in a later part of the subsection. The problem in Equation 8.89 is simplified to

$$\text{Minimize}: \left(\frac{3}{N}\right)\sum_{c=1}^{N} P_c, \tag{8.90}$$

$$\text{Subject to}: \text{ Modified peak power constraints.} \tag{8.91}$$

Using the decision variables, the above LP objective function is formulated as

$$\text{Minimize}: \sum_{l}\sum_{i\in F_{k,L_V}}\sum_{L_V} y_{i,L_V,l,(l+L_{i,L_V}-1)} \times \left(\frac{3}{N}\right) P(C_{swi},L_V,L_{f_{clk}}), \tag{8.92}$$

$$\text{Subject to}: \text{ Modified peak power constraints,} \tag{8.93}$$

where f_{clk} is the operating frequency level of the data path circuit in multicycling mode. This is further simplified to

$$\text{Minimize}: \sum_{l}\sum_{i\in F_{k,L_V}}\sum_{L_V} y_{i,L_V,l,(l+L_{i,L_V}-1)} \times P^*(C_{swi},L_V,L_{f_{clk}}), \tag{8.94}$$

$$\text{Subject to}: \text{ Modified peak power constraints,} \tag{8.95}$$

where $P^*(C_{swi},L_V,L_{f_{clk}}) = \left(\frac{3}{N}\right) \times P(C_{swi},L_V,L_{f_{clk}})$ are modified power values.

Uniqueness Constraints: These constraints ensure that every vertex v_i is scheduled in appropriate control steps within the mobility range $(C_S[i], C_L[i])$ with a particular supply voltage. Depending on the supply voltage, it may be operated at more than one clock cycle. We represent them as $\forall i, 1 \leq i \leq N_v$,

$$\sum_{L_V}\sum_{l=C_S[i]}^{C_S[i]+C_L[i]+1-L_{i,L_V}} y_{i,L_V,l,(l+L_{i,L_V}-1)} = 1. \tag{8.96}$$

When the operators are computed at the highest voltage, they are scheduled in one unique control step; whereas when they are to be operated at lower voltages, they need more than one clock cycle for completion. Thus, for lower voltage, the mobility is restricted.

Precedence Constraints: These constraints guarantee that for an operation o_i, all its predecessors are scheduled in earlier control steps and its successors are scheduled in later control steps. These constraints should also take care of the multicycling operations. These are modeled as $\forall i, j, v_i \in \text{Pred}_{v_j}$,

$$\sum_{L_V}\sum_{l=S_i}^{E_i}(l+L_{i,L_V}-1) \times y_{i,L_V,l,(l+L_{i,L_V}-1)}$$

$$-\sum_{L_V}\sum_{l=S_j}^{E_j} l \times y_{j,L_V,l,(l+L_{j,L_V}-1)} \leq -1. \tag{8.97}$$

Resource Constraints: These constraints make sure that no control step contains more than F_{k,L_V} operations of type k operating at voltage level L_V. These can be enforced as $\forall L_V$ and $\forall l, 1 \leq l \leq N$,

$$\sum_{i \in F_{k,L_V}} \sum_{l} y_{i,L_V,l,(l+L_{i,L_V}-1)} \leq M_{k,L_V}. \tag{8.98}$$

Peak Power Constraints: As discussed earlier with reference to Equations 8.87 and 8.89, these constraints are enforced to eliminate the fractional non-linearity of the objective function. They are enforced as follows: $\forall l, 1 \leq l \leq N$,

$$\sum_{i \in F_{k,L_V}} \sum_{L_V} y_{i,L_V,l,(l+L_{i,L_V}-1)} \times P(C_{\mathrm{sw}i}, L_V, L_{f_{\mathrm{clk}}}) \leq P_{\mathrm{peak}}. \tag{8.99}$$

Modified Peak Power Constraints: These constraints are introduced to eliminate the absolute function non-linearity of the objective function. These constraints can be enforced as $\forall l, 1 \leq l \leq N$,

$$\frac{1}{N} \sum_{l} \sum_{i \in F_{k,L_V}} \sum_{L_V} y_{i,L_V,l,(l+L_{i,L_V}-1)} \times P(C_{\mathrm{sw}i}, L_V, L_{f_{\mathrm{clk}}})$$
$$- \sum_{i \in F_{k,L_V}} \sum_{L_V} y_{i,L_V,l,(l+L_{i,L_V}-1)} \times P(C_{\mathrm{sw}i}, L_V, L_{f_{\mathrm{clk}}}) \leq P_{\mathrm{peak}}^*, \tag{8.100}$$

where P_{peak}^* is the modified peak power constraint, which is also minimized as a part of the objective function.

8.7 ILP-Based Scheduling Algorithm for CPF* Minimization

8.7.1 Introduction

In this section, the solutions for the ILP formulations obtained in the previous section are discussed through the development of scheduling algorithms for both MVDFC and MVMC schemes. The same target architecture model as in the previous sections is assumed for the scheduling schemes. The inputs to the algorithm are an UDFG, the resource constraints, the number of allowable voltage levels (L_V), the number of allowable frequencies (L_f), the delay of each resource such as functional unit (d_{FU}), multiplexor (d_{Mux}), and register (d_{Reg}) at different voltage levels. The delays of level converters (d_{Conv}) are represented in the form of a matrix that shows the delay in converting one signal at voltage level V_i to another at voltage level V_j (where both $V_i, V_j \in V_{L_V}$). The resource constraint includes the number of

252 8 Transient Power Reduction

ALUs and multipliers at different voltage levels V_i (where, $V_i \in V_{L_V}$). The scheduling algorithm determines f_{base}, cfi_c time stamp for each operation and voltage level such that the function CPF* (Equation 8.44) is minimum.

8.7.2 Algorithm

The ILP-based scheduler which minimizes the CPF* of the DFG is outlined in Algorithm 31. In step 1, the scheduler constructs a LUT for effective switching capacitance for known values of the average switching activity pair. In step 2, the scheduler determines the switching activities at the inputs of each node by using behavioral simulation of DFG. It should be noted that if the LUT (in step 1) does not have the switching capacitance for an average switching activity value (in step 2), then the scheduler uses interpolation techniques to estimate it. The third step is to determine the ASAP time stamp of each operation. The fourth step is the determination of the ALAP time stamp of each vertex for the DFG. The ASAP time stamp is the start time and the ALAP time stamp is the finish time of each operation. These two time stamps provide the mobility of an operation, and the operation must be scheduled within this mobility range. This mobility graph needs to be modified for the MVMC scheme. The ILP formulations are constructed based on the models described in the previous sections. The scheduler uses the modeling language AMPL to model the ILP formulations. At this step, the power consumption of the FUs are calculated as follows. The operational delay of a functional unit is assumed as $(d_{\text{FU}} + d_{\text{Mux}} + d_{\text{Reg}} + d_{\text{Conv}})$. For the MVMC scheme, the operating frequency is the frequency corresponding to the operational delay at the highest operating voltage of the multiplier unit. On the other hand, for the MVDFC scheme, the operating frequency of an FU is calculated based on these operational delays. It is assumed to be the inverse of the operational delay of an FU at a corresponding supply voltage. We obtain the switching capacitance from step 1 and step 2, and the power values are calculated whenever necessary for different operating voltages and frequencies. The scheduled DFG is obtained after the ILP formulation is solved using LP-Solve.

Algorithm 31 ILP-Based Scheduling for CPF* Minimization

1: Construct a LUT for effective switching capacitance.
2: Calculate the switching activities at each node through behavioral simulation.
3: Find the ASAP and ALAP schedules for the unscheduled data flow graph (UDFG).
4: Determine the mobility graph of each node.
5: Modify the mobility graph for MVMC.
6: Model the ILP formulations of the DFG using AMPL.
7: Solve the ILP formulations using LP-Solve.
8: Find the scheduled DFG.
9: Determine the cycle frequencies (f_c), f_{base} and cfi_c for MVDFC scheme.
10: Estimate the power and energy consumptions of the scheduled DFG.

Then, the scheduler determines f_{base}, cfi_c and the cycle frequency (f_c) based on the delay of each cycle. Finally, the power consumption, the energy consumption and the EDP of the scheduled DFG are calculated.

8.7.2.1 CPF-MVDFC Scheduling Scheme

The solution for the ILP formulation in the MVDFC case is illustrated with the help of the DFG shown in Fig. 8.5. From the unconstrained ASAP and ALAP schedules, the mobility graph, Fig. 8.5(a), is obtained. The ILP formulation is obtained using this mobility graph. The formulation is solved using LP-Solve; based on the results, the scheduled DFG shown in Fig. 8.5(b) is obtained for the resource constraint (RC5), two multipliers at 2.4 V and one ALU operating at 3.3 V. Similarly, other schedules can be obtained for different resource constraints.

8.7.2.2 CPF-MVMC Scheduling Scheme

The solution for the ILP formulations of the MVMC case is illustrated using the DFG shown in Fig. 8.6. From the unconstrained ASAP and the ALAP schedules, the mobility graph shown in Fig. 8.6(a) is obtained. This mobility graph is different

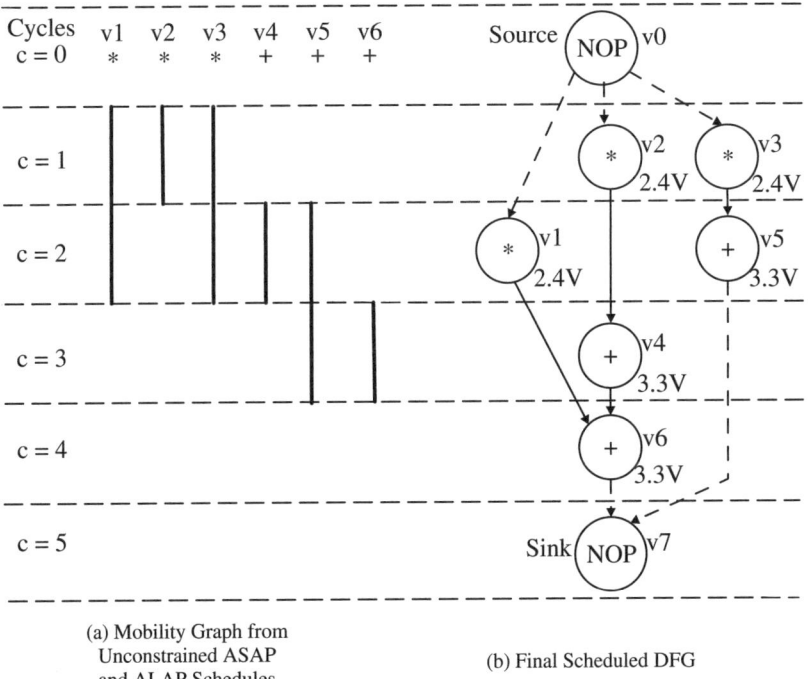

(a) Mobility Graph from
Unconstrained ASAP
and ALAP Schedules

(b) Final Scheduled DFG

Fig. 8.5 Mobility graph and final schedule for the example of DFG for RC5 using MVDFC

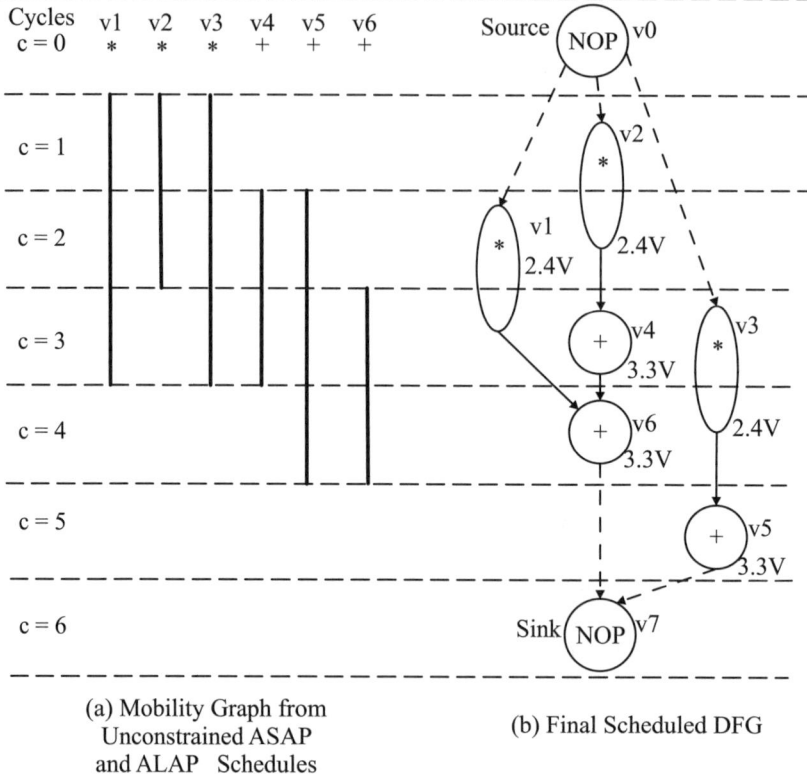

Cycles
c = 0

	v1	v2	v3	v4	v5	v6
	*	*	*	+	+	+

(a) Mobility Graph from
    Unconstrained ASAP
    and ALAP   Schedules

(b) Final Scheduled DFG

Fig. 8.6 Mobility graph and final schedule for the example of DFG for RC5 using MVMC

from that shown in Fig. 8.5(a). In the MVMC case, the mobility graph considers the multicycle operations. In this illustration, it is assumed that there are two operating voltage levels; and when the multipliers are operated at the lower voltage, they take two clock cycles. It should be noted that the mobility graph depends on the number of operating voltages and the assumed operating frequency. The ILP formulation is solved using LP-Solve; and based on the results, the scheduled DFG shown is Fig. 8.6(b) is obtained for the resource constraint (RC5), two multipliers at 2.4 V and one ALU operating at 3.3 V.

8.7.3 Experimental Results

The ILP-based CPF-MVDFC and CPF-MVMC scheduling algorithms were tested with several benchmark circuits, such as Example circuit (EXP), FIR filter, HAL differential equation solver, IIR filter and ARF. The LUT method is used for average switching capacitance calculation. Both scheduling algorithms, CPF-MVDFC and CPF-MVMC, were tested using five different sets of resource constraints (RC1, RC2, RC3, RC4 and RC5) [210, 228]. The reason behind choosing these sets of resource constraints is that they cover a good representation of types of resources at different operating voltages. The number of allowable voltage levels is two (2.4 V, 3.3 V), and the maximum number of allowable frequencies is three. The power/energy estimation includes the power consumption of the overhead, such as level converters. The results are reported for two supply voltages. In the case of CPF-MVDFC scheduling, the frequencies calculated are 4.5 MHz, 9 MHz and 18 MHz. For CPF-MVMC scheduling scheme, the operating frequency (f_{clk}) is 9 MHz.

The average experimental results are plotted in Figs. 8.7 and 8.8 to provide a visual picture. The figures show the average reductions for different benchmarks averaged over all resource constraints. It is obvious from the figure that the

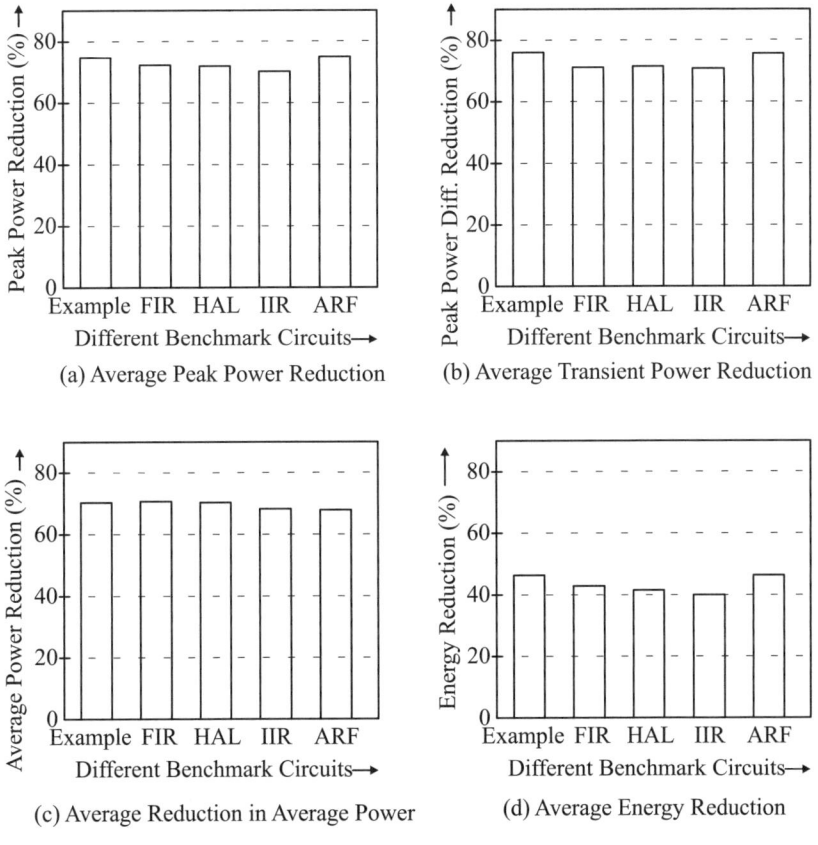

(a) Average Peak Power Reduction

(b) Average Transient Power Reduction

(c) Average Reduction in Average Power

(d) Average Energy Reduction

Fig. 8.7 Average percentage reductions for benchmarks using CPF-MVDFC

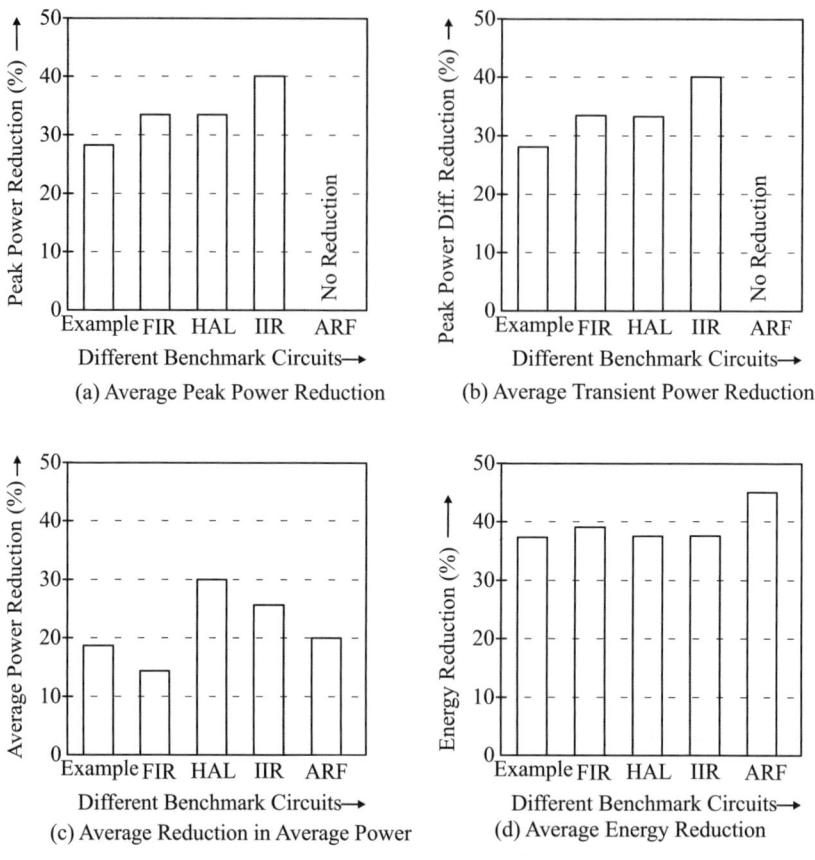

Fig. 8.8 Average percentage reductions for benchmarks using CPF-MVMC

reductions are significant. It is also noted that the reduction for the MVDFC scheme is better than the MVMC scheme. The CPF-MVDFC scheme works effectively for all resource constraints and all benchmarks, whereas the CPF-MVMC scheme does not produce good results for the ARF benchmark.

In order to study the power consumption per cycle, the power profile for different benchmarks is plotted over all the control steps (clock steps). Figures 8.9(a), 8.9(b), 8.10(a), and 8.10(b) show the power profile for benchmarks for resource constraints RC1, RC2, RC3, and RC4, respectively. The curves labeled as "SVSF" correspond to the profile when the schedule is operated at a single frequency (which is the maximum frequency of the slower operator, multiplier) and single voltage. The profiles labeled as "MVDFC" correspond to the case when dynamic clocking and multiple voltage scheme are used. Similarly, the profiles labeled as "MVMC" is for the MVMC scheme. The effectiveness of the proposed scheduling schemes is obvious from the figures.

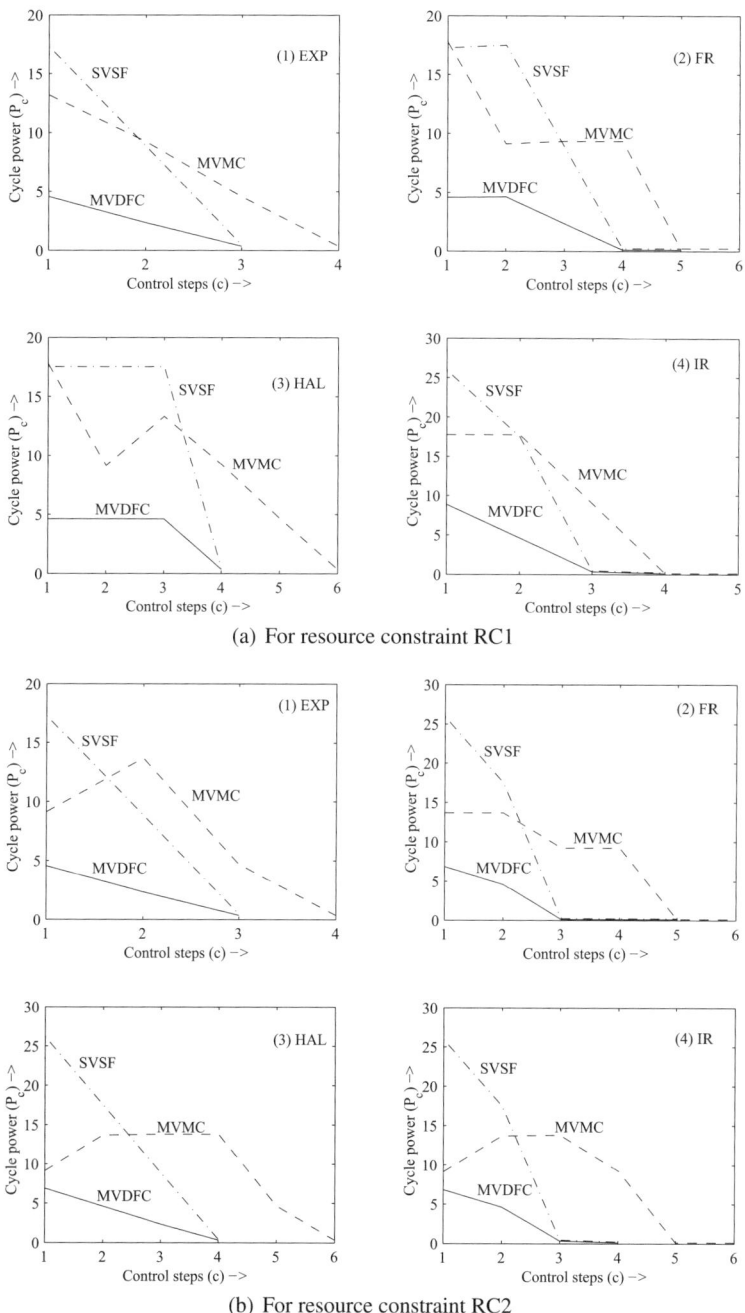

(a) For resource constraint RC1

(b) For resource constraint RC2

Fig. 8.9 Power profile for benchmark for different resource constraints, RC1 and RC2

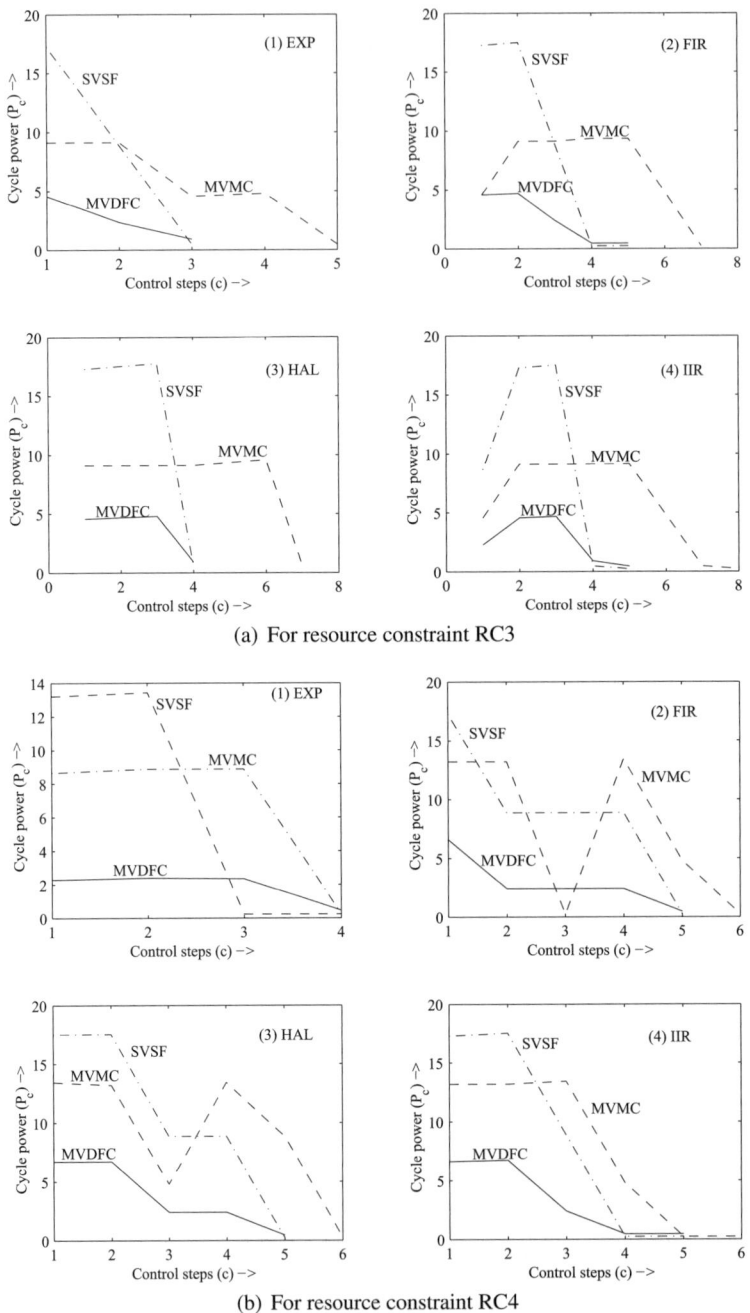

(a) For resource constraint RC3

(b) For resource constraint RC4

Fig. 8.10 Power profile for benchmark for different resource constraints, RC3 and RC4

8.7.4 Conclusions

The modified-CPF model defined and used in this section essentially facilitates simultaneous optimization using ILP formulations. The optimization is performed by using the MVDFC and MVMC schemes. The data path scheduling algorithm described in this chapter is particularly useful for synthesizing data-intensive ASICs. The algorithm attempts to optimize energy and power while maintaining performance. The scheduling algorithm assumes a number of different types of resources at each voltage level (both CPF-MVDFC and CPF-MVMC) and the number of allowable frequencies (CPF-MVMC scheme) as resource constraints. The EDP for both CPF-MVDFC and CPF-MVMC scheduling scenarios was estimated to keep track of the effect of scheduling algorithms on circuit performance. The CPF-MVDFC scheduling resulted in reduction of EDP for all benchmarks and all resource constraints, which shows its effectiveness. On the other hand, the CPF-MVMC scheme resulted in improvement in EDP in almost all cases, except for a few cases, where there was no improvement. *The results clearly indicate that the MVDFC scheme yields better power and energy minimization than the MVMC scheme.* The effectiveness of the scheduling schemes in the context of pipelined data path and control-intensive applications needs to be investigated.

8.8 Data Monitoring for Transient Power Minimization

In [291], Raghunathan et al. present a high-level synthesis approach for transient power management. The power optimization includes the peak power and peak power differential. The authors advocate the need for judicious choice of a transient power metric to avoid area and performance overhead. The authors propose the use of data monitor operations for simultaneous reduction of peak power and peak power differential. The presented scheduling algorithm considers constraints on power characteristics in addition to conventional resource and time constraints. In this scheme, peak power reduction in the range of 17%–32% has been obtained. The reduction in the peak power differential is in the range of 25%–58%.

8.9 Summary and Conclusions

Transient power (or power fluctuation) reduction is needed to improve battery life and efficiency. Only a few research works are reported in this direction – they propose some metrics related to the power profile and the transient power. Only a few approaches are available in this area that use voltage or frequency scaling for the power reduction. These use heuristic or ILP-based optimizations. Moreover, here a data-monitoring technique has also been proposed for transient power management.

Chapter 9
Leakage Power Reduction

9.1 Introduction

As was described in Chapter 3, in a short-channel nano-CMOS transistor, several forms of leakage current exist: reverse-biased diode leakage, subthreshold leakage, gate-oxide tunneling current, hot-carrier gate current, gate-induced drain leakage and channel punch-through current [308]. Of all these leakage mechanisms, gate-oxide (direct) tunneling current that flows during both active and sleep modes of a device is the most significant component for low-end nano-CMOS technology of 45 nm and below. Thus, the major sources of power dissipation in a nano-CMOS circuit can be summarized as dynamic and short-circuit current along with subthreshold and gate-oxide leakage [211]. In this chapter, reduction of gate-oxide leakage and subthreshold leakage of a CMOS data path circuit during high-level synthesis is discussed.

Why Gate-Oxide or Subthreshold Leakage Reduction During High-Level Synthesis? To make high-level synthesis suitable for nano-CMOS circuits, the aim is to develop models to capture gate leakage and optimize it during the high-level synthesis process. The behavioral level is not as highly abstracted as the system level nor as lowly abstracted as the gate or transistor level. Hence, at the behavioral level there is a balanced degree of freedom to explore power reduction mechanisms; and it can help in investigating lower-power design alternatives prior to circuit layout in actual silicon. Moreover, correct design decisions at early phases of circuit abstraction (like high level) will ensure that design errors are not propagated to lower levels, which may be costly to rectify. In a similar philosophy, subthreshold leakage reduction is beneficial if performed during behavioral synthesis.

S.P. Mohanty et al., *Low-Power High-Level Synthesis for Nanoscale CMOS Circuits*,
DOI: 10.1007/978-0-387-76474-0_9, © Springer Science+Business Media, LLC 2008

9.2 Gate-Oxide Leakage Reduction

9.2.1 Dual-T_{ox} Technique

In this section, reduction of total gate-oxide leakage of a CMOS data path circuit based on a dual-T_{ox} technique during high-level synthesis is presented [216, 217]. The algorithm incorporates time constraints as a performance (or delay) trade-off factor and offers the user the choice of predetermining the performance of a circuit, vis-à-vis, power requirements. The algorithms consider an unscheduled DFG schedule for each of their nodes at appropriate control steps, and simultaneously binds them to the best available resource while considering resource constraints so as to achieve the desired performance with the minimum gate-oxide leakage.

9.2.1.1 Motivation

As the gate-oxide thickness decreases, the gate-oxide leakage current increases exponentially and delay decreases. It is believed that using higher gate-oxide thickness resources in off-critical paths and lower gate-oxide thickness resources in critical paths can obviate this effect. The extent to which the reduction can take place depends on both resource and time constraints. Using a larger number of higher-thickness resources implies added silicon cost. Thus, three important elements used to support design decisions are gate-oxide leakage power (current) dissipation, circuit performance (or critical path delay) and silicon cost (number of resources in the data path circuit to be generated).

The following can be concluded from the related background literature:

- Low-power behavioral synthesis research works have mostly considered dynamic power reduction. Only few behavioral synthesis works address subthreshold leakage. No high-level synthesis work is available in the current literature to optimize gate leakage. This calls for a new behavioral synthesis approach considering tunneling current for the low-end nanoscale technology (45 nm and below) domain.
- Behavioral synthesis works either focus on operation scheduling or on resource binding only. Behavioral scheduling algorithms minimize subthreshold leakage current without considering resource constraint (i.e., ignoring the silicon cost), whereas resource-binding algorithms take an already scheduled graph (ignoring the circuit delay).

Thus, simultaneous scheduling and binding are necessary to minimize the gate-oxide leakage current of the overall data path circuit while keeping performance degradation under control. Although performance can be incorporated as time constraints in the form of a delay trade-off factor (or implicitly through a current-delay product), the cost of silicon is taken into account as resource constraints.

Why Multi-T_{ox} Approach During High-Level Synthesis?: The probability of tunneling is a strong function of the barrier height (i.e., the voltage drop across the gate oxide) and barrier thickness. For supply voltage V_{DD} and effective gate-oxide thickness T_{ox}, gate-oxide leakage (tunneling current) dissipation in a CMOS can be described as follows [62, 217, 211]:

$$I_{ox} \propto \left(\frac{V_{DD}}{T_{ox}} \right)^2 \exp \left(-\gamma \frac{T_{ox}}{V_{DD}} \right), \tag{9.1}$$

where γ is an experimentally derived factor. Based on Equation (9.1), the following possible options for reduction of gate leakage are (i) decreasing supply voltage, (ii) increasing gate-oxide thickness and (iii) decreasing gate width since the gate leakage current is proportional to the area of the gate. Decreasing power supply voltage is used as a popular option to reduce dynamic power consumption [320, 221] which will play its role in the reduction of leakage power as well. It may not be sufficient, however, to control the exponential growth of gate leakage. Increase in the gate-oxide (SiO_2) thickness leads to increase in propagation delay. Moreover, reduction of gate width may not be an attractive option because gate leakage current is only linearly dependent on it. Thus, it can be concluded that use of multiple gate-oxide thicknesses can serve as a leakage power (current) and performance (delay) trade-off. In this chapter, the multiple thickness (multi-T_{ox}) approach for reduction of direct tunneling gate current during behavioral synthesis is explored. The usage of resources of multiple gate-oxide thicknesses (multi-T_{ox}) during behavioral synthesis for gate-oxide leakage and performance trade-off is presented. It may be noted that the multi-T_{ox} technique can be used along with any of the other available techniques, such as multi-V_{DD}, multi-V_{Th} or clock gating to provide a complete low-power solution for 45 nm and below CMOS technology data path circuits.

9.2.1.2 Problem Definition

Given an unscheduled DFG (UDFG) $G(V,E)$, it is required to find the SDFG with appropriate resource binding such that the total gate-oxide leakage current is minimized and resource constraints (silicon cost) and latency constraints (circuit performance) are satisfied.

The above can be stated as an optimization problem as follows. Let V be the set of all vertices and V_{CP} be the set of vertices in the critical path from the source S of the DFG to the output or sink node D. For simplicity, it is assumed that the DFG has a single-source node and a single-sink node. However, the formulation can be extended to multiple-source, multiple-sink graphs, whereby we will have corresponding response times for each source and sink pair. The tunneling current minimization problem driven by cost (resource constrained) and performance (latency constrained) can thus be formulated as follows:

$$\text{Minimize} \sum_{v_i \in V} I_{ox}(v_i), \tag{9.2}$$

where $I_{ox}(v_i)$ is the tunneling current consumed per sample node v_i of the DFG, such that the following resource and latency constraints, respectively, are satisfied:

$$\sum_{v_i \in V_{cp}} D_i(v_i) \le D_{cp} \quad \forall \, , v_i \in V_{cp}, \tag{9.3}$$

$$\text{Allocated } (\text{FU}_i(k, T_{ox})) \le \text{Available } (\text{FU}_i(k, T_{ox})). \tag{9.4}$$

The constraints in Equation (9.3) ensure that the summation of all delays $D_i(v_i)$ is less than the critical path delay D_{cp}. The resource allocation is summarized in Equation (9.4), where the total allocation of the ith resources (functional units) of type k and made up of transistors of oxide thickness T_{ox} denoted as $(\text{FU}_i(k, T_{ox}))$ should be less than the total number of corresponding resources available.

9.2.1.3 High-Level Synthesis Flow for Gate Leakage Reduction

The high-level synthesis flow for gate-oxide leakage minimization is shown in Fig. 9.1. Various steps involved in the behavioral synthesis are compilation, transformation, data path scheduling, resource (or FU) allocation, operation binding, connection allocation and architecture generation. Scheduling and binding are the major phases of low-power behavioral synthesis. During the compilation and transformation phases, behavioral VHDL is compiled to structural VHDL to obtain a DFG. Resource- or time-constrained scheduling time stamps the variables and operations in the DFG so that the operations in the same group can be executed concurrently. Whereas allocation fixes the number and types of resources to be used in the data path circuit, the binding process involves attaching operations to FUs and variables

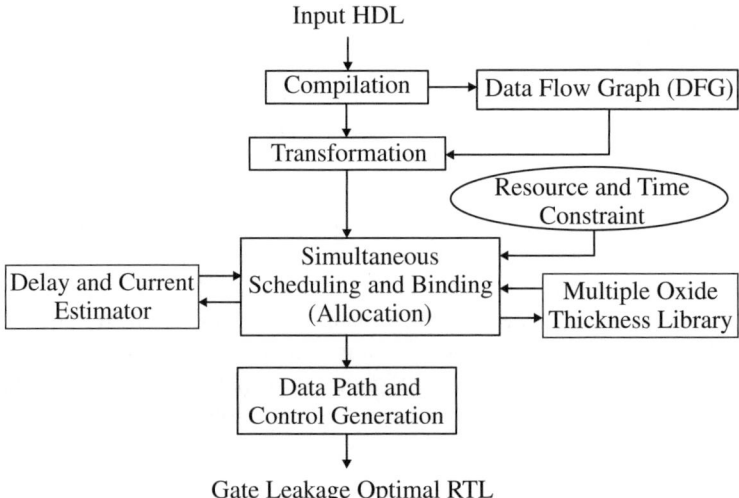

Fig. 9.1 High-level synthesis framework with simultaneous scheduling and binding

to memory units. With a conventional scheduling and binding approach, we lose flexibility while binding is performed. More flexibility is achieved with a simultaneous scheduling and binding approach while binding resources to operations. When the proposed behavioral scheduling–binding algorithm is used along with gate-oxide leakage (direct tunneling) and propagation delay estimators, the system generates a circuit that dissipates minimal gate-oxide leakage power. The delay-current estimator uses a pre-characterized multi-T_{ox} data path library. It also calculates the total gate-oxide leakage current and critical path delay of the circuits when a scheduled DFG is given to it. The connection allocation step determines types or number of buses, buffers and multiplexers for the communication between resources. Finally, RTL descriptions for data path and control circuits are generated.

9.2.1.4 Dual-T_{ox}-Based Algorithm for Gate Leakage Optimization

It is assumed that the data path is specified as a sequencing DFG, which is a directed acyclic DFG. Each vertex of the DFG represents an operation, and each edge represents a dependency. Each vertex also has attributes that specify the operation type. The delay of a control step is dependent on the delays of the FU, the multiplexer and the register. It is assumed that each node connected to the primary input is assigned two registers and one multiplexer, and the inner nodes of the DFG have one register and one multiplexer. It is also assumed that the inputs to the algorithm are an unscheduled DFG, a resource constraint matrix and a delay trade-off factor (T_{F}); and the output is an RTL description that satisfies the resource constraint and trade-off factor and minimizes the tunneling current.

The overall algorithm flow is presented in Fig. 9.2, and the notations used for its description are defined in Table 9.1. The algorithm considers time constraints explicitly and can allow flexibility to the designer to provide time constraints as an input. The scheduling–binding algorithms generate various outputs, such as scheduled DFG with appropriate data path FU assignment to an operation and estimates of gate-oxide leakage and delay. The algorithm assumes that different FUs are characterized for gate-oxide leakage current and propagation delay for various gate-oxide thicknesses and are available in the data path component library. All the transistors inside the same resource have the same gate-oxide thickness, and the transistor gate-oxide thickness may differ for various FUs.

The proposed time–resource-constrained heuristic algorithm that considers time constraints explicitly is presented in Algorithm 32. The algorithm first performs a resource-constrained ASAP schedule C_{S} and a resource-constrained ALAP schedule C_{L} and identifies the set of nodes with zero (V_{Z}) and non-zero (V_{NZ}) mobility. With the available resources (R_{Avl}), maximum gate-oxide leakage reduction can be achieved when resources with higher thickness ($T_{\text{ox}_{\text{H}}}$) are assigned to as many nodes as possible. Since the algorithm has no prior knowledge of the critical path, it initially assigns $T_{\text{ox}_{\text{H}}}$ resources to zero mobility nodes. For each allowable clock cycle of the non-zero mobility nodes, the algorithm identifies the clock cycle that can allocate a $T_{\text{ox}_{\text{H}}}$ resource and schedules the node to that clock cycle and binds its

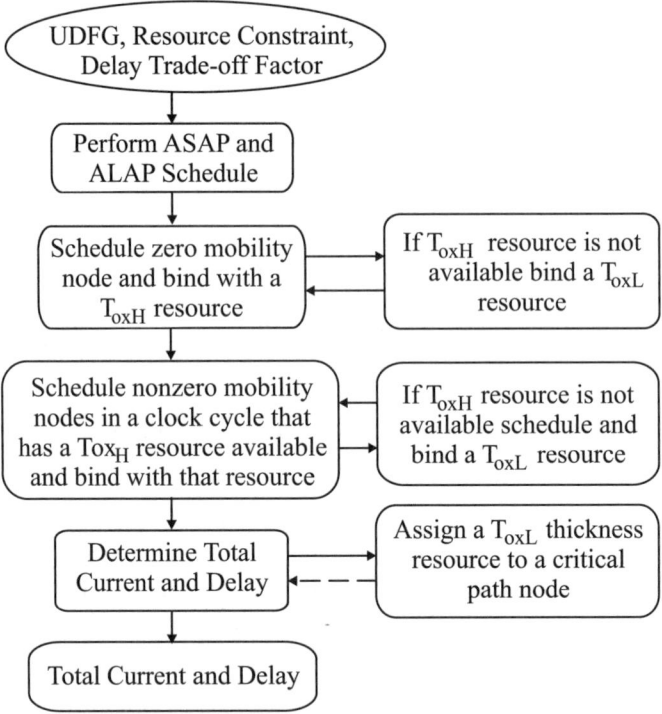

Fig. 9.2 Flow of the simultaneous scheduling and binding algorithm

operation with that T_{ox_H} resource. If such a clock cycle is not available, then it allocates a T_{ox_L} resource in the ASAP clock cycle and schedules it to that clock cycle. Once the scheduling, allocation and binding are done, the critical path delay is calculated. If the critical path delay violates the trade-off factor then the algorithm tries to assign T_{ox_L} resources by swapping the resources of the critical path node and the node containing T_{ox_L} resource in that clock cycle. If there is no T_{ox_L} resource in that clock cycle, the algorithm assigns a T_{ox_L} resource from the available resources.

The algorithm is now demonstrated with the help of an example. Consider an unscheduled DFG of an IIR filter: the resource-constrained ASAP and ALAP schedules are shown in Fig. 9.3(a) and 9.3(b), respectively. Nodes 3, 4, 5, 7, 8 and 9 have zero mobility; and nodes 1, 2 and 6 have non-zero mobility. Since the schedule of zero mobility nodes is fixed, nodes 3, 4, 5, 7, 8 and 9 are scheduled to ASAP time stamp, which tries to bind these operations to T_{ox_H} resources. Therefore, nodes 3, 4, 5, 7, 8 and 9 are assigned T_{ox_H} resources. For the non-zero mobility node 1, there are two possible clock cycles: either clock cycle 1 or 2. Since a T_{ox_H} multiplier can be assigned in clock cycle 2, node 1 is scheduled in clock cycle 2. Next, for node 2, both possible clock cycles do not have a T_{ox_H} multiplier. Therefore, a T_{ox_L} multiplier is assigned, and the node is scheduled in clock cycle 1. This process is continued until all the nodes are scheduled and bound. The DFG after scheduling and binding

Table 9.1 Notations used in algorithm description

$G(V,E)$	Unscheduled DFG (DFG)
$V\{v_i\}$	Set of nodes or vertices in $G(V,E)$
E	Set of edges in $G(V,E)$
V_Z	Set of zero mobility vertices in $G(V,E)$
V_{NZ}	Set of non-zero mobility vertices in $G(V,E)$
$C_S[v_i]$	ASAP time stamp of a vertex v_i
$C_L[v_i]$	ASAP time stamp of a vertex v_i
$C[v_i]$	Final time stamp of a vertex v_i
$FU_j(k,T_{ox})$	jth resource of type k and thickness T_{ox}
T_{ox_H}	Higher gate-oxide thickness
T_{ox_L}	Lower gate-oxide thickness
$V_{Bin}[k][T_{ox}][r_j]$	Binding matrix; r_j is any jth resource
$R_{Avl}[c][k][T_{ox}]$	Availability matrix; c is any clock cycle
$T_{cp_{ST}}$	Critical path delay for single-oxide thickness
$T_{cp_{DT}}$	Critical path delay for dual-oxide thickness
T_F	Performance or delay trade-off factor
$I_{ox_{ST}}$	Tunneling current for single-oxide thickness
$I_{ox_{DT}}$	Tunneling current for dual-oxide thickness
N_c	Number of control steps
n_{FU_c}	Number of resources active in a control step c
$I_{ox_{FU}}(c,r)$	Gate-oxide leakage current of the r-th FU active in the step c
$T_{pd_{FU}}(c,r)$	Propagation delay of the r-th FU active in the control step c

with maximum utilization of available T_{ox_H} resources is shown in Fig. 9.3(c). Now, the algorithm identifies 3, 7, 8 and 9 and 4, 7, 8 and 9 as critical paths; and since the algorithm runs in ASAP order, 3-7-8-9 is chosen. Then, the algorithm calculates the critical path delay. For $T_F = 1$, all the critical path nodes need to be assigned T_{ox_L} resources. For node 3 there is a T_{ox_L} multiplier available in its clock cycle, therefore, node 2 is assigned T_{ox_H} multiplier and node 3 is assigned T_{ox_L} multiplier. The same is the case with node 8. But for nodes 9 and 7, no T_{ox_L} adders are available in their respective clock cycles; therefore, a T_{ox_L} adder can be assigned from the available adders. The final DFG after all scheduling and binding satisfying resource and time constraints with a trade-off factor $T_F = 1.0$ is shown in Fig. 9.3(d).

A special case of the above time–resource-constrained algorithm is the case of time-constrained approach when the constraints on the number of resources are relaxed. Here, the algorithm gives an RTL that has the least tunneling current while satisfying a time constraint. With this approach, maximum tunneling current reduction can be achieved by assigning T_{ox_H} resources to all non-critical path nodes. The algorithm in this case produces the minimal number of resources required for the execution of the scheduled DFG.

9.2.1.5 Experimental Results

The algorithms have been exhaustively tested with several behavioral-level benchmark circuits for several constraints. We present the experimental results in this

Algorithm 32 Time and Resource-Constrained Simultaneous Scheduling and Binding Heuristic

1: Get resource-constrained ASAP schedule C_S of the DFG.
2: Get resource-constrained ALAP schedule C_L of the DFG.
3: Fix the number of control steps as maximum of the ASAP and ALAP schedules.
4: Classify the vertices in V to zero mobile and non-zero mobile vertices.
5: **for all** {Zero mobile vertex v_i in ASAP order} **do**
6:     v_i is ASAP time stamped.
7:     **if** {$FU_j(k, T_{ox_H})$ is available for control step $C[v_i]$} **then**
8:         **return** Assign $FU_j(k, T_{ox_H})$ resource vertex v_i.
9:     **else**
10:         **return** Assign $FU_j(k, T_{ox_L})$ resource vertex v_i.
11:     **end if**
12: **end for**
13: **for all** {Non-zero vertex in ASAP order} **do**
14:     **for all** {Possible clock cycle $c : C_S[v_i] \rightarrow C_L[v_i]$} **do**
15:         **if** {$FU_j(k, T_{ox_H})$ is available for c} **then**
16:             **return** Schedule v_i in control step c and Assign $FU_j(k, T_{ox_H})$ resource vertex v_i.
17:         **else if** {$FU_j(k, T_{ox_L})$ is available for c and v_i is not already scheduled} **then**
18:             **return** Schedule v_i in control step c and Assign $FU_j(k, T_{ox_H})$ resource vertex v_i.
19:         **end if**
20:     **end for**
21: **end for**
22: Call Floyd's algorithm to determine V_{CP}.
23: Calculate critical path delays $T_{cp_{ST}}$ and $T_{cp_{DT}}$.
24: **for all** {Control step c in the available schedule} **do**
25:     **if** {If time constraint not met, i.e., $T_{cp_{DT}} > T_F \times T_{cp_{ST}}$} **then**
26:         **return** Find all vertices in control step c.
27:         **if** {v_i is in the critical path and using T_{ox_H} resource} **then**
28:             **return**
29:             **if** {v_j is not in critical path and using T_{ox_L} resource} **then**
30:                 **return** Swap the resources of v_i and v_j.
31:             **else**
32:                 **return** Assign T_{ox_L} resource to v_i.
33:             **end if**
34:         **end if**
35:     **end if**
36: **end for**

section for a selected set of benchmarks and constraints. While calculating the gate-leakage current for single-oxide thickness, a nominal 1.4 nm oxide thickness is used. We considered resources of three dual-oxide thickness pairs of (i) 1.4–1.5 nm, (ii) 1.4–1.6 nm, and (iii) 1.4–1.7 nm. For each benchmark and for each pair of dual thicknesses, four sets of experiments were performed. In the first set of experiments, a smaller number of T_{ox_H} resources and a higher number of T_{ox_L} resources are used. In the second set of experiments, a higher number of T_{ox_H} resources, as compared to the first set of experiments, are used. In the third set of experiments, a higher number of T_{ox_H} resources, as compared to the second set of experiments

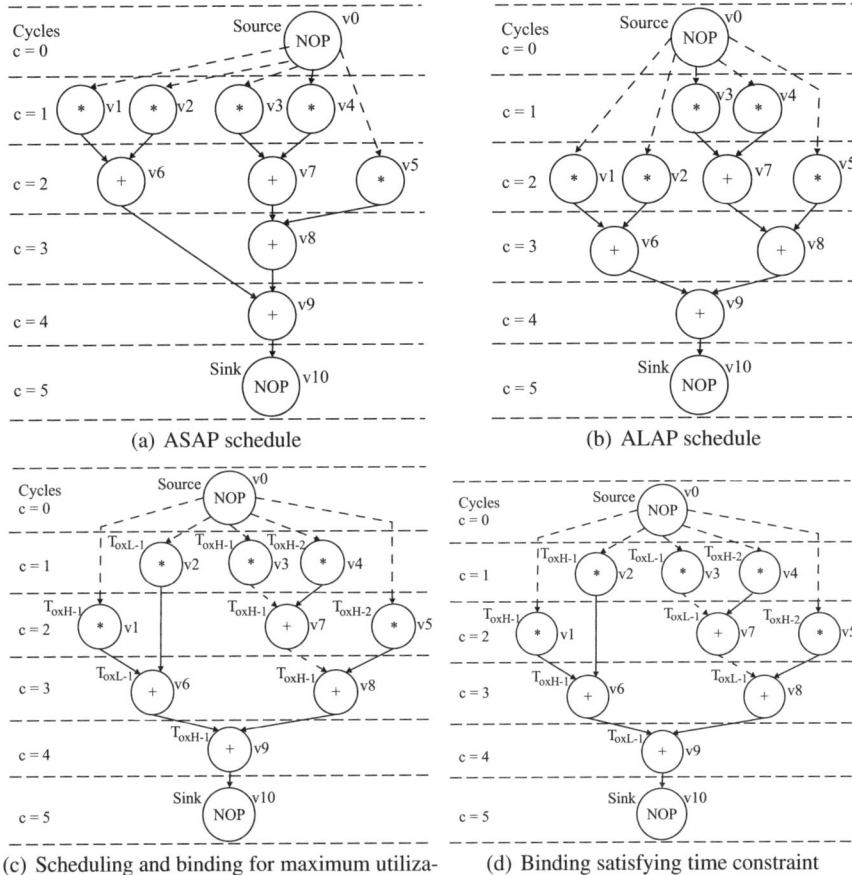

(a) ASAP schedule   (b) ALAP schedule

(c) Scheduling and binding for maximum utiliza-   (d) Binding satisfying time constraint
tion of higher oxide thickness resources

Fig. 9.3 Illustration of algorithm 1 for simultaneous scheduling and binding for tunneling current reduction for IIR benchmark for time constraint $T_F = 1.0$. The resource constraint is assumed to be 2 T_{ox_H} multipliers, 2 T_{ox_L} multipliers, 1 T_{ox_H} adder and 1 T_{ox_L} adder

are used. In the fourth set of experiments, we relaxed the resource constraints to study the time-constrained approach only. The resource constraints represent the FU of different oxide thicknesses available to the behavioral scheduling–binding algorithms. The sets of resource constraints were chosen so as to cover FUs consisting of different oxide thicknesses. They are representatives of various forms of the corresponding RTL representation. The experimental results consider the tunneling current and propagation delay of FUs and storage units present in the data path circuit. The percentage reduction in gate-oxide leakage current is calculated as $\Delta I = \left(\frac{I_{ox_{ST}} - I_{ox_{DT}}}{I_{ox_{ST}}} \right) \times 100\%$. We estimate the critical path delay of the circuit as the sum of the delays of the vertices in the longest path of the DFG. The delay penalty is calculated as $\left(\frac{T_{pd_{MT}} - T_{pd_{ST}}}{T_{pd_{ST}}} \right) \times 100\%$.

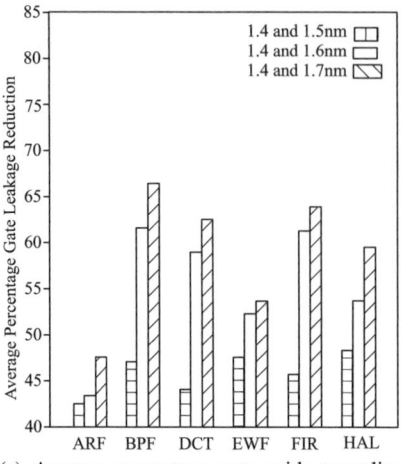
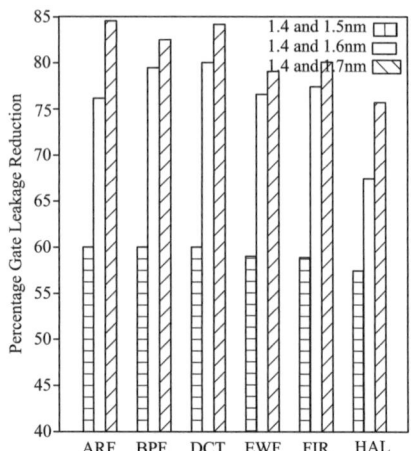

(a) Average percentage gate-oxide tunneling current reduction for $T_F = 1.5$ with resource constraints

(b) Percentage gate-oxide tunneling current reduction for time-constrained approach for $T_F = 1.5$ without resource constraint

Fig. 9.4 Bar charts showing percentage reduction in gate-oxide tunneling current for three cases: (i) 1.4–1.5 nm, (ii) 1.4–1.6 nm and (iii) 1.4–1.7 nm for algorithm 1

The bar charts in Fig. 9.4 represent the experimental results for all the thickness pairs under consideration for a trade-off factor of 1.5. The bar chart in Fig. 9.4(a) shows the variation of average gate-oxide tunneling current reduction for all the benchmarks in the case of the combination of 1.4 nm thickness gate-oxide CMOS devices with 1.5, 1.6 and 1.7 nm gate-oxide thickness devices, respectively. This shows that for a particular performance requirement with resource constraint in effect, there is a definite gain in the leakage reduction with increase in the thickness of the gate oxide. In Fig. 9.4(b), a time-constrained perspective of the result is presented. Here, with all resource constraints relieved, we see an appreciable reduction taking place for all the benchmarks tested. When compared with the resource-constrained approach, this method shows an average reduction of almost 15% across all benchmarks.

From the results for various benchmark circuits for the dual-thickness technique for 1.4–1.7 nm, it is observed that the reduction in gate-oxide leakage across all benchmarks ranges from 11.24 to 75.83% for different trade-off factors considered in the experiment and all three dual-thickness pairs. The results also show that the reduction in tunneling current is in the range of 34.8–84.8% with a tradeoff factor of $T_F = 1.5$. The average reduction for each benchmark improves with the size of the circuit. The tunneling current decreases drastically as the number of available T_{oxH} resources increases. It can be seen that the reduction in gate-oxide leakage is maximum for the DCT and EWF benchmarks and minimum for the ARF benchmark. For time-constrained-only scheduling and binding without the resource constraints assuming unlimited resource constraint, it is observed that a reduction in the tunneling current in the range 58.45–60.05% for dual thickness of 1.4 and 1.5 nm, 67.64–80.24% for 1.4 and 1.6 nm thicknesses and 75.79–84.80% for 1.4 and 1.7 nm dual-thickness pair with a trade-off factor of 1.5.

9.2.1.6 Conclusions

An algorithm has been presented for simultaneous scheduling and binding of behavioral-level elements utilizing FUs of dual-oxide thickness. As can be seen from the results of the experiments performed in this context, the dual-oxide thickness methodology is highly effective in reducing gate leakage. The choice of FUs is being made during scheduling; and we are in the process of evaluating its impact on area, capacitance and dynamic power. We are also exploring the option of replacing the heuristic-based approach with more advanced optimization techniques. Finally, we aim to extend this work on gate leakage to a holistic solution to the entire spectrum of power dissipation at the behavioral level, considering multi-V_{DD}, multi-V_{Th} and multi-T_{ox}.

9.2.2 Dual-κ Technique

9.2.2.1 Introduction

As a result of aggressive scaling, gate-oxide direct tunneling current (gate leakage) has become a major component of total power dissipation. Use of dielectrics of higher permittivity (dual-κ) or use of silicon dioxide of higher thicknesses (dual-T) are being considered as methods for its reduction. Recently, the "dual dielectric" approach was introduced for logic-level gate leakage reduction, where SiO_2 is selectively replaced with high-κ materials such as SiON, Si_3N_4 [240]. This section presents a dual dielectric for gate leakage reduction in a behavioral synthesis perspective. An algorithm is presented for the gate leakage current reduction that does simultaneous scheduling, allocation and binding during behavioral synthesis accounting for process variations [234].

9.2.2.2 A Dual-κ-Based Algorithm for Gate Leakage Optimization

In this section, we present a simulated annealing-based algorithm that performs simultaneous scheduling, allocation and binding and minimizes gate leakage current. Given a reduction mechanism of dual dielectric constants for the gate material (dual-κ) and time constraints, it is needed to determine an RTL implementation that has minimum leakage current. In both approaches, in order to maximize the leakage reduction, we need to ensure that every node can be scheduled in such a way that a less leaky resource can be assigned to non-critical resources so that the total delay is not affected. Analogous to an annealing process, the mobility of nodes in a DFG is dependent on the total available resources. Here, the nodes of a DFG are analogous to atoms, and temperature is analogous to the total number of available resources. The mobility of the nodes (chance of assigning a higher thickness or dielectric resource) is dependent on the total number of available less leaky resources. The annealing principle is applied to the problem and used to explore the trade-offs among power, performance and area.

The input to this algorithm is a DFG, input data streams, gate leakage reduction mechanism (either dual-K or dual-T), and the output is an RTL description with gate leakage. The pseudocode is presented in Algorithm 33, and the notations used in the algorithm are presented in Table 9.1. The algorithm starts with the ASAP schedule and assigns RL (leaky resources, lower thickness resources for dual-T approach and lower dielectric for dual-κ approach) to all the operations with the function Allocate_Bind. The total leakage is determined as the sum of leakages of all the allocated resources, so the minimum number of resources required for the schedule is determined and allocated. Once the execution in a clock cycle is finished, all the resources are assumed to be in the ready state before the next clock cycle.

Algorithm 33 Simulated Annealing Algorithm for Minimizing Gate Leakage

---

1: Initialize the Temperature; i.e., $t \leftarrow t_o$.
2: Initialize Available Resources as unlimited; i.e., $\leftarrow \infty$.
3: **while** (There exists a schedule with available resources) **do**
4:     Assign the interaction counter as i = number of iterations.
5:     Perform resource-constrained as soon as possible (ASAP) scheduling.
6:     Perform resource-constrained as late as possible (ALAP) scheduling.
7:     Assign the initial solution S to ASAP schedule.
8:     Perform binding and allocation following standard approach.
9:     Assign gate leakage cost as Gate-Leakage-Cost(S).
10:     **while** $(i > 0)$ **do**
11:         Generate a random thicknesses in range of $(T_{ox} - \delta T_{ox}, T_{ox} + \delta T_{ox})$.
12:         Generate random transition from S to S^*.
13:         ΔGate-Leakage-Cost $\leftarrow$ Gate-Leakage-Cost(S) $-$ Gate-Leakage-Cost(S^*).
14:         **if** (ΔGate-Leakage-Cost > 0) **then**
15:             **return** $S \leftarrow S^*$
16:         **else**
17:             **if** $\left\{ \left[exp \left(\dfrac{\Delta\text{Gate-Leakage-Cost}}{t} \right) \right] > random[0,1) \right\}$ **then**
18:                 **return** $S \leftarrow S^*$
19:             **end if**
20:         **end if**
21:         $i \leftarrow i - 1$.
22:     **end while**
23:     Decrement available resources.
24:     $t \leftarrow \alpha \times t$.
25: **end while**
26: return solution S.

In the outer loop during each iteration, the number of less leaky resources (higher-thickness resources for dual-T approach and higher dielectric constant for dual-K approach) is decreased, which restricts the mobility of the nodes. The algorithm attempts to find an RTL that has minimum leakage for a given number of available resources. In the inner loop during each iteration, a neighborhood solution is generated. If this solution has less leakage than the current solution, the neighborhood solution is made the current solution. This way, the algorithm converges to a solution that has minimum leakage. In generating a neighborhood solution, a node is randomly selected and checked if a less leaky resource can be assigned in all possible clock cycles that satisfy a time constraint. Each time a different resource is assigned a random thickness in the range of $(T_{ox} - \delta T_{ox}, T_{ox} + \delta T_{ox})$ is generated to take into account process variation. Assuming a monolayer misplacement of SiO_2, δT_{ox} is approximately 15%.

For calculating the total delay of the circuit, the critical path delay for a single-cycle case is used. While generating a random neighborhood solution for a dual dielectric approach, the algorithm ensures that the nodes in the critical path are not assigned a higher K resource. In the case of dual thickness, the algorithm ensures that the nodes in the critical path are not assigned a higher T_{ox} resource. For multicycling, the total delay of the circuit is calculated as the product of total number of control steps and the maximum delay of any resource in the circuit. Assigning higher thickness/dielectric resources will increase the delay, which can be compensated using chaining and multicycling. Using multicycling or chaining alone may not be enough. While multicycling increases the number of control steps, only a few operations for chaining can be implemented. The idea behind using both multicycling and chaining is to ensure that the execution of any operation that is ready (all its predecessors finished execution) has a resource available to start execution.

9.2.2.3 Experimental Results

The algorithm was experimented with various benchmark circuits. In the dual dielectric approach, the base case tunneling current is calculated using SiO_2 as the dielectric. A dual dielectric pair $SiO_2(\kappa = 3.9)$–$Si_3N_4(\kappa = 7)$ is considered for the dual-κ approach. To start with, it is assumed that there is an infinite number of RL and RH resources; and during each iteration, the number of RL resources are decreased. The experiments were performed for both multicycling and chaining-based data path as well as single-cycle data path circuits.

The results consider gate leakage, area and propagation delay of FUs, interconnect units and storage units present in the data path circuit. The reduction in direct tunneling current is calculated as $\Delta I = \left(\frac{I_{gate_{SK}} - I_{gate_{DK}}}{I_{gate_{SK}}} \right) \times 100\%$. The critical path delay of the circuit is estimated as the sum of the delays of the vertices in the longest path of the DFG for the single-cycle case and the number of control steps times the slowest delay resource for multicycling-chaining case. The delay trade-off factor (DTF) is used to provide various time constraints for the experiments. T_{cp} is used to denote the critical path delay of the circuit.

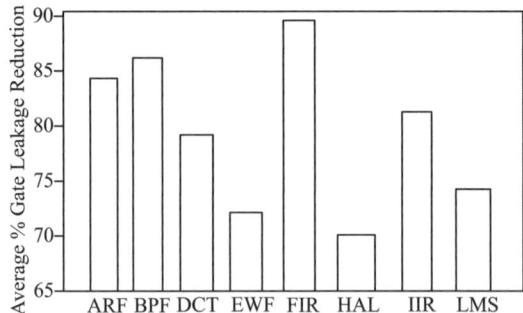

Fig. 9.5 Average percentage reduction in gate leakage using dual-K approach for various benchmarks circuits

The average percentage reduction for the benchmark circuits is shown in Fig. 9.5. The reduction in tunneling current for all the benchmarks ranges from 57.9 to 91.5% for an area penalty ranging from 6.6 to 34.8% for different DTFs considered in the experiment. For the dual dielectric approach, it ranges from 61.5 to 95.7%. It can be seen from the results table that the area penalty increases with tunneling current reduction for a dual thickness approach. The variations in area with the dual-κ approach is due to the effect of allocating a higher number of resources. For multicycling and chaining data path with a dual-T approach, tunneling current reduction ranging from 30.5 to 91.1% is observed for an area penalty ranging from 4.80 to 28.67% for different DTFs. One significant observation is that there is a drastic reduction in delay compared to single-cycle operation. The design space exploration for this case is very much similar to the single-cycle case.

9.2.2.4 Conclusions

In this section, a dual dielectric approach for reduction of tunneling current at behavioral level using simultaneous scheduling and binding of FU is presented. The results achieved with the proposed method outperform other behavioral-level leakage reduction works available in the literature in terms of percentage reduction. Further exploration of these techniques in addition to the use of dual-κ along with dual-T_{ox} is being investigated for future implementations. The ultimate objective is to extend the work on tunneling current to provide a broader solution to the problem of power dissipation in all its forms at the behavioral level.

9.3 Subthreshold Leakage Reduction

9.3.1 Prioritization Algorithm for Dual-V_{Th}-Based Optimization

An algorithm for subthreshold leakage power analysis and reduction during behavioral synthesis using a dual-threshold (dual-V_{Th}) approach is given in [154, 155].

The algorithm targets the least used modules as candidates for leakage minimization by using dual-threshold voltages. The candidate for leakage power reduction can be computed by using a priority function. The priority function can consider the potential of an FU to reduce leakage and the possibility that FU results are needed in the next cycle. One observation that can be made is that there is a synergy between leakage power reduction and switching power reduction. Performing leakage reduction in many cases reveals potential solutions for switching power reduction, which may not have been previously encountered. Also in some cases, the same benchmark yields more leakage at a lower technology and feature size. The algorithm results in higher percentage reduction as technology scales from 350 to 70 nm. The average reduction in this approach is observed to be 58.0% [155].

9.3.2 MTCMOS-Based Clique Partitioning for Subthreshold Leakage Reduction

An MTCMOS-based binding approach for reduction of subthreshold current during high-level synthesis is given in [107]. The binding algorithm can consider trade-offs for power, delay and area. The algorithm is based on clique partitioning and performs resource allocation and binding. The clique partitioning is modified by addition of two different types of information, the idle time of resources and the number of transitions of resources. Resources with higher idle times are bound to MTCMOS instances. The first parameter ensures that the MTCMOS resources are idle for longer consecutive clock cycles so that they can be turned off. The second parameter ensures that the transitions between idle and active states of a resource are minimal to reduce short-circuit power dissipation. Average reduction in leakage power using this approach is found to be 22.4% [107].

9.3.3 MTCMOS-Based Knapsack Binding for Subthreshold Leakage Reduction

A binding approach, based on 0-1 knapsack algorithm, which selectively binds MTCMOS (Multi-threshold CMOS) is given in [106]. The binding is done in such a way that leakage reduction is performed under area constraints. Experiments demonstrate that reductions as high as 30% are possible. An important observation from this research is that the higher the number of MTCMOS resources, the higher the area and the higher the leakage reduction. However, beyond a certain area, the leakage reduction does not increase; rather it decreases, signifying the law of diminishing returns.

9.3.4 Power Island Technique for Subthreshold Leakage Reduction

A high-level synthesis technique, Power Island, is used to minimize leakage in a circuit by partitioning it into islands [87, 85]. In this technique, maximally compatible circuit components, the components with maximally overlapping lifetimes, are identified and clustered together. Each cluster is then placed into a separate island. Hence, each island can be turned off when the logic within it is idle. Thus, leakage is eliminated during the power-down mode of the islands. The experiments show 41–80% leakage improvement for 70 nm technology node.

9.3.5 Maximum Weight-Independent Set (MWIS) Problem Heuristic for Dual-V_{Th}-Based Optimization

A heuristic algorithm for leakage optimization based on a *maximum weight-independent set* (MWIS) problem that uses dual-V_{Th} is given in [343]. The heuristic solves the high-level synthesis problem being formulated as an MWIS problem. The approach allows the time slack dependency among operation nodes and module instances to be analyzed leading to an average of 70.9% leakage power reduction.

9.4 Summary and Conclusions

Leakage power due to various sources such as subthreshold conduction and gate-oxide tunneling are prominent sources of power dissipation in nano-CMOS circuits. The research works in optimization of leakage during high-level synthesis are few and are based on primarily dual-V_{Th} technology. However, recently, dual-T_{ox} and dual-κ technologies have been developed. The algorithms proposed are primarily heuristic, and there is a need for new high-level synthesis approaches for leakage reduction.

Chapter 10
Conclusions and Future Directions

The relentless drive for miniaturization has pushed CMOS technology deeply into the nanometer regime. As a consequence, a host of physical effects, previously unimportant, have come into play to produce sources of power dissipation in nano-CMOS devices that can dramatically affect their performance. New leakage mechanisms such as direct tunneling current through the gate oxide, in particular, along with subthreshold leakage, have come to dominate power dissipation. They will be the major challenge in the integration of digital and analog components into an SoC. Armed with accurate models, the designer can explore and characterize exhaustively the performance of the building block of digital circuits, the logic gate. Even the smallest universal gate, a two-input NAND, can exhibit surprisingly complex behavior, with regard to leakage, depending on its state. This has significant consequences for the characterization and power profiling of the next level of hierarchy, the FU. Closely related to the problem of gate and FU characterization is the issue of process and design parameter variation. Nanoscale processes have intrinsic variations that are large because of small dimensions involved. Both on-chip and off-chip power distribution also include variability. To guarantee the functionality of a design and to account for process and design variation according to DFM principles, the designer uses statistical techniques such as corners analysis and MCM.

This text has emphasized the importance of grounding the higher level estimation and optimization decisions on the underlying physics and the characteristics of the technology used in building a circuit. A stark example of the sort of technology breakthrough is high-κ metal gate solution that shows a fundamental shift in the leakage characteristics that could not have been simply extrapolated from the available data from previous generations (65 nm and earlier), in order to estimate leakage and make design decisions for a new generation (45 nm) processor. The technology innovations are in processing technologies, materials, and chemicals – stable and process-compatible high-κ materials, work-function-appropriate metal gate and "atomic layer deposition" technique – that yielded a ten-fold decrease in leakage for the same drive current. In fact, it allows a process-level trade-off from 25% increase in performance (drive current) at the same leakage level all the way to a 10X decrease in leakage at the same performance level. These quantum changes

S.P. Mohanty et al., *Low-Power High-Level Synthesis for Nanoscale CMOS Circuits*,
DOI: 10.1007/978-0-387-76474-0_10, © Springer Science+Business Media, LLC 2008

in trends need a same-generation, early circuit study to properly tune the estimation and optimization models for use during design refinement and implementation.

Various techniques for power estimation at the abstract RTL level and their role in early design exploration were presented throughout this book. Power estimation is integral to optimization and designed for reliability, and high-level estimation enables power versus performance trade-offs at the most effective level of abstraction. Estimation techniques for specific, important architectural blocks such as register files and caches, and synthesizable control logic were discussed. The effect of variation and the interplay of power with temperature were also examined. The importance of performing technology readiness that predicts characteristics of major design structures and the use of significant circuit and process techniques in power optimization were also discussed. The objective is that high-level power modeling and estimation can anticipate these trends to be successful and sufficiently accurate. To make traditional simulators more accurate, correlating the simulator directly to silicon measurement is important. But this approach has a deficiency: a single-chip-wide power value is used for correlation in such an approach. Consequently, this approach loses the level of granularity that can be accomplished via correlating a high-level power simulator to a design team's internal microarchitecture simulator. A good power simulator should include such microarchitectural features as (i) the front-end prefetch and micro-instruction translation engine that translates macro-instructions into the component micro-ops; (2) the trace cache and control; (3) the out-of-order logic that performs allocation, reorder, register renaming and scheduling; (4) the execution engine, which includes integer, floating point and multimedia units; (5) the memory subsystem which provides memory access logic and buffers; and (6) the system interface unit defining the buses, queues, arbiters and contention logic.

High-level synthesis is important for several fundamental reasons. In particular, reduction of design cycle time and rapid design space exploration at the higher level of abstraction will ensure that wrong decisions are not propagated to the lower levels of design abstraction, which will be difficult and costly to correct. High-level synthesis involves several important steps, such as scheduling, allocation and binding. Several graph-theory-based algorithms are available that can perform optimization while doing these tasks. The high-level synthesis tasks need estimation of power, leakage and performance using models and simulators for generating low-power circuits. Both data path and control synthesis need to be performed to ensure that synthesis covers all types of real-life circuits. Industrial interest for the development of high-level synthesis tools has increased recently. There are existing tools that perform high-level synthesis explicitly, and some tools perform the behavioral to RTL compilation as an intermediate process.

Low-power design is also important for many reasons. The prominent one is the significant demand of portable appliances, in which the battery life is an important factor. Power optimization can be performed at different levels of design abstraction. Different methods have been proposed at each level of design abstraction for power reduction. Power optimization during high-level synthesis can significantly reduce power, and the optimization process is much faster. Low-power, high-level synthesis techniques, such as multiple V_{DD}, multiple V_{Th} and multiple T_{ox}, have been used. At the same time, technology-independent techniques, such as pipelining, parallelization, precomputation and shutting down, are used for high-level power optimization. Both leakage and dynamic power reduction are important for nano-CMOS circuits, and methods are being investigated for their reduction. While device power is important and has received major attention, reduction of interconnect power dissipation also needs attention.

Energy or average power reduction is important because it has a direct impact on battery life and energy bill. Techniques for energy reduction are proposed that work during different high-level synthesis tasks, such as scheduling and binding. These techniques rely on voltage reduction, frequency reduction or switching activity reduction to optimize power consumption. Heuristic, ILP and GTS algorithms are proposed in the literature. The algorithms provide time- and resource-constrained approaches for design space exploration. Peak power reduction, which is the maximum instantaneous power dissipation in a circuit, is becoming important. This reduction will improve battery efficiency and circuit reliability. The current literature is sparse in this area of research. Only few approaches are available that use voltage or frequency scaling for peak power reduction. They use ILP-based optimization to achieve this goal. Transient power reduction or power fluctuation reduction is also needed to improve battery life and efficiency. There are only a few research works in this area. They propose metrics to capture the power profile and transient power. Only a few approaches are available that use voltage or frequency scaling for its reduction. They use heuristic- or ILP-based optimization techniques. A data-monitoring technique is proposed for transient power management. Leakage power due to various sources such as subthreshold conduction and gate-oxide tunneling are prominent sources of power dissipation in nano-CMOS circuits. Research in optimization of leakage during high-level synthesis is rare. The existing approaches are based on primarily dual-V_{Th} technology. However, recently, dual-T_{ox} and dual-K technologies have been developed. The algorithms proposed are primarily heuristic, and there is a need for new high-level synthesis approaches for leakage reduction.

The future directions of low-power high-level synthesis are (i) high-level synthesis and optimization for analog and mixed-signal circuits, (ii) battery-aware synthesis and optimization, (iii) high-level synthesis for emerging technology and nanotechnology and (iii) high-level synthesis for nano-electro-mechanical-systems (NEMS).

References

1. http://www.energybulletin.net/
2. Agility Compiler. http://www.celoxica.com/products/agility/default.asp
3. Berkely BSIM4 Homepage. http://www-device.eecs.berkeley.edu/~bsim3/bsim4.html
4. Cascade. http://www.criticalblue.com/products/index.html
5. CyberWorkBench. http://www.cyberworkbench.com/
6. Cynthesizer and High-Level Design. http://www.forteds.com/products/index.asp
7. eBook: Design of VLSI Systems. http://lsiwww.epfl.ch/LSI2001/teaching/webcourse/ch01/ch01.html (15 Nov 2005)
8. ESL Synthesis Extensions to SystemC (ESE). http://www.bluespec.com/products/ ESLSynthesisExtensions.htm
9. eXCite. http://www.yxi.com/products3.html
10. Express: High-Level Synthesis Benchmarks. http://express.ece.ucsb.edu/benchmark/
11. High Level Synthesis – Moving Beyond Hand-Coded RTL Methodologies. http://www.mentor.com/products/esl/high_level_synthesis/
12. Incisive verification platform products. http://www.cadence.com/products/product_index.aspx?platform=FunctionalVer
13. ISE Foundation. http://www.xilinx.com/ise/logic_design_prod/foundation.htm
14. lp-solve, Solve (mixed integer) linear programming problems. http://lpsolve.sourceforge.net/5.5/
15. PICO Express. http://www.synfora.com/products/picoexpress.html
16. Quartus II. http://www.altera.com/products/software/products/quartus2/design/qts-design_flow_pld.html
17. Semiconductor Industry Association, International Technology Roadmap for Semiconductors. http://public.itrs.net
18. Synplify DSP. http://www.synplicity.com/products/dsp_solutions.html
19. The Senate Energy & Natural Resources Committee Report. http://energy.senate.gov/ public/_files/TheEnergyPolicyActof2005anniversaryreport5.doc
20. VCS: Comprehensive RTL Verification Solution. http://www.synopsys.com/products/ simulation/simulation.html
21. Intel Technology Journal **7**(2), 47 (2003)
22. Abdel-Kader, R.F.: Resource-constrained loop scheduling in high-level synthesis. In: Proceedings of the 43rd ACM Annual Southeast Regional Conference, pp. 195–200 (2005)
23. Acquaviva, A., Benini, L., Ricco, B.: Processor frequency setting for energy minimization of streaming multimedia application. In: Proceedings of the 9th International Symposium on Hardware/Software Codesign, pp. 249–253 (2001)
24. Ahmad, I., Dhodhi, M.K., Ali, F.M.: TLS: A tabu search based scheduling algorithm for behavioral synthesis of functional pipelines. The Computer Journal **43**(2), 152–166 (2000)
25. Aho, R.V., Ullman, J.: Compilers: Principles, Techniques and Tools. Addison-Wesley (1997)

26. Al-Turki, U., Fedjki, C., Andijani, A.: Tabu search for a class of single-machine scheduling problems. Computers & Operations Research **28**(12), 1223–1230 (2001)
27. Alidina, M., Monteiro, J.C., Devadas, S., Ghosh, A., Papaefthymiou, M.C.: Precomputation-based sequential logic optimization for low power. In: Proceedings of the IEEE/ACM International Conference on Computer-Aided Design (ICCAD), pp. 74–81 (1994)
28. Alupoaei, S., Katkoori, S.: Ant colony system application to macrocell overlap removal. IEEE Transaction of VLSI Systems **12**(10), 1118–1123 (2004)
29. Ameliford, B., Fallah, F., Pedram, M.: Reducing the sub-threshold and gate-tunneling leakage of SRAM cells using dual-V_t and dual-T_{ox} assignment. In: Proceedings of Design, Automation and Test in Europe, pp. 1–6 (2006)
30. Amella, S., Kaminska, B.: Scheduling of a control and data flow graph. In: Proceedings of the IEEE International Symposium on Circuits and Systems (ISCAS), pp. 1666–1669 (1993)
31. Antola, A., Ferrandi, F., Piuri, V., Sami, M.: Semiconcurrent error detection in data paths. IEEE Transactions on Computers **50**(5), 449–465 (2001)
32. Antola, A., Piuri, V., Sami, M.: A low-redundancy approach to semi-concurrent error detection in datapaths. In: Proceedings of the Design Automation and Test in Europe, pp. 266–272 (1998)
33. Arato, P., Visegrady, T., Jankovits, I.: High Level Synthesis of Pipelined Datapaths. Wiley (1997)
34. Arvind, Nikhil, R.S., Rosenband, D.L., Dave, N.: High-level synthesis: an essential ingredient for designing complex ASICs. In: Proceedings of the International Conference on Computer-Aided Design (ICCAD), pp. 775–782 (2004)
35. Benini, L., Bogliolo, A., Micheli, G.D.: A survey of design techniques for system-level dynamic power management. IEEE Transactions on Very Large Scale Integration (VLSI) Systems **8**(3), 299–316 (2000)
36. Benini, L., Castelli, G., Macii, A., Macii, E., Poncino, M., Scarsi, R.: Discrete-time battery models for system-level low-power design. IEEE Transactions on VLSI Systems **9**(5), 630–640 (2001)
37. Benini, L., Casterlli, G., Macii, A., Scarsi, R.: Battery-driven dynamic power management. IEEE Design and Test of Computers **13**(2), 53–60 (2001)
38. Benini, L., Macii, E., Pnocino, M., Micheli, G.D.: Telescopic units : a new paradigm for performance optimization of VLSI design. IEEE Transactions on Computer-Aided Design of Integrated Circuits and Systems **17**(3), 220–232 (1998)
39. Benini, L., Micheli, G.D.: System-level power optimization: techniques and tools. ACM Transactions on Design Automation of Electronic Systems **5**(2), 115–192 (2000)
40. Benini, L., Micheli, G.D., Lioy, A., Macii, E., Odasso, G., Poncino, M.: Automatic synthesis of large telescopic units based on near-minimum timed supersetting. IEEE Transactions on Computers **48**(8), 769–779 (1999)
41. Benini, L., Michelli, G.D., Macii, A.: Designing low-power circuits : practical recipes. IEEE Circuits and Systems Magazine **1**(1), 6–25 (2001)
42. Bernstein, K., Frank, D.J., Gattiker, A.E., Haensch, W., Ji, B.L., Nassif, S.R., Nowak, E.J., Pearson, D.J., Rohrer, N.J.: High-performance CMOS variability in the 65-nm regime and beyond. IBM Journal of Research and Development **50**(4/5), 433–449 (2006)
43. Berrebi, E., Kission, P., Vernalde, S., Troch, S.D., Herluison, J.C., Fréhel, J., Jerraya, A.A., Bolsens, I.: Combined control flow dominated and data flow dominated high-level synthesis. In: Proceedings of the 33rd Annual Conference on Design Automation, pp. 573–578 (1996)
44. Bhatia, S., Jha, N.K.: Behavioral synthesis for hierarchical testability of controller/datapath circuit with conditional branches. In: Proceedings of the International Conference on Computer Design (1994)
45. Blumrich, M. et al.: A holistic approach to system reliability in blue gene. In: Proceedings of the International Workshop on Innovative Architecture for Future Generation High Performance Processors and Systems (IWIA), pp. 3–12 (2006)
46. Bohr, M.T., Chau, R.S., Ghani, T., Mistry, K.: The high-κ solution. IEEE Spectrum **44**(10), 29–35 (2007)

47. Borkar, S.: Design challenges of technology scaling. IEEE Micro **19**(4), 23–29 (1999)
48. Brooks, D., Tiwari, V., Martonosi, M.: Wattch: a framework for architectural-level power analysis and optimizations. In: Proceedings of the International Symposium on Computer Architecture, pp. 83–94 (2000)
49. Brynjolfson, I., Zilic, Z.: Dynamic clock management for low power applications in FPGAs. In: Proceedings of the IEEE Custom Integrated Circuits Conference, pp. 139–142 (2000)
50. Brynjolfson, I., Zilic, Z.: FPGA clock management for low power applications. In: Proceedings of the International Symposium on FPGAs, pp. 219–219 (2000)
51. Burd, T., Brodersen, R.W.: Design issues for dynamic voltage scaling. In: Proceedings of the International Symposium on Low Power Electronics and Design, pp. 9–14 (2000)
52. Burd, T., Pering, T., Stratakos, A., Brodersen, R.W.: A dynamic voltage scaled microprocessor system. In: Proceedings of the IEEE International Solid-State Circuits Conference, pp. 294–295 (2000)
53. Burd, T., Pering, T.A., Stratakos, A.J., Brodersen, R.W.: A dynamic voltage scaled microprocessor system. IEEE Journal of Solid-State Circuits **35**(11), 1571–1580 (2000)
54. Burger, D., Austin, T.M.: The SimpleScalar Tool Set, Version 2.0. Computer Architecture News, pp. 13–25 (1997)
55. Butts, J.A., Sohi, G.S.: A static power model for architects. In: Proceedings of the 33rd Annual IEEE/ACM International Symposium on Microarchitecture (MICRO-33), pp. 191–201 (2000)
56. Buyuksahin, K.M., Najm, F.N.: Early power estimation for vlsi circuits. IEEE Transactions on CAD of Integrated Circuits and Systems **24**(7), 1076–1088 (2005)
57. Büyüksahin, K.M., Patra, P., Najm, F.N.: ESTIMA: an architectural-level power estimator for multi-ported pipelined register files. In: Proceedings of the 2003 International Symposium on Low Power Electronics and Design, pp. 294–297 (2003)
58. Camposano, R.: Path-Based Scheduling for Synthesis. IEEE Transactions on Computer-Aided Design of Integrated Circuits and Systems **10**(1), 85–93 (1991)
59. Cerny, V.: A thermodynamical approach to the travelling salesman problem: an efficient simulation algorithm. Journal of Optimization Theory and Applications **45**(1), 41–51 (1985)
60. Cesário, W.O., Sugar, Z., Moussa, I., Jerraya, A.A.: Efficient integration of behavioral synthesis within existing design flows. In: Proceedings of the 13th International Symposium on System Synthesis (ISSS), pp. 85–90 (2000)
61. Chandra, S., Lahiri, K., Raghunathan, A., Dey, S.: Considering process variations during system-level power analysis. In: Proceedings of the International Symposium on Low Power Electronics and Design, pp. 342–345 (2006)
62. Chandrakasan, A., Bowhill, W., Fox, F.: Design of High-Performance Microprocessor Circuits. IEEE Press (2001)
63. Chandrakasan, A., Brodersen, R.W.: Low-Power CMOS Design. Wiley-IEEE Press (1998)
64. Chandrakasan, A., Potkonjak, M., Mehra, R., Rabaey, J., Brodersen, R.W.: Optimizing power using transformations. IEEE Transactions on Computer-Aided Design of Integrated Circuits and Systems **14**(1), 12–31 (1995)
65. Chandrakasan, A., Potkonjak, M., Rabaey, J., Brodersen, R.W.: HYPER-LP: a system for power minimization using architectural transformations. In: Proceedings of the International Conference on Computer-Aided Design, pp. 300–303 (1992)
66. Chandrakasan, A., Sheng, S., Brodersen, R.W.: Low-power CMOS digital design. IEEE Journal of Solid-State Circuits **27**(4), 473–483 (1992)
67. Chandrakasan, A.P., Brodersen, R.: Minimizing power consumption in digital CMOS circuits. Proceedings of the IEEE **83**(4), 498–523 (1996)
68. Chang, C.T., Rose, K., Walker, R.A.: High-level DSP synthesis using the COMET design system. In: Proceedings of the Sixth Annual IEEE International ASIC Conference and Exhibit, pp. 408–411 (1993)
69. Chang, J.M., Pedram, M.: Energy minimization using multiple supply voltages. In: Proceedings of the International Symposium on Low Power Electronics and Design, pp. 157–162 (1996)

70. Chang, J.M., Pedram, M.: Energy minimization using multiple supply voltages. IEEE Transactions on VLSI Systems **5**(4), 436–443 (1997)
71. Chang, J.M., Pedram, M.: Power optimization and synthesis at behavioral and system levels using formal methods. Kluwer Academic Publishers (1999)
72. Chappell, B., Wang, X., Patra, P., Saxena, P., Vendrell, J., Gupta, S., Varadarajan, S., Gomes, W., Hussain, S., Krishnamurthy, H., Venkateshmurthy, M., Jain, S.: A system-level solution to domino synthesis with 2 GHz application. In: Proceedings of the IEEE International Conference on Computer Design (ICCD), pp. 164–171 (2002)
73. Chappidi, S.K.: Peak Power Minimization Through Datapath Scheduling using ILP Based Models. Master's Thesis, University of South Florida (Spring, 2003)
74. Chau, R., Datta, S., Doczy, M., Doyle, B., Kavalieros, J., Metz, M.: High-κ/metal-gate stack and its MOSFET characteristics. IEEE Electron Device Letters **25**(6), 408–410 (2004)
75. Chaudhuri, S., Walker, R.A.: ILP-based scheduling with time and resource constraints in high level synthesis. In: Proceedings of the International Conference on VLSI Design, pp. 17–20 (1994)
76. Chaudhuri, S., Walker, R.A., Mitchell, J.E.: Analyzing and exploiting the structure of the constraints in the ILP approach to the scheduling problem. IEEE Transactions on Very Large Scale Integration (VLSI) Systems **2**(4), 456–471 (1994)
77. Chen, C.T., Küçükçakar, K.: High-level scheduling model and control synthesis for a broad range of design applications. In: Proceedings of the International Conference on Computer Aided Design (ICCAD), pp. 236–243 (1997)
78. Cherabuddi, R.V., Bayoumi, M.A.: A low power based partitioning and binding technique for single chip application specific DSP architectures. In: Proceedings of the Second Annual IEEE International Conference on Innovative Systems in Silicon, pp. 350–361 (1997)
79. Cherabuddi, R.V., Bayoumi, M.A., Krishnamurthy, H.: A low power based system partitioning and binding technique for multi-chip module architectures. In: Proceedings of the 7th Great Lakes Symposium on VLSI, pp. 156–162 (1997)
80. Chiou, L.Y., Muhammand, K., Roy, K.: DSP data path synthesis for low-power applications. In: Proceedings of the International Conference on Acoustics, Speech, and Signal Processing, pp. 1165–1168 (2001)
81. Composano, R., Wolf, W.: High-Level VLSI Synthesis. Kluwer Academic Publishers (1991)
82. Cong, J., Ma, T., Bolsens, I., Moorby, P., Rabaey, J.M., Sanguinetti, J., Wakabayashi, K., Watanabe, Y.: Are we ready for system-level synthesis? In: Proceedings of the ASP-DAC (2005)
83. Cormen, T.H., Leiserson, C.E., Rivest, R.L., Stein, C.: Introduction to Algorithms, second edition. The MIT Press (2001)
84. Daalder, J., Eklund, P.W., Ohmori, K.: High-level synthesis optimization with genetic algorithms. In: Proceedings of the 4th Pacific Rim International Conference on Artificial Intelligence, pp. 276–287 (1996)
85. Dal, D., Kutagulla, D., Nunez, A., Mansouri, N.: Power islands: a high-level synthesis technique for reducing spurious switching activity and leakage. In: Proceedings of the 48th Midwest Symposium on Circuits and Systems, pp. 1875–1879 (2005)
86. Dal, D., Mansouri, N.: A high-level register optimization technique for minimizing leakage and dynamic power. In: Proceedings of the 17th Great Lakes Symposium on VLSI (GLSVLSI), pp. 517–520 (2007)
87. Dal, D., Nunez, A., Mansouri, N.: Power islands: a high-level technique for counteracting leakage in deep sub-micron. In: Proceedings of the 7th International Symposium on Quality Electronic Design (2006)
88. Dave, G., Jha, N.: Cosyn: hardware–software cosynthesis of heterogeneous distributed embedded systems. IEEE Transactions on VLSI Systems **7**(1), 92–104 (1999)
89. De, V., Borkar, S.: Technology and design challenges for low power and high performance [microprocessors]. In: Proceedings of the International Symposium on Low Power Electronics and Design, pp. 163–168 (1999)
90. Devadas, S., Newton, A.R.: Algorithms for allocation in datapath synthesis. IEEE Transactions on Computer-Aided Design of Integrated Circuits and Systems **8**(7), 768–781 (1989)

91. Dick, R.P., Rhodes, D.L., Wolf, W.: TGFF: task graphs for free. In: Proceedings of the 6th International Workshop on Hardware/Software Codesign (CODES), pp. 97–101 (1998)

92. Ding, L., Mazumder, P.: On circuit techniques to improve noise immunity of CMOS dynamic logic. IEEE Transactions on VLSI Systems **12**(9), 910–925 (2004)

93. Dorigo, M., Gambardella, L.M.: Ant colony system: a cooperative learning approach to the traveling salesman problem. IEEE Transactions on Evolutionary Computation **1**(1), 53–66 (1997)

94. Dorigo, M., Maniezzo, V., Colorni, A.: The ant system: optimization by a colony of cooperating agents. IEEE Transactions on Systems, Man, Cybernatics **26**(1), 29–41 (1996)

95. Drennan, P.G., McAndrew, C.C.: Understanding MOSFET mismatch for analog design. IEEE Journal of Solid-State Circuits **38**(3), 450–456 (2003)

96. Dunga, M.V., Yang, W., Xi, X., He, J., Liu, W., Cao, M., Jin, X., Ou, J., Chan, M., Niknejad, A.M., Hu, C.: Bsim 4.6.1 Mosfet Model – User's Manual. Tech. Rep., EECS Department, University of California, Berkeley (2007)

97. Elgamel, M.A., Bayoumi, M.: On low-power high-level synthesis using genetic algorithms. In: Proceedings of the 9th International Conference on Electronics, Circuits and Systems, pp. 725–728 (2002)

98. Elgamel, M.A., Bayoumi, M.A.: On low power high level synthesis using genetic algorithms. In: Proceedings of the International Workshop on Logic and Synthesis (IWLS), pp. 37–40 (2002)

99. Ferré, A., Figueras, J.: On estimating leakage power consumption for submicron CMOS digital circuits. In: Proceedings of the 7th International Workshop on Power, Timing, Modeling, Optimization and Simulation (PATMOS) (1997)

100. Fetweis, G., Chiu, J., Fraenkel, B.: A low-complexity bit-serial DCT/IDCT architecture. In: Proceedings of the IEEE International Conference on Communications, pp. 217–221 (1993)

101. Fourer, R., Gay, D., Kernighan, B.: AMPL: A Modeling Language for Mathematical Programming. Thomson Brooks Cole (2003)

102. Gajski, D., Dutt, N.: High-Level Synthesis: Introduction to Chip and System Design. Kluwer Academic Publishers (1992)

103. Gerez, S.H.: Algorithms for VLSI Design Automation. Wiley (2004)

104. Gold, S.: A PSPICE macromodel for lithium-ion batteries. In: Proceedings of the 12th Annual Battery Conference on Applications and Advances, pp. 215–222 (1997)

105. Goldberg, D.E.: Genetic Algorithms in Search Optimization and Machine Learning. Addison-Wesley (1989)

106. Gopalakrishnan, C., Katkoori, S.: Knapbind: an area-efficient binding algorithm for low-leakage datapaths. In: Proceedings of the 21st International Conference on Computer Design, pp. 430–435 (2003)

107. Gopalakrishnan, C., Katkoori, S.: Resource allocation and binding approach for low leakage power. In: Proceedings of 16th International Conference on VLSI Design, pp. 297–302 (2003)

108. Gopalakrishnan, C., Katkoori, S.: Tabu search based behavioral synthesis of low leakage datapaths. In: IEEE Computer Society Annual Symposium on VLSI (ISVLSI), pp. 260–261 (2004)

109. Gopalakrishnan, C., Katkoori, S., Gupta, S.: Power optimisation of combinational circuits by input transformations. IEE Proceedings – Computers and Digital Techniques **150**(3), 133–142 (2003)

110. Gopalan, R., Gopalakrishnan, C., Katkoori, S.: Leakage power driven behavioral synthesis of pipelined datapaths. In: IEEE Computer Society Annual Symposium on VLSI (ISVLSI), pp. 167–172 (2005)

111. Grewal, G.W., Wilson, T.C.: An enhanced genetic algorithm for solving the high-level synthesis problems of scheduling, allocation, and binding. International Journal of Computational Intelligence and Applications **1**(1) (2001)

112. Gu, Z.P., Yang, Y., Wang, J., Dick, R.P., Shang, L.: TAPHS: thermal-aware unified physical-level and high-level synthesis. In: Proceedings of the ASP-DAC, pp. 879–885 (2006)

113. Gupta, S., Dutt, N.D., Gupta, R.K., Nicolau, A.: SPARK: a high-level synthesis framework for applying parallelizing compiler transformations. In: Proceedings of the International Conference on VLSI Design, pp. 461–466 (2003)

114. Gupta, S., Katkoori, S.: Force-directed scheduling for dynamic power optimization. In: Proceedings of the IEEE Computer Society Annual Symposium on VLSI, pp. 68–73 (2002)

115. Gupta, S., Najm, F.N.: Energy-per-cycle estimation at RTL. In: IEEE International Symposium on Low Power Electronics and Design (ISLPED), pp. 121–126 (1999)

116. Gupta, S., Najm, F.N.: Energy and peak-current per-cycle estimation at RTL. IEEE Transactions on VLSI Systems 11(4), 525–537 (2003)

117. Gupta, S., Savoiu, N., Kim, S., Dutt, N.D., Gupta, R.K., Nicolau, A.: Speculation techniques for high level synthesis of control intensive designs. In: Proceedings of the 38th Design Automation Conference (DAC), pp. 269–272 (2001)

118. Hamada, M., Takahashi, M., Arakida, H., Chiba, A., Terazawa, T., Ishikawa, T., Kanazawa, M., Igarashi, M., Usami, K., Kuroda, T.: A top-down low power design technique using clustered voltage scaling with variable supply-voltage scheme. In: Proceedings of the 1998 IEEE Custom Integrated Circuits Conference, pp. 495–498 (1998)

119. Hanchate, N., Ranganathan, N.: LECTOR: a technique for leakage reduction in CMOS circuits. IEEE Transactions on VLSI Systems 12(2), 196–205 (2004)

120. Hansen, J.G.: Design of CMOS Cell Libraries for Minimal Leakage Currents. Master's thesis, Department of Informatics and Mathematical Modelling, Computer Science and Engineering Technical University of Denmark (Fall, 2004)

121. Haynal, S., Brewer, F.: Automata-based symbolic scheduling for looping DFGs. IEEE Transactions on Computers 50(3), 250–267 (2001)

122. He, J., Xi, X., Wan, H., Dunga, M., Chan, M., Niknejad, A.M.: BSIM5: an advanced charge-based MOSFET model for nanoscale VLSI circuit simulation. Solid-State Electronics 51(3), 433–444 (2007)

123. Heijligers, M.J.M.: The Applications of Genetic Algorithms to High-Level Synthesis. Ph.D. thesis, Technische Universiteit Eindhoven (1996)

124. Heijligers, M.J.M., Cluitmans, L.J.M., Jess, J.A.G.: High-level synthesis scheduling and allocation using genetic algorithms. In: Proceedings of the 28th Design Automation Conference, pp. 61–66 (1991)

125. Heijligers, M.J.M., Cluitmans, L.J.M., Jess, J.A.G.: High-level synthesis scheduling and allocation using genetic algorithms. In: Proceedings of the Asia Pacific Design Automation Conference (ASP-DAC) (1995)

126. Henning, R., Chakrabarti, C.: Activity models for use in low power, high-level synthesis. In: Proceedings of the International Conference on Acoustics, Speech, and Signal Processing, pp. 1881–1884 (1999)

127. Henning, R., Chakrabarti, C.: An approach to switching activity consideration during high-level, low-power design space exploration. IEEE Transactions on Circuits and Systems II: Analog and Digital Signal Processing 49(5), 339–351 (2002)

128. Hertz, A., de Werra, D.: The tabu search metaheuristic: how we used it. Annals of Mathematics and Artificial Intelligence 1 (1990)

129. Hsiao, M.S.: Peak power estimation using genetic spot optimization for large VLSI circuits. In: Proceedings of the Design Automation and Test in Europe (DATE), pp. 175–179 (1999)

130. Hsu, C.H., Kremer, U., Hsiao, M.: Compiler-directed dynamic frequency and voltage scheduling. In: Proceedings of the Workshop on Power-Aware Computer Systems, pp. 65–81 (2000)

131. Hsu, C.H., Kremer, U., Hsiao, M.: Compiler-directed dynamic voltage/frequency scheduling for energy reduction in microprocessor. In: Proceedings of the International Symposium on Low Power Electronics and Design, pp. 275–278 (2001)

132. Hu, J., Deng, Y., Marculescu, R.: System-level point-to-point communication synthesis using floorplanning information. In: Proceedings of the ASP-DAC, pp. 573–579 (2002)

133. Huff, H.R. et al.: Integration of high-k gate stack systems into planar CMOS process flows. In: International Workshop on Gate Insulator, pp. 2–11 (2001)

134. Hwang, C.T., Lee, J.H., Hsu, Y.C.: A formal approach to the scheduling problem in high level synthesis. IEEE Transactions on Computer-Aided Design of Integrated Circuits and Systems 10(4), 85–93 (1991)

135. Igarashi, M., Usami, K., Nogami, K., Minami, F., Kawasaki, Y., Aoki, T., Takano, M., Sonoda, S., Ichida, M., Hatanaka, N.: A low-power design method using multiple supply voltages. In: Proceedings of the International Symposium on Low Power Electronics and Design, pp. 18–20 (1997)

136. Iman, S., Pedram, M.: Logic Synthesis for Low Power VLSI Designs. Kluwer Academic Publishers (1998)

137. Ishihara, F., Sheikh, F.: Level conversion for dual supply systems. IEEE Transactions on VLSI Systems 12(2), 185–195 (2004)

138. Jain, R., Majumdar, A., Sharma, A., Wang, H.: Empirical evaluation of some high-level synthesis scheduling heuristics. In: Proceedings of the 28th Design Automation Conference, pp. 210–215 (1991)

139. Jain, R., Panda, P.R.: An efficient pipelined vlsi architecture for lifting-based 2d-discrete wavelet transform. In: Proceedings of the International Symposium on Circuits and Systems (ISCAS), pp. 1377–1380 (2007)

140. Jha, N.K.: Batteries for low power electronics. Proceedings of the IEEE 83(4), 687–693 (1995)

141. Jha, N.K.: Low power system scheduling and synthesis. In: Proceedings of the International Conference on Computer-Aided Design, pp. 259–263 (2001)

142. Jha, N.K.: Low-power system scheduling, synthesis and displays. IEE Proceedings – Computers and Digital Techniques 152(3), 344–352 (2005)

143. Johnson, M., Roy, K.: Optimal selection of supply voltages and level conversions during datapath scheduing under resource constraints. In: Proceedings of the International Conference on Computer Design, pp. 72–77 (1996)

144. Johnson, M., Roy, K.: Datapath scheduling with multiple supply voltages and level converters. ACM Transactions on Design Automation of Electronic Systems 2(3), 227–248 (1997)

145. Johnson, M., Roy, K.: Scheduling and optimal voltage selection for low power multiple-voltage DSP datapaths. In: Proceedings of the IEEE Symposium on Circuits and Systems, pp. 2152–2155 (1997)

146. Kamble, M.B., Ghose, K.: Analytical energy dissipation models for low power caches. In: Proceedings of the International Symposium on Low Power Electronic Design, pp. 143–148 (1997)

147. Kanno, Y., Mizuno, H., Tanaka, K., Watanabe, T.: Level converters with high immunity to power-supply bouncing for high-speed sub-1-V LSIs. In: Proceedings of the Symposium on VLSI Circuits Digest of Technical Papers, pp. 202–203 (2000)

148. Karnik, T., Ye, Y., Tschanz, J., Wei, L., Burns, S., Govindarajulu, V., De, V., Borkar, S.: Total power optimization by simultaneous dual-vt allocation and device sizing in high performance microprocessors. In: Proceedings of the 39th Conference on Design automation, pp. 486–491 (2002)

149. Katkoori, S., Kumar, N., Rader, L., Vemuri, R.: A profile driven approach for low power synthesis. In: Proceedings of the International Conference on Asian and South Pacific Design Automation Conference (ASP-DAC), pp. 759–765 (1995)

150. Katkoori, S., Roy, J., Vemuri, R.: A hierarchical register optimization algorithm for behavioral synthesis. In: Proceedings of the 9th International Conference on VLSI Design, pp. 120–132 (1996)

151. Katkoori, S., Vemuri, R.: Scheduling for low power under resource and latency constraints. In: Proceedings of the IEEE International Symposium on Circuits and Systems (ISCAS), pp. 53–56 (2000)

152. Kerntopf, P.: A new heuristic algorithm for reversible logic synthesis. In: Proceedings of the Design Automation Conference (DAC), pp. 834–837 (2004)

153. Khargharia, B., Hariri, S., Szidarovszky, F., Houri, M., El-Rewini, H., Khan, S.U., Ahmad, I., Yousif, M.S.: Autonomic power and performance management for large-scale data centers.

In: Proceedings of the 21st International Parallel and Distributed Processing Symposium (IPDPS), pp. 1–8 (2007)

154. Khouri, K.S., Jha, N.K.: Leakage power analysis and reduction during behavioral synthesis. In: Proceedings of International Conference on Computer Design, pp. 561–564 (2000)

155. Khouri, K.S., Jha, N.K.: Leakage power analysis and reduction during behavioral synthesis. IEEE Transactions on VLSI Systems **10**(6), 876–885 (2002)

156. Khouri, K.S., Lakshminarayana, G., Jha, N.K.: High-level synthesis of low-power control-flow intensive circuits. IEEE Transactions on Computer-Aided Design of Integrated Circuits and Systems **18**(12), 1715–1729 (1999)

157. Kim, J.M., Chae, S.I.: New MPEG2 decoder architecture using frequency scaling. In: Proceedings of the IEEE International Symposium on Circuits and Systems, pp. 253–256 (1996)

158. Kim, J.T., Shin, D.R.: New efficient clique partitioning algorithms for register-transfer synthesis of data paths. Journal of the Korean Physical Society **40**(4), 754–758 (2002)

159. Kim, N.S., Austin, T., Blaauw, D., Mudge, T., Flautner, K., Hu, J.S., Irwin, M.J., Kandemir, M., Vijaykrishnan, N.: Leakage current – Moore's law meets static power. IEEE Computer, pp. 68–75 (2003)

160. Kim, N.S., Kgil, T., Bertacco, V., Austin, T., Mudge, T.: Microarchitectural power modeling techniques for deep sub-micron microprocessors. In: ISLPED '04: Proceedings of the 2004 International Symposium on Low Power Electronics and Design, pp. 212–217 (2004)

161. Kirkpatrick, S., Gelatt, C.D., Vecchi, M.P.: Optimization by simulated annealing. Science **220**(4598), 671–680 (1983)

162. Kollig, P., Al-Hashimi, B.M.: Simultaneous scheduling, allocation and binding in high level synthesis. IEE Electronics Letters **33**(18), 1516–1518 (1997)

163. Kopuri, S., Mansouri, N.: Enhancing scheduling solutions through ant colony optimization. In: Proceedings of the International Symposium on Circuits and Systems (ISCAS), pp. 257–260 (2004)

164. Kougianos, E., Mohanty, S.P.: Effective tunneling capacitance: a new metric to quantify transient gate leakage current. In: Proceedings of the IEEE International Symposium on Circuits and Systems, pp. 2937–2940 (2006)

165. Kougianos, E., Mohanty, S.P.: Metrics to quantify steady and transient gate leakage in nanoscale transistors: NMOS vs. PMOS perspective. In: Proceedings of the 16th International Conference on VLSI Design, pp. 195–200 (2007)

166. Kountouris, A.A., Wolinski, C.: Hierarchical conditional dependency graphs as a unifying design representation in the CODESIS high-level synthesis system. In: Proceedings of the 13th International Symposium on System Synthesis, pp. 66–71 (2000)

167. Kountouris, A.A., Wolinski, C.: Efficient scheduling of conditional behaviors for high-level synthesis. ACM Transactions on Design Automation of Electronic Systems **7**(3), 380–412 (2002)

168. Krishna, V., Ranganathan, N., Vijaykrishnan, N.: Energy efficient datapath synthesis using dynamic frequency clocking and multiple voltages. In: Proceedings of the International Conference on VLSI, pp. 440–445 (1999)

169. Krishnan, V., Katkoori, S.: A genetic algorithm for the design space exploration of datapaths during high-level synthesis. IEEE Transactions on Evolutionary Computation **10**(3), 213–229 (2006)

170. Krishnan, V., Katkoori, S.: Design space exploration of RTL datapaths using rent parameter based stochastic wirelength estimation. In: Proceedings of the 7th International Symposium on Quality of Electronic Design (ISQED), pp. 364–369 (2006)

171. Kulkarni, S.H., Sylvester, D.: Fast and energy-efficient asynchronous level converters for multi-VDD design. In: Proceedings of the IEEE International Systems-on-Chip Conference, pp. 169–172 (2003)

172. Kulkarni, S.H., Sylvester, D.: High performance level conversion for dual VDD design. IEEE Transactions on VLSI Systems **12**(9), 926–936 (2004)

173. Kumar, A., Bayoumi, M.: A novel scheduling-based CAD methodology for exploring the design space of ASICs for low power. In: Proceedings of the 11th Annual IEEE International ASIC Conference, pp. 115–118 (1998)

174. Kumar, A., Bayoumi, M.: A novel scheduling-based CAD methodology for exploring the design space of ASICs for low power. In: Proceedings of the 1998 IEEE Asia-Pacific Conference on Circuits and Systems, pp. 391–394 (1998)

175. Kumar, A., Bayoumi, M.: Multiple voltage-based scheduling methodology for low power in the high level synthesis. In: Proceedings of the International Symposium on Circuits and Systems (ISCAS), pp. 371–379 (1999)

176. Kumar, N., Katkoori, S., Rader, L., Vemuri, R.: Profile-driven behavioral synthesis for low power VLSI system. IEEE Design and Test of Computers 12(3), 70–84 (1995)

177. Kurdahi, F.J., Parker, A.C.: REAL: a program for REgister ALlocation. In: Proceedings of the Design Automation Conference (DAC), pp. 210–215 (1987)

178. Kuroda, T. et al.: A 0.9-V, 150-MHz, 10-mW, 4 mm^2, 2-D discrete cosine transform core processor with variable threshold-voltage (VT) scheme. IEEE Journal of Solid-State Circuits 31(11), 1770–1779 (1996)

179. Lackey, D.E., Zuchowski, P.S., Koehl, J.: Designing mega-ASICs in nanogate technologies. In: Proceedings of the Design Automation Conference, pp. 770–775 (2003)

180. Landman, P.E., Rabaey, J.M.: Architectural power analysis : the dual bit type method. IEEE Transactions on VLSI Systems 3(2), 173–187 (1995)

181. Lee, C., Potkonjak, M., Mangione-Smith, W.H.: MediaBench: A tool for evaluating and synthesizing multimedia and communications systems. In: Proceedings of the 30th Annual ACM/IEEE International Symposium on Microarchitecture (MICRO-30), pp. 330–335 (1997)

182. Lee, C., Potkonjak, M., Mangione-Smith, W.H.: A detailed analysis of MediaBench. In: Proceedings of the IEEE Workshop on Signal Processing Systems, pp. 448–455 (1999)

183. Lee, J.S., Lee, H.D., Park, C.W., Hwang, S.Y.: Power-conscious scheduling algorithm for performance-driven datapath synthesis. IEE Electronics Letters 32(17), 1574–1576 (1996)

184. Lee, Y.H., Krishna, C.M.: Voltage-clock scaling for low energy consumption in real-time embedded systems. In: Proceedings of the 6th International Conference on Real-Time Computing Systems and Applications, pp. 272–279 (1999)

185. Lin, Y.L.: Recent developments in high-level synthesis. ACM Transactions on Design Automation of Electronic Systems 2(1), 2–21 (1997)

186. Lin, Y.R., Hwang, C.T., Wu, A.C.H.: Scheduling techniques for variable voltage low power design. ACM Transactions on Design Automation of Electronic Systems 2(2), 81–97 (1997)

187. Lis, J.S., Gajski, D.D.: Synthesis from VHDL. In: Proceedings of the International Conference on Computer Design, pp. 378–381 (1988)

188. Liu, M., Wang, W.S., Orshansky, M.: Leakage power reduction by dual-V_{th} designs under probabilistic analysis of V_{th} variation. In: Proceedings of International Symposium on Low Power Electronics and Design, pp. 2–7 (2004)

189. Liu, Y., Yang, H., Dick, R.P., Wang, H., Shang, L.: Thermal vs energy optimization for DVFS-enabled processors in embedded systems. In: Proceedings of the 8th International Symposium on Quality of Electronic Design (ISQED), pp. 204–209 (2007)

190. Lu, Y.H., Benini, L., Micheli, G.D.: Dynamic frequency scaling with buffer insertion for mixed workloads. IEEE Transactions on Computer-Aided Design of Integrated Circuits and System 21(11), 1284–1305 (2002)

191. Luo, J., Jha, N.K.: Static and dynamic variable voltage scheduling algorithms for real-time heterogeneous distributed embedded systems. In: Proceedings of the 15th International Conference on VLSI Design, pp. 719–726 (2002)

192. Luo, J., Jha, N.K.: Power-profile driven variable voltage scaling for heterogeneous distributed real-time embedded systems. In: Proceedings of the 16th International Conference on VLSI Design, pp. 369–375 (2003)

193. Macii, E.: Ultra Low-Power Electronics and Design. Kluwer Academic Publishers (2004)

194. Macii, E., Pedram, M., Somenzi, F.: High-level power modeling, estimation, and optimization. In: Proceedings of the 34th Annual Conference on Design Automation, pp. 504–511. ACM Press, New York (1997). DOI http://doi.acm.org/10.1145/266021.266268

195. Mamidipaka, M., Khouri, K., Dutt, N., Abadir, M.: Analytical models for leakage power esti-
     mation of memory array structures. In: Proceedings of the 2nd IEEE/ACM/IFIP International
     Conference on Hardware/Software Codesign and System Synthesis, pp. 146–151 (2004)
196. Manchanda, L., Busch, B., Green, M.L., Morris, M., van Dover, R.B., Kwo, R.,
     Aravamudhan, S.: High K gate dielectrics for the silicon industry. In: International Workshop
     on Gate Insulator, pp. 56–60 (2001)
197. Mandal, C.A., Zimmer, R.M.: A genetic algorithm for the synthesis of structured data paths.
     In: Proceedings of the 13th International Conference on VLSI Design, pp. 206–211 (2000)
198. Manzak, A., Chakrabarti, C.: A low power scheduling scheme with resources operating at
     multiple voltages. In: Proceedings of the 1999 IEEE International Symposium on Circuits
     and Systems, pp. 354–357 (1999)
199. Manzak, A., Chakrabarti, C.: A low power scheduling scheme with resources operating at
     multiple voltages. IEEE Transactions on VLSI Systems 10(1), 6–14 (2002)
200. Martin, R.S., Knight, J.P.: PASSOS: a different approach for assignment and scheduling for
     power, area and speed optimization in high-level synthesis. In: Proceedings of the 37th
     Midwest Symposium on Circuits and System, pp. 339–342 (1994)
201. Martin, R.S., Knight, J.P.: Optimizing power in ASIC behavioral synthesis. IEEE Design
     and Test of Computers 13(2), 58–70 (1996)
202. Martin, R.S., Knight, J.P.: Using spice and behavioral synthesis tools to optimize ASICs'
     peak power consumption. In: Proceedings of the 38th Midwest Symposium on Circuits and
     Systems, pp. 1209–1212 (1996)
203. Martin, T.L., Siewiorek, D.P.: Non-ideal battery properties and low power operation in wear-
     able computing. In: Proceedings of the 3rd International Symposium on Wearable Comput-
     ers, pp. 101–106 (1999)
204. Martin, T.L., Siewiorek, D.P.: Nonideal battery and main memory effects on CPU speed-
     setting for low power. IEEE Transactions on VLSI Systems 9(1), 29–34 (2001)
205. McCarl, B.A., Spreen, T.H.: Applied Mathematical Programming using Algebraic Systems.
     Online book at http://agecon.tamu.edu/faculty/mccarl/regbook.htm (1997)
206. McFarland, M.C., Parker, A.C., Camposano, R.: The high-level synthesis of digital systems.
     Proceedings of the IEEE 78(2), 301–318 (1990)
207. McKelvey, R.D., McLennan, A., Turocy, T.: Gambit: Software Tools for Game Theory. Tech.
     Rep., California Institute of Technology, University of Minnesota and Texas A & M Univer-
     sity, http://econweb.tamu.edu/gambit/ (2002)
208. Metropolis, N., Rosenbluth, A., Teller, A., Teller, E.: Equation of state calculations by fast
     computing machines. Journal of Chemical Physics 21, 1087 (1953)
209. Micheli, G.D.: Synthesis and Optimization of Digital Circuits. McGraw-Hill. (1994)
210. Mohanty, S.P.: Energy and Transient Power Minimization During Behavioral Synthesis.
     Ph.D. thesis, University of South Florida (Fall, 2003)
211. Mohanty, S.P., Kougianos, E.: Modeling and reduction of gate leakage during behavioural
     synthesis of NanoCMOS circuits. In: Proceedings of International Conference on VLSI
     Design (2006)
212. Mohanty, S.P., Kougianos, E.: Steady and transient state analysis of gate leakage current in
     nanoscale CMOS logic gates. In: Proceedings of IEEE International Conference on Com-
     puter Design, pp. 210–215 (2006)
213. Mohanty, S.P., Kougianos, E.: Simultaneous power fluctuation and average power minimiza-
     tion during nano-CMOS behavioural synthesis. In: Proceedings of the 20th IEEE Interna-
     tional Conference on VLSI Design, pp. 577–582 (2007)
214. Mohanty, S.P., Kougianos, E., Ghai, D., Patra, P.: Interdependency study of process and
     design parameter scaling for power optimization of nano-cmos circuits under process varia-
     tion. In: Proceedings of the 16th ACM/IEEE International Workshop on Logic and Synthesis
     (IWLS), pp. 207–213 (2007)
215. Mohanty, S.P., Kougianos, E., Mahapatra, R.N.: A comparative analysis of gate leakage and
     performance of high-k nanoscale cmos logic gates. In: Proceedings of the 16th ACM/IEEE
     International Workshop on Logic and Synthesis (IWLS), pp. 31–38 (2007)

216. Mohanty, S.P., Kougianos, E., Velagapudi, R., Mukherjee, V.: Scheduling and binding for low gate leakage NanoCMOS datapath circuit synthesis. In: Proceedings of the 38th IEEE International Symposium on Circuits and Systems (ISCAS), pp. 5291–5294 (2006)
217. Mohanty, S.P., Mukherjee, V., Velagapudi, R.: Analytical modeling and reduction of direct tunneling current during behavioral synthesis of nanometer CMOS circuits. In: Proceedings of the 14th ACM/IEEE International Workshop on Logic and Synthesis (IWLS), pp. 249–256 (2005)
218. Mohanty, S.P., Ranganathan, N.: A framework for energy and transient power reduction during behavioral synthesis. In: Proceedings of the International Conference on VLSI Design, pp. 539–545 (2003)
219. Mohanty, S.P., Ranganathan, N.: Energy efficient scheduling for datapath synthesis. In: Proceedings of the International Conference on VLSI Design, pp. 446–451 (2003)
220. Mohanty, S.P., Ranganathan, N.: A framework for energy and transient power reduction during behavioral synthesis. IEEE Transactions on VLSI Systems 12(6), 562–572 (2004)
221. Mohanty, S.P., Ranganathan, N.: Energy efficient datapath scheduling using multiple voltages and dynamic clocking. ACM Transactions on Design Automation of Electronic Systems (TODAES) 10(2), 330–353 (2005)
222. Mohanty, S.P., Ranganathan, N.: Simultaneous peak and average power minimization during datapath scheduling. IEEE Transactions on Circuits and Systems Part I (TCAS-I) 52(6), 1157–1165 (2005)
223. Mohanty, S.P., Ranganathan, N., Balakrishnan, K.: Design of a low power image watermarking encoder using dual voltage and frequency. In: Proceedings of the 18th International Conference on VLSI Design (VLSID), pp. 153–158 (2005)
224. Mohanty, S.P., Ranganathan, N., Balakrishnan, K.: A dual voltage-frequency VLSI chip for image watermarking in DCT domain. IEEE Transactions on Circuits and Systems II (TCAS-II) 53(5), 394–398 (2006)
225. Mohanty, S.P., Ranganathan, N., Chappidi, S.K.: Peak power minimization through datapath scheduling. In: Proceedings of the IEEE Computer Society Annual Symposium on VLSI, pp. 121–126 (2003)
226. Mohanty, S.P., Ranganathan, N., Chappidi, S.K.: Power fluctuation minimization during behavioral synthesis using ILP-based datapath scheduling. In: Proceedings of the IEEE International Conference on Computer Design, pp. 441–443 (2003)
227. Mohanty, S.P., Ranganathan, N., Chappidi, S.K.: Simultaneous peak and average power minimization during datapath scheduling for DSP processors. In: Proceedings of the ACM Great Lakes Symposium on VLSI, pp. 215–220 (2003)
228. Mohanty, S.P., Ranganathan, N., Chappidi, S.K.: ILP models for simultaneous energy and transient power minimization during behavioral synthesis. ACM Transaction on Design Automation of Electronic Systems 11(1), 186–212 (2006)
229. Mohanty, S.P., Ranganathan, N., Krishna, V.: Datapath scheduling using dynamic frequency clocking. In: Proceedings of the IEEE Computer Society Annual Symposium on VLSI, pp. 65–70 (2002)
230. Mohanty, S.P., Rangnathan, N., Chappidi, S.K.: An ILP-based scheduling scheme for energy efficient high performance datapath synthesis. In: Proceedings of the International Symposium on Circuits and Systems (ISCAS), pp. 313–316 (2003)
231. Mohanty, S.P., Rangnathan, N., Chappidi, S.K.: Transient power minimization through datapath scheduling in multiple supply voltage environment. In: Proceedings of the 10th IEEE International Conference on Electronics, Circuits and Systems, pp. 300–303 (2003)
232. Mohanty, S.P., Rangnathan, N., Chappidi, S.K.: ILP models for energy and transient power minimization during behavioral synthesis. In: Proceedings of the 17th International Conference on VLSI Design, pp. 745–748 (2004)
233. Mohanty, S.P., Vadlamudi, S.T., Kougianos, E.: A universal voltage level converter for multi-Vdd based low-power nano-CMOS Systems-on-Chips (SoCs). In: Proceedings of the 13th NASA Symposium on VLSI Design, p. 2.2 (2007)

234. Mohanty, S.P., Velagapudi, R., Kougianos, E.: Dual-K versus dual-T technique for gate leak-age reduction : a comparative perspective. In: Proceedings of International Symposium on Quality Electronic Design (ISQED), pp. 564–569 (2006)

235. Mohanty, S.P., Velagapudi, R., Kougianos, E.: Physical-aware simulated annealing optimization of gate leakage in nanoscale datapath circuits. In: Proceedings of the Conference on Design, Automation and Test in Europe (DATE), pp. 1191–1196 (2006)

236. Monteiro, J., Devadas, S., Ashar, P., Mauskar, A.: Scheduling techniques to enable power management. In: Proceedings of the ACM/IEEE Design Automation Conference, pp. 349–352 (1996)

237. Montgomery, D.C., Runger, G.C.: Applied Statistics and Probability for Engineers, 3rd edition Wiley (2006)

238. Mooney, E.L., Rardin, R.L.: Tabu search for a class of scheduling problems. Annals of Operations Research **41**(3), 253–278 (1993)

239. Mudge, T.N.: Power: a first class design constraint for future architecture and automation. In: Proceedings of the International Conference on High Performance Computing, pp. 215–224 (2000)

240. Mukherjee, V., Mohanty, S.P., Kougianos, E.: A dual dielectric approach for performance aware gate tunneling reduction in combinational circuits. In: Proceedings of the IEEE International Conference on Computer Design (ICCD), pp. 441–443 (2005)

241. Mukhopadhay, S., Roy, K.: Modeling and estimation of total leakage current in nano-scaled CMOS devices considering the effect of parameter variation. In: Proceedings of the IEEE International Symposium on Low Power Design, pp. 172–175 (2003)

242. Mukhopadhyay, S., Raychowdhury, A., Roy, K.: Accurate estimation of total leakage in nanometer-scale bulk CMOS circuits based on device geometry and doping profile. IEEE Transactions on Computer-Aided Design of Integrated Circuits and Systems **24**(3), 363–381 (2005)

243. Murugavel, A., Ranganathan, N.: A game theoretic approach for binding in behavioral synthesis. In: Proceedings of the International Conference on VLSI Design, pp. 452–458 (2003)

244. Murugavel, A.K., Ranganathan, N.: A game theoretic approach for power optimization during behavioral synthesis. IEEE Transactions on Very Large Scale Integration (VLSI) Systems **11**(6), 1031–1043 (2003)

245. Murugavel, A.K., Ranganathan, N.: Game theoretic modeling of voltage and frequency scaling during behavioral synthesis. In: Proceedings of the International Conference on VLSI Design, pp. 670–673 (2004)

246. Musoll, E., Cortadella, J.: Scheduling and resource binding for low power. In: Proceedings of the 8th International Symposium on System Synthesis, pp. 104–109 (1995)

247. Mutoh, S., Douseki, T., Matsuya, Y., Aoki, T., Shigematsu, S., Yamada, J.: 1-V power supply high-speed digital circuit technology with multithreshold-voltage CMOS. IEEE Journal of Solid-State Circuits **30**(8), 847–854 (1995)

248. Nahar, S., Sahni, S., Shragowitz, E.: Simulated annealing and combinatorial optimization. In: Proceedings of the 23rd ACM/IEEE conference on Design automation, pp. 293–299 (1986)

249. Namballa, R., Ranganathan, N., Ejnioui, A.: Control and data flow graph extraction for high-level synthesis. In: Proceedings of the IEEE Computer Society Annual Symposium on VLSI (ISVLSI), p. 192 (2004)

250. Namballa, R.K.: CHESS: A tool for CDFG extraction and high-level synthesis of VLSI systems. Master's thesis, University of South Florida (Spring, 2003)

251. Narasimhan, M., Ramanujam, J.: On lower bounds for scheduling problems in high-level synthesis. In: Proceedings of the Design Automation Conference (DAC), pp. 546–551 (2000)

252. Narendra, S., De, V., Borkar, S., Antoniadis, D., Chandrakasan, A.: Full-chip subthreshold leakage power prediction and reduction techniques for sub-0.18μm CMOS. IEEE Journal on Solid-State Circuits **39**(2), 501–510 (2004)

253. Nash., J.F.: Non-cooperative games. Annals of Mathematics **54**(2), 286–295 (1951)

254. Nielsen, S.F., Madsen, J.: Power constrained high-level synthesis of battery powered digital systems. In: Proceedings of the Conference on Design, Automation and Test in Europe (2003)

255. Nikara, J., Takala, J., Akopian, D., Astola, J., Saarinen, J.: Pipeline architecture for 8×8 discrete cosine transform Acoustics, Speech, and Signal Processing (ICASSP). In: Proceedings of the IEEE International Conference on, pp. 3303 – 3306 (2000)

256. Nikara, J., Takala, J., Akopian, D., Saarinen, J.: Pipeline architecture for DCT/IDCT. In: Proceedings of the International Symposium on Circuits and Systems (ISCAS), pp. 902–905 (2001)

257. Nowka, K.J., Carpenter, G.D., MacDonald, E.W., Ngo, H.C., Brock, B.C., Ishii, K.I., Nguyen, T.Y., Burns, J.L.: A 32-bit powerPC system-on-a-chip with support for dynamic voltage scaling and dynamic frequency scaling. IEEE Journal of Solid-State Circuits **37**(11), 1441–1447 (2002)

258. Ortiz, A.G., Lichtenau, C., Pfluger, T.: Technological and architectural power optimizations for advance microprocessors. In: Proceedings of the International Symposium on Signals, Circuits and Systems (ISSCS), pp. 11–14 (2005)

259. Pandini, D., Pileggi, L.T., Strojwas, A.J.: Bounding the efforts on congestion optimization for physical synthesis. In: Proceedings of the ACM Great Lakes Symposium on VLSI, pp. 7–10 (2003)

260. Panik, M.J.: Linear Programming: Mathematics, Theory and Practice. Kluwer Academic Publishers (1996)

261. Pant, P., Roy, R.K., Chattejee, A.: Dual-threshold voltage assignment with transistor sizing for low power CMOS circuits. IEEE Transactions on VLSI Systems **9**(2), 390–394 (2001)

262. Papa, G., Silc, J.: Multi-objective genetic scheduling algorithm with respect to allocation in high-level synthesis. In: Proceedings of the 26th EUROMICRO 2000 Conference, Informatics: Inventing the Future, pp. 339–346 (2000)

263. Papachristou, C., Spining, M., Nourani, M.: a multiple clocking scheme for low power RTL design. IEEE Transactions on VLSI Systems **7**(2), 266–276 (1999)

264. Papachristou, C.A., Konuk, H.: A linear program driven scheduling and allocation method. In: Proceedings of the 27th ACM/IEEE Design Automation Conference, pp. 77–83 (1990)

265. Papadimitriou, C.: Algorithms, games, and the internet. In: Proceedings of the 33rd Annual ACM Symposium on Theory of computing (STOC), pp. 749–753 (2001)

266. Park, C.: Task Scheduling in High Level Synthesis. Ph.D. Thesis, University of Illinois at Urbana-Champaign (1996)

267. Park, I.C., Kyung, C.M.: Fast and near optimal scheduling in automatic data path synthesis. In: Proceedings of the 28th Design Automation Conference, pp. 680–685 (1991)

268. Patra, P., Narayanan, U.: Automated phase assignment for the synthesis of low power domino circuits. In: Proceedings of the 36th ACM/IEEE Conference on Design Automation (1999)

269. Patterson, D.A., Hennessy, J.L.: Computer Organization and Design: The Hardware/Software Interface. Morgan Kaufmann (1997)

270. Paul, B.C., Agarwal, A., Roy, K.: Low-power design techniques for scaled technologies. Integration. The VLSI Journal **39**(2), 64–89 (2006)

271. Paulin, P.G., Knight, J.P.: Force-directed scheduling in automatic data path synthesis. In: Proceedings of the Design Automation Conference (DAC), pp. 195–202 (1987)

272. Paulin, P.G., Knight, J.P.: Force directed scheduling for the behavioral synthesis of ASICs. IEEE Transactions on Computer-Aided Design of Integrated Circuits and Systems **8**(6), 661–679 (1989)

273. Paulin, P.G., Knight, J.P.: Scheduling and binding algorithms for high-level synthesis. In: Proceedings of the 26th ACM/IEEE Design Automation Conference, pp. 1–6 (1989)

274. Paulin, P.G., Knight, J.P.: Algorithms for high-level synthesis. IEEE Design and Test of Computers **6**(6), 18–31 (1999)

275. Pedram, M.: Power minimization in IC design: principles and applications. ACM Transactions on Design Automation of Electronic Systems **1**(1), 3–56 (1996)

276. Pedram, M., Rabaey, J.M.: Power aware design methodologies. Kluwer Academic Publishers (2002)

277. Pering, T., Burd, T., Brodersen, R.W.: Dynamic voltage scaling and the design of a low-power microprocessor system. In: Proceedings of the Workshop on Power Driven Microarchitecture (1998)

278. Pering, T., Burd, T., Brodersen, R.W.: Voltage scheduling in the lpARM microprocessor system. In: Proceedings of the International Symposium on Low Power Electronics and Design, pp. 96–101 (2000)

279. Ping-Yuan, C., Chien-Cheng, Y.: A voltage level converter circuit design with low power consumption. In: Proceedings of the 6th International Conference on ASIC, pp. 358–359 (2005)

280. Pouwelse, J., Langendoen, K., Sips, H.: Dynamic voltage scaling on a low-power microprocessor. In: Proceedings of the 7th International Conference on Mobile Computing Network (2001)

281. Prabhakaran, P., Bannerjee, P., Crenshaw, J., Sarrafzadeh, M.: Simultaneous scheduling, allocation and floorplanning for interconnect power optimization. In: Proceedings of the 12th International Conference on VLSI Design, pp. 423–427 (1999)

282. Pu, G., He, J., Qiu, Z.: An optimal lower-bound algorithm for the high-level synthesis scheduling problem. In: Proceedings of the 9th IEEE Workshop on Design and Diagnostics of Electronic Circuits and Systems (DDECS 2006), pp. 151–152 (2006)

283. Puri, R., Stok, L., Cohn, J., Sylvester, D., Srivastava, A., Kung, D., Pan, D., Kulkarni, S.: Pushing ASIC performance in a power envelope. In: Proceedings of the Design Automation Conference, pp. 788–793 (2003)

284. Rabaey, J., Chu, C., Hoang, P., Potkonjak, M.: Fast prototyping of datapath-intensive architectures. IEEE Design and Test of Computer **8**(2), 40–51 (1991)

285. Rabaey, J.M., Chandrakasan, A., Nikolic, B.: Digital Integrated Circuits: A Design Perspective. Prentice Hall (2003)

286. Raghunathan, A., Dey, S., Jha, N.K.: High-level macro-modeling and estimation techniques for switching activity and power consumption. IEEE Transactions on Very Large Scale Integration System **11**(4), 538–557 (2003)

287. Raghunathan, A., Jha, N.: Behavioral synthesis for low power. In: Proceedings of the International Conference on Computer Design, pp. 318–322 (1994)

288. Raghunathan, A., Jha, N.K.: SCALP: an iterative-improvement based low-power datapath synthesis system. IEEE Transactions on CAD of Integrated Circuits and Systems **16**(11), 1260–1277 (1997)

289. Raghunathan, A., Jha, N.K., Dey, S.: High-Level Power Analysis and Optimization. Kluwer Academic Publishers (1998)

290. Raghunathan, V., Ravi, S., Lakshminarayana, G.: High-level synthesis with variable-latency components. In: Proceedings of the International Conference on VLSI Design, pp. 220–227 (2000)

291. Raghunathan, V., Ravi, S., Raghunathan, A., Lakshminarayana, G.: Transient power management through high level synthesis. In: Proceedings of the International Conference on Computer-Aided Design, pp. 545–552 (2001)

292. Rahman, H., Chakrabarti, C.: A leakage estimation and reduction technique for scaled CMOS logic circuits considering gate-leakage. In: Proceedings of the IEEE International Symposium on Circuits and Systems (ISCAS), pp. 297–300 (2004)

293. Raje, S., Bergamaschi, R.A.: Generalized resource sharing. In: Proceedings of the Design Automation Conference (DAC), pp. 326–332 (1997)

294. Raje, S., Sarrafzadeh, M.: GEM: A geometric algorithm for scheduling. In: Proceedings of the IEEE International Symposium on Circuits and Systems (Vol. 3), pp. 1991–1994 (1993)

295. Rakhmatov, D., Vrudhula, S., Wallach, D.A.: A model for battery lifetime analysis for organizing applications on a pocket computer. IEEE Transactions on VLSI Systems **11**(6), 1019–1030 (2003)

296. Rakhmatov, D.N., Vrudhula, S.B.K.: An analytical high-level battery model for use in energy management of portable electronic systems. In: Proceedings of the International Conference on Computer Aided Design, pp. 488–493 (2001)

297. Ramprasad, S., Shanbhag, N.R., Hajj, I.N.: Analytical estimation of signal transition activity from word-level statistics. IEEE Transactions on CAD of Integrated Circuits and Systems **16**(7), 718–733 (1997)

298. Ranganathan, N., Murugavel, A.K.: A low power scheduler using game theory. In: Proceedings of the 1st IEEE/ACM/IFIP International Conference on Hardware/Software Codesign and System Synthesis, CODES+ISSS, pp. 126–131 (2003)

299. Ranganathan, N., Namballa, R., Hanchate, N.: CHESS: a comprehensive tool for CDFG extraction and synthesis of low power designs from VHDL. In: Proceedings of the IEEE Computer Society Annual Symposium on VLSI (ISVLSI), pp. 329–334 (2006)

300. Ranganathan, N., Vijaykrishnan, N., Bhavanishankar, N.: A linear array processor with dynamic frequency clocking for image processing applications. IEEE Transactions on Circuits and Systems for Video Technology **8**(4), 435–445 (1998)

301. Rao, R., Vrudhula, S.B.K., Rakhmatov, D.N.: Battery modeling for energy-aware system design. IEEE Computer **36**(12), 77–87 (2003)

302. Rao, S.S.: Engineering Optimization: Theory and Practice. Addison-Wesley (1996)

303. Rawat, S., Camposano, R., Kahng, A., Sawicki, J., Gianfagna, M., Zafar, N., Sharan, A.: DFM: where's the proof of value? In: Proceedings of the Design Automation Conference (DAC), pp. 1061–1062 (2006)

304. Rim, M., Jain, R., Leone, R.D.: Optimal allocation and binding in high-level synthesis. In: Proceedings of the Design Automation Conference (DAC), pp. 120–123 (1992)

305. Rim, M., Mujumdar, A., Jain, R., de Leone, R.: Optimal and heuristic algorithms for solving the binding problem. IEEE Transactions on Very Large Scale Integration (VLSI) Systems **2**(2), 211–225 (1994)

306. Rong, P., Pedram, M.: An analytical model for predicting the remaining battery capacity of lithium-ion batteries. In: Proceedings of the Design Automation and Test in Europe Conference, pp. 1148–1149 (2003)

307. Rosien, M., Smit, G., Krol, T.: Generating a CDFG from C/C++ code. In: Proceedings of the 3rd Progress Workshop on Embedded Systems, pp. 200–202 (2002)

308. Roy, K., Mukhopadhyay, S., Meimand, H.M.: Leakage current mechanisms and leakage reduction techniques in deep-submicrometer CMOS circuits. Proceedings of the IEEE **91**(2), 305–327 (2003)

309. Roy, K., Prasad, S.C.: Low Power CMOS VLSI Circuits. Wiley (2000)

310. Sadeghi, K., Emadi, M., Farbiz, F.: Using level restoring method for dual supply voltage. In: Proceedings of the 19th International Conference on VLSI Design, pp. 601–605 (2006)

311. Sarrafzadeh, M., Raje, S.: Scheduling with multiple voltages under resource constraints. In: Proceedings of the IEEE Symposium on Circuits and Systems, pp. 350–353 (1999)

312. Satyanarayan, J.H., Parhi, K.K.: Theoretical analysis of word-level switching activity in the presence of glitch and correlation. IEEE Transactions on VLSI Systems **8**(2), 148–159 (2000)

313. Schuegraf, K., Hu, C.: Hole injection SiO_2 breakdown model for very low voltage lifetime extrapolation. IEEE Transactions on Electron Devices **41**(5), 761–767 (1994)

314. Sherwani, N.A., Mack, S.L., Alexanian, A., Buch, P., Guardiani, C., Lehon, H., Rabkin, P., Sharan, A.: Dfm rules! In: Proceedings of the Design Automation Conference (DAC), pp. 168–169 (2005)

315. Shin, M.K., Lin, C.H.: An efficient resource allocation algorithm with minimal power consumption. In: Proceedings of the IEEE Region 10 International Conference on Electrical and Electronic Technology, pp. 703–706 (2001)

316. Shiue, W.T.: High level synthesis for peak power minimization using ILP. In: Proceedings of the IEEE International Conference on Application Specific Systems, Architectures and Processors, pp. 103–112 (2000)

317. Shiue, W.T.: Peak power minimization using novel scheduling algorithm based on an ILP model. In: Proceedings of the 10th NASA Symposium on VLSI Design (2002)

318. Shiue, W.T., Chakrabarti, C.: Low power scheduling with resources operating at multiple voltages. In: Proceedings of the 9th International Symposium on Circuits and Systems, pp. 437–440 (1998)

319. Shiue, W.T., Chakrabarti, C.: ILP based scheme for low power scheduling and resource binding. In: Proceedings of the IEEE International Symposium on Circuits and Systems, pp. 279–282 (2000)

320. Shiue, W.T., Chakrabarti, C.: Low-power scheduling with resources operating at multiple voltages. IEEE Transactions on Circuits and Systems-II: Analog and Digital Signal Processing **47**(6), 536–543 (2000)

321. Shiue, W.T., Denison, J., Horak, A.: A novel scheduler for low power real time systems. In: Proceedings of the 43rd Midwest Symposium on Circuits and Systems, pp. 312–315 (2000)

322. Shivakumar, P., Jouppi, N.P.: CACTI 3.0: An Integrated Cache Timing, Power and Area Model. Tech. Rep., Compaq Western Research Laboratory, 250 University Avenue Palo Alto, California 94301, USA (2001)

323. Shukla, S.K., Gupta, R.K.: A model checking approach to evaluating system level dynamic power management policies for embedded systems. In: Proceedings of the International High-level Validation and Test Workshop, pp. 53–57 (2001)

324. Sicard, E., Bendhia, S.D.: Deep-Submicron CMOS Circuit Design (Simulator in Hands). Brooks/Coles (2003)

325. Simunic, T., Benini, L., Acquaviva, A., Glynn, P., Micheli, G.D.: Dynamic voltage scaling and power management for portable systems. In: Proceedings of the ACM/IEEE Design Automation Conference, pp. 524–529 (2001)

326. Singh, D., Rabaey, J.M., Pedram, M., Catthoor, F., Rajgopal, S., Sehgal, N., Mozdzen, T.J.: Power conscious CAD tools and methodologies: a perspective. Proceedings of the IEEE **83**, 570–594 (1995)

327. Singhal, K., Visvanathan, V.: Statistical device models from worst case files and electrical test data. IEEE Transactions on Semiconductor Manufacturing **12**(4), 470–484 (1999)

328. Sirisantana, N., Roy, K.: Low-power design using multiple channel lengths and oxide thicknesses. IEEE Design and Test of Computers **21**(1), 56–63 (2004)

329. Sirisantana, N., Wei, L., Roy, K.: High-performance low-power CMOS circuits using multiple channel length and multiple oxide thickness. In: Proceedings of the IEEE International Conference on Computer Design, pp. 227–232 (2000)

330. Sllame, A.M., Drábek, V.: An efficient list-based scheduling algorithm for high-level synthesis. In: Proceedings of the Euromicro Symposium on Digital Systems Design (DSD), pp. 316–323 (2002)

331. Small, C.: Shrinking devices put the squeeze on system packaging. Electronic Design News **39**, 41–46 (1994)

332. Smith, D.: What is logic synthesis. In: VLSI Design and Test (1988)

333. Smith, M.J.S.: Application-Specific Integrated Circuits (ASICs... the book). Addison-Wesley (1997)

334. Springer, D.L., Thomas, D.E.: Exploiting the special structure of conflict and compatibility graphs in high-level synthesis. IEEE Transactions on CAD of Integrated Circuits and Systems **13**(7), 843–856 (1994)

335. Su, F., Chakrabarty, K.: Unified high-level synthesis and module placement for defect-tolerant microfluidic biochips. In: Proceedings of the 42nd Annual Conference on Design automation, pp. 825–830 (2005)

336. Sullivan, C., Wilson, A., Chappell, S.: Using C based logic synthesis to bridge the productivity gap. In: Proceedings of the ASP-DAC, pp. 349–354 (2004)

337. Sultania, A.K., Sylvester, D., Sapatnekar, S.S.: Tradeoffs between gate oxide leakage and delay for dual T_{ox} circuits. In: Proceedings of Design Automation Conference, pp. 761–766 (2004)

338. Sultania, A.K., Sylvester, D., Sapatnekar, S.S.: Transistor and pin reordering for gate oxide leakage reduction in dual T_{ox} circuits. In: Proceedings of IEEE International Conference on Computer Design, pp. 228–233 (2004)

339. Sultania, A.K., Sylvester, D., Sapatnekar, S.S.: Gate oxide leakage and delay tradeoffs for dual-T_{ox} circuits. IEEE Transactions on Very Large Scale Integration (VLSI) Systems **13**(12), 1362–1375 (2005)

340. Sylvester, D., Kaul, H.: Future performance challenges in nanometer design. In: Proceedings of the 38th Design Automation Conference, pp. 3–8 (2001)

341. Sylvester, D., Kaul, H.: Power-driven challenges in nanometer design. IEEE Design Test Computers **13**, 12–21 (2001)

342. Takahara, S., Miyamoto, S.: An adaptive tabu search (ATS) and other metaheuristics for a class of optimal allocation problems. Journal of Advanced Computational Intelligence and Intelligent Informatics (JACIII) **3**(1), 21–27 (1999)

343. Tang, X., Zhou, H., Banerjee, P.: Leakage power optimization with dual-v_{th} library in high-level synthesis. In: DAC '05: Proceedings of the 42nd Annual Conference on Design Automation, pp. 202–207 (2005)

344. Taur, Y., Ning, T.H.: Fundamentals of Modern VLSI Devices. Cambridge University Press (1998)

345. Teich, J., Blickle, T., Thiele, L.: An evolutionary approach to system-level synthesis. In: Proceedings of the 5th International Workshop on Hardware/Software Codesign, pp. 167–172 (1997)

346. Thakkar, S.T.: Battery life challenges on future mobile notebook platforms. In: Proceedings of the International Symposium on Low Power Electronics and Design (ISLPED), p. 187 (2004)

347. Tiwari, V., Malik, S., Ashar, P.: Guarded evaluation: pushing power management to logic synthesis/design. IEEE Transactions on CAD of Integrated Circuits and Systems **17**(10), 1051–1060 (1998)

348. Tiwari, V., Singh, D., Rajgopal, S., Mehta, G., Patel, R., Baez, F.: Reducing power in high-performance microprocessors. In: Proceedings of the ACM/IEEE Design Automation Conference, pp. 732–737 (1998)

349. Tseng, C.J., Siewiorek, D.P.: Automatic synthesis of data path on digital systems. IEEE Transactions on CAD of Integrated Circuits and Systems **5**(3), 379–395 (1986)

350. Tsividis, Y.: Operation and Modeling of the MOS Transistor, 2nd edition. Oxford University Press (2003)

351. Tugsinavisut, S., Su, R., Beerel, P.A.: High-level synthesis for highly concurrent hardware systems. In: Proceedings of the International Conference on Application of Concurrency to System Design (ACSD), pp. 79–90 (2006)

352. Usami, K., Igarashi, M., Minami, F., Ishikawa, T., Kanzawa, M., Ichida, M., Nogami, K.: Automated low-power technique exploiting multiple supply voltages applied to a media processor. IEEE Journal of Solid-State Circuits **33**(3), 463–472 (1998)

353. Vallerio, K.S., Jha, N.K.: Task graph extraction for embedded system synthesis. In: Proceedings of the International Conference on VLSI Design, pp. 480–485 (2003)

354. Veendrick, H.J.M.: Short circuit dissipation of static CMOS circuitry and its impact on the design of buffer circuits. IEEE Journal of Solid State Circuits **sc-19**(4), 468–473 (1984)

355. Verhaegh, W.F.J., Lippens, P.E.R., Aarts, E.H.L., Korst, J.H.M., van Meerbergen, J.L., van der Werf, A.: Improved force-directed scheduling in high-throughput digital signal processing. IEEE Transactions on CAD of Integrated Circuits and Systems **14**(8), 945–960 (1995)

356. Vijaykrishnan, N., Kandemir, M.T., Irwin, M.J., Kim, H.S., Ye, W.: Energy-driven integrated hardware–software optimizations using simplepower. In: Proceedings of the International Symposium on Computer Architecture (ISCA), pp. 95–106 (2000)

357. Voldman, S.H., Brachitta, J.A., Fitzgerald, D.J.: Band-to-band tunneling and thermal generation gate-induced drain leakage. IEEE Transactions on Electron Devices **35**(12), 2433 (1988)

358. Wakerly, J.F.: Leakage in Nanometer CMOS Technologies. Springer (2005)

359. Wakerly, J.F.: Digital Design Principles and Practices. Pearson Education (2006)

360. Walker, R.A., Chaudhuri, S.: Introduction to the scheduling problems. IEEE Design and Test of Computers **12**(2), 60–69 (1995)

361. Wang, G., Gong, W., DeRenzi, B., Kastner, R.: Design space exploration using time and resource duality with the ant colony optimization. In: Proceedings of the 43rd Design Automation Conference (DAC), pp. 451–454 (2006)

362. Wang, G., Gong, W., Kastner, R.: Instruction scheduling using *MAX-MIN* ant system optimization. In: Proceedings of the 15th ACM Great Lakes Symposium on VLSI (GLSVLSI), pp. 44–49 (2005)

363. Wang, L., Jiang, Y., Selvaraj, H.: Synthesis scheme for low power designs with multiple supply voltages by tabu search. In: Proceedings of the International Symposium on Circuits and Systems (ISCAS), pp. 261–264 (2004)

364. Weste, N.H., Harris, D.: Principles of CMOS VLSI Design: A Systems Perspective. Addison-Wesley (2004)

365. Weste, N.H.E., Harris, D.: CMOS VLSI Design: A Circuit and Systems Perspective. Addison-Wesley (2005)

366. Williams, A.C., Brown, A.D., Zwolinski, M.: Simultaneous optimization of dynamic power, area, and delay in behavorial synthesis. IEE Proceedings on Computer and Digital Techniques **147**(6), 383–390 (2000)

367. Wilson, T.C., Grewal, G.W., Banerji, D.K.: An ILP solution for simultaneous scheduling, allocation, and binding in multiple block synthesis. In: Proceedings of IEEE International Conference on Computer Design: VLSI in Computer and Processors (ICCD), pp. 581–586 (1994)

368. Wilson, T.C., Mukherjee, N., Garg, M.K., Banerji, D.K.: An integrated and accelerated ILP solution for scheduling, module allocation, and binding in datapath synthesis. In: Proceedings of the International Conference on VLSI Design, pp. 192–197 (1993)

369. Wolf, S.: The Submicron MOSFET. Lattice Press (1995)

370. Wolf, S., Tauber, R.N.: Silicon Processing for the VLSI Era, Vol. 1: Process Technology. Lattice Press (1999)

371. Woods, M.: MOS VLSI reliability and yield trends. Proceedings of the IEEE **74**(2), pp. 1715–1729 (1986)

372. Ye, T.T., Chaudhuri, S., Huang, F., Savoj, H., Micheli, G.D.: Physical synthesis for ASIC datapath circuits. In: Proceedings of the IEEE International Conference on Circuits and Systems (ISCAS), pp. 365–368 (2002)

373. Yeo, K.S., Roy, K.: Low-Voltage, Low-Power VLSI Subsystems. McGraw Hill (2005)

374. Yu, C.C., Wang, W.P., Liu, B.D.: A new level converter for low power applications. In: Proceedings of the IEEE International Symposium on Circuits and Systems, pp. 113–116 (2001)

375. Yun, H.S., Kim, J.: Power-aware modulo scheduling for high-performance VLIW processors. In: Proceedings of the International Symposium on Low Power Electronics and Design, pp. 40–45 (2001)

376. Zhong, L., Jha, N.K.: Interconnect-aware low-power high-level synthesis. IEEE Transactions on CAD of Integrated Circuits and Systems **24**(3), 336–351 (2005)

377. Zhong, L., Luo, J., Fei, Y., Jha, N.K.: Register binding based power management for high-level synthesis of control-flow intensive behaviors. In: International Conference on Computer Design (ICCD), pp. 391–394 (2002)

378. Zhu, J., Gajski, D.D.: Soft scheduling in high level synthesis. In: Proceedings of the 36th Design Automation Conference, pp. 219–224 (1994)

379. Zyuban, V., Kogge, P.: The energy complexity of register files. In: Proceedings of the International Symposium on Low Power Electronics and Design, pp. 305–310 (1998)

Index

Printed in the United States of America